ISBN 978-0-428-99231-6
PIBN 10273102

The

Kentucky Geological Survey

WILLARD ROUSE JILLSON
DIRECTOR AND STATE GEOLOGIST

SERIES SIX
VOLUME THREE

Oil Field Stratigraphy
of Kentucky

1922

THE MOST PRODUCTIVE OIL "SAND" IN KENTUCKY.

Detail of the outcrop of the Onondaga limestone (Corniferous "sand") on the L & N. R. R., Northwest of Irvine.
The cherty "hornstone" characteristic is at once apparent.

OIL FIELD
STRATIGRAPHY OF KENTUCKY

A Systematic Presentation of the Several Oil Sands of
the State as Interpreted from Twelve Hundred
New and Detailed Well Records

By

WILLARD ROUSE JILLSON
DIRECTOR AND STATE GEOLOGIST

AUTHOR OF

OIL AND GAS RESOURCES OF KENTUCKY
CONTRIBUTIONS TO KENTUCKY GEOLOGY
ECONOMIC PAPERS ON KENTUCKY GEOLOGY
PRODUCTION OF EASTERN KENTUCKY CRUDE OILS
ETC.

*Illustrated with 35 Photographs
Maps and Diagrams*

First Edition
1500 Copies

THE KENTUCKY GEOLOGICAL SURVEY
FRANKFORT, KY.
1922

Exchange

THE STATE JOURNAL COMPANY
Printer to the Commonwealth
Frankfort, Ky.

PREFACE

The oil resources of the State of Kentucky have demonstrated their great importance to the economic prosperity and growth of the Commonwealth. In 1918 Kentucky produced 4,308,893 barrels of crude oil valued at $11,128,421. This was more than doubled in 1919, when 9,226,473 barrels were produced valued at $24,459,017. In 1920 due to steadily increased prices paid for crude oil, the total value of Kentucky crude again jumped and reached the large figure of value of $33,-525,210; while the volume slowly declined to 8,546,027 barrels. The loss during the year 1920 was therefore nearly 700,000 barrels. Recently completed figures of petroleum production for Kentucky for the year 1921 again show an increase of 534, 818 barrels over the year 1920, or at a total petroleum production for 1921 of 9,080,845 barrels valued at $16,674,969. This increase, which is regarded as temporary, has been due entirely to the new pools found in Johnson, Magoffin, Lawrence and Warren Counties, for the main producing pools of Lee County have declined steadily.

In view of this critical condition of the oil producing industry in Kentucky, it has been regarded as worth while to present an uptodate study of the oil sands of the State as interpreted from representative records selected from the large amount of recent drilling. The idea had its inception in the minds of a number of practical operators who have anticipated the value of such a report to drillers generally, and especially those working in "wildcat" localities. It is hoped that the practical values so earnestly predicted may be realized, and that this volume containg over 1,200 new and, here to fore unpublished Kentucky well records, may be the source of much general information as well as assistance in staying the declining oil production of this State.

W. R. Gillson

Director and State Geologist
Kentucky Geological Survey

OLD CAPITOL
Frankfort, Ky., Jan. 1, 1922.

494213

CONTENTS

ILLUSTRATIONS

OIL FIELD STRATIGRAPHY OF KENTUCKY

OIL AND GAS SANDS.

Historical.

The first serious attempt to correlate the strata in which oil and gas occur in Kentucky was made in 1888-1889 by Edward Orten, the celebrated Ohio geologist, who made a personal reconnaissance of the western portion of this State for the Second Kentucky Geological Survey.* Excellent results attended this early petroleum investigation, though it followed closely upon the opening of the first oil and gas fields of Pennsylvania, Ohio and Indiana. Among the fundamental points established was the productive nature of several "sands" in Kentucky. Chief among

BLOCKS OF "BEAVER SAND."
These weathered fragments of a ledge of "Beaver Sand" (New Providence Limestone) well illustrate the effects of differential weathering. This outcrop, which is four feet thick, occurs surrounded by the characteristic blue-green shales on Beaver Creek in Wayne County.

these were: (1) the Trenton (Middle Ordovician) limestone series in southern Kentucky; (2) the Clinton (Silurian) limestones of Barren County; (3) the Devonian black shale in Meade County; (4) the lower

---

*Report on the occurrence of Petroleum, Natural Gas, and Asphalt Rock in Western Kentucky. Edward Orten. Kentucky Geological Survey, Series II, 233 pp., 1891.

Mississippian limestones and shales in Allen, Barren, Warren and Breckinridge counties; and (5) the Pottsville and Allegheny (Pennsylvanian) "sands" in the Western coal field.

Today all of these determinations still stand correct, except the second, which is now known to be the Niagaran (Silurian), instead of the Clinton. Yet this early report had something of incompleteness about it, for the Corniferous (Devonian) and Niagaran (Silurian) limestones, which have been responsible when taken together for the greater portion of the oil produced in western Kentucky, were not recognized as oil producing horizons. Furthermore, the Berea and Wier (Mississippian) now so well and favorably known in eastern Kentucky were at this time quite unsuspected as large petroliferous sources.

In 1904, Joseph B. Hoeing, later the sixth State Geologist of Kentucky, prepared the second report on the oil and gas sands of Kentucky.[†] Coming at a much later date, and after a sixteen-year period of state-wide prospecting, Mr. Hoeing's report added much to the conclusions already presented by Dr. Orten. Hoeing's statement of the order and sequence of the various oil and gas producing strata was, with one or two exceptions in reference to the Ordovician, quite correct, and stood for many years to follow. He added to Orten's list a number of important sands, among which are the following:

(1) Knox Dolomite (Cambro-Ordovician), southern Kentucky.
(2) Calciferous (Ordovician), state-wide.
(3) Sunnybrook (Ordovician), southern Kentucky.
(4) Clinton (Silurian), of Morgan County, Kentucky.
(5) Niagaran (Silurian), of Barren County, Kentucky.
(6) Corniferous (Devonian), of Estill, Menifee, and Bath counties, Kentucky.
(7) The Beaver (Mississippian), of Wayne County, Kentucky.
(8) Oil City (Mississippian), of Barren County, Kentucky.
(9) Berea (Mississippian), of Eastern Kentucky.
(10) Big Injun (Mississippian), of eastern Kentucky.
(11) Big Lime (Mississippian), of eastern Kentucky, and
(12) The four separate sands of the Pottsville (Pennsylvanian) in eastern Kentucky.

The contribution to the knowledge of the petroleum geology of Kentucky made by Hoeing was notable; yet, in the light of the great advances made in the drilling of this State in the decade and a half which followed, it finally came to be considered incomplete. In 1919, fifteen years after the preparation of the "Oil and Gas Sands of Kentucky," the writer presented a new discussion* on the oil stratigraphy of Kentucky which elaborated considerably upon the work of both Orten and Hoeing.

---

†The Oil & Gas Sands of Kentucky, J. B. Hoeing, Ky. Geological Survey, Series III, Bull. I, 233 pp., 1904.
*Oil and Gas Resources of Kentucky. W. R. Jillson, Ky. Geol. Survey, Series V, Bull. I, 630 pp., 1st and 2d Eds. 1919, 3d Ed. 1920.

AN EASTERN KENTUCKY DRILLING.

The J. C. Hunter No. 1, drilled by the Ohio Oil Co. near Sandy Hook, Elliott County, in 1921. The rig is of the portable type but the derrick floor has been roofed over as a protection against bad weather.

At the present time there may be added to the list of "sands" enumerated the following "sands" which appear to have a rather important bearing on the oil industry in Kentucky:

(1) "Deep" sand (Niagaran-Silurian) of Allen, Simpson, Edmonson, Butler, and Warren counties; (2) "Wier" sand (Mississippian) of eastern Kentucky; (3) "Shallow" sand (St. Louis-Mississippian) of Warren, Logan, Butler, Simpson and Edmonson counties; (4) "Maxton"* (Mauch Chunk-Mississippian) of eastern Kentucky; (5) "Sebree" (Allegheny-Pennsylvanian) of Union and Henderson counties; (6) the "Penrod" (Chester-Mississippian) of Muhlenberg County; (7) "Pellville" sand (Chester-Mississippian) of southern Hancock and northern Ohio counties; (8) "Shallow" sand (Niagaran-Silurian) of Olympia, Bath County and Stanton, Powell County, and other eastern Kentucky oil pools. The producing possibilities of the Warren County "shallow sand" is now well known but the ultimate possibilities of the "Wier" and "Maxton" (Mississippian) in eastern Kentucky, and the "Sebree" (Pennsylvanian) and "Penrod" (Mississippian) in western Kentucky, all of which with the exception of perhaps the last are true silica sands, cannot be estimated at this time.

---

*Frequently corrupted into "Maxon."

Oil and Gas Sands of Kentucky.

Since it is simply the purpose of this work to set forth the sequence of
oil and gas sands as recognized by the oil driller in Kentucky for the use
of all who may be interested in the oil and gas industry in this State,
and especially the practical man, no attempt will be made to present
the extreme fullness of detail descriptive matter available.

A GROUP OF OIL SCOUTS.

Important test wells drilled in possible new oil territory generally
draw the attention of oil men about the time the "Sand" is reached.
These men are watching a new well in Martha district of Lawrence
County.

While the experienced oil and gas operator recognizes that no well
record can ever be presented on a printed page in a form more accurate
than that in which it is prepared by the driller, and that inaccuracies of
one form or another are inherent undoubtedly in every log, he knows
that the best record available is the best information on which to base
further drilling. Stratigraphers who regard as basic the law of changing
measurements of outcrop sections within the same series and even short
distances, have been slow to realize that the same law applies to sub-
surface stratigraphy—the stratigraphy of oil and gas wells. There has been
too much of an attempt on the part of many to try and harmonize well
records with known surface measurements, a practice which while it gives
an air of finish to a report, cannot assist the practical man at all. It is a
bit of square peg and round hole labor that does not produce oil. The important thing for both the professional and practical man working in oil

field stratigraphy to do, is to recognize the several horizons penetrated by the bit, and learn their lithologic character, and their productive or non-productive measurements.

In the light of these considerations, which are fundamental, this book preempts a special geologic field to itself, and does not compete with the standard works on the stratigraphy of Kentucky or adjoining States. The following statements are principally based, therefore, upon an intelligent and practical interpretation of many well logs. All descriptions and conclusions have been condensed as much as possible in the interest of the practical oil man, who is justifiably more desirous of securing an adequate summary of the nature of the producing oil and gas sands of Kentucky than he is in a detailed account of its stratigraphy on outcrop.

Knox Dolomite.

Considered in ascending order, the lowest and oldest formation referred to as a possible oil producing "sand" about which anything is known in Kentucky is a "sand" which occurs at a depth ranging from 1,350 to 1,385 feet below the base of the Chattanooga (Devonian) black shale in the Beech Bottom section of southeastern Clinton County, Kentucky. This "sand" is regarded, by some, as occurring below the Trenton and the Calciferous, and has been referred* to the Knox Dolomite (Cambro-Ordovician) though the ultimate decision is yet in question, due to lack of sufficient detailed information of a paleontologic and lithologic character. In this region the Devonian and Silurian limestones are regarded as missing, and no outcrops as low in the Paleozoic section occur nearer than Jacksboro, Campbell County, northeastern Tennessee, a point forty miles distant in an air line. Further investigations will undoubtedly lead to a definite decision concerning this deep "sand" which is either of Lower Ordovician or Upper Cambrian age. In eastern Tennessee the Knox Dolomite (Cambro-Ordovician) attains a maximum thickness of about 3,500 feet. It is a light to dark magnesean limestone with many chert nodules, and with this description the cuttings from the Beech Bottom sand seem to agree. In the Beech Bottom region of Clinton County oil was found in the Geo. Smith No. 1 (lessor) well at a depth of 1,728 feet, and produced about five barrels of high gravity green oil from 1,770 to 1,780 feet in depth. Chemically it was a magnesian limestone, fine in texture, and the sand as cut from the bit resembled very fine beech sand. In the Pickett and Fentress counties at the south and southeast in Tennessee a similar if not indeed the same "sand" is recognized at a depth of from 1,562 to 1,617 feet, which is at a "top" depth of 1,335 feet below the Chattanooga (Devonian) black shale.

The Calciferous.

At some distance above the Knox Dolomite, possibly of Beekman-

---

*Administrative report of the State Geologist, Wilbur A. Nelson, Nashville, Tennessee, 1920. Tenn. Geol. Survey, Bulletin No. 25, page 57.

town age, occurs a hard, sandy limestone which has been correlated with the St. Peters' Sandstone, and is known as the ''Calciferous'' (Lower Ordovician). This "sand" produced commercial quantities of gas in Estill and Hardin counties some years ago, but has been unproductive elsewhere in the State so far as known. While a remote gas producing possibility in deep wells, it does not hold forth much prospect as an oil producer in this State, as the large number of costly tests which have penetrated it at widely separated points clearly indicate.

OILY LITTERAL FORK.
A drilling of the Carter oil on the, head of Litteral Fork, Magoffin County. This well sprayed oil over the derrick as may be noted on the darkened beam.

The Trenton.

Widely prospected in Ohio, Indiana, and Illinois, where it is a large producer of both oil and gas, the Trenton "sand" (Middle Ordovician) can only be regarded as one of the minor oil sands of Kentucky. A shaly and somewhat cherty limestone series of considerable thickness on its outcrop in central Kentucky, the Trenton appears to be of very similar lithologic character in the productive regions of Wayne, Clinton, Russell, Cumberland, Barren and Monroe counties in southern Kentucky. The oil produced is a fairly high though not uniform gravity, and the wells are characteristically small, ten to fifteen barrels being about the maximum settled yield. Productive horizons are of irregular occurrence, depth, and thickness. The lower Sunnybrook sand of Wayne County is undoubtedly of Trenton age.

Upper Sunnybrook.

Overlying the Trenton series, in a limestone group, occurs the Upper Sunnybrook (Upper Ordovician) "sand," which is a correlative of the Caney sand, the Barren County "Deep" sand, and the Cumberland County "Shallow" sand. Like the underlying Trenton, the series in which the Sunnybrook is found is mainly calcareous, but exhibits some intercallated blue calcareous shales. The oil production secured is generally small, though of medium high grade. The series itself is thick, ranging from 400 to 700 feet. This "sand" has been tested generally without success over a very wide area in Kentucky, and for this reason its future is not regarded as very promising from an oil producing standpoint.

Niagaran.

Omitting the Clinton (Middle Silurian), which is practically of no importance in Kentucky, we come to the Niagaran group of limestones and shales which have recently come to take a rather important producing position in Allen, Barren, Warren, Simpson, Butler and Edmonson counties in western Kentucky, and in Bath, Rowan, Powell, Breathitt and other counties in eastern Kentucky. In Warren County this sand is known among the drillers as the "Deep" sand, and is penetrated by the bit west and northwest of Bowling Green at from 60 to 75 feet below the Chattanooga (Devonian) black shale. In productive regions it generally carries one or two oil pays. The Niagaran oil is dark green and of excellent grade. Wells producing 50 to 100 barrels in a test are not uncommon, and the "staying" qualities of the "sand" has been the cause of much meritorious comment. Though not generally recognized as such, the Niagaran is very probably one of the contributing though possibly small sources of the oil contained in the Corniferous in the larger fields located along the Pottsville outcrop in eastern Kentucky. It also shows commercial oil and gas "pays" of its own over a large part of eastern Kentucky.

The Corniferous.

The chief producing sand of Kentucky is the Corniferous (Devonian) limestone. This formation is probably of Onondaga age and is responsible for the oil secured in eastern Kentucky in the counties of Estill, Lee, Powell, Wolfe, Menifee, Jackson, Bath, Rowan and Morgan, where the production is secured just below the Chattanooga (Devonian) black shale. This "sand" is a magnesian limestone of a very irregularly cherty or "hornstone" characteristic, somewhat creviced, and ranging in thickness One, two, and sometimes three pay sands are encountered inside of this limestone, depending upon the locality. The Corniferous limestone is the chief oil "sand" of the Big Sinking pool, the largest producing pool in Kentucky. Its areal distribution is somewhat limited, since it does not cover the Blue Grass in Kentucky at all, and is absent in the Cumberland River Valley of southern central Kentucky, and is also probably absent under cover in some of the central southern Kentucky counties.

THE DEVONIAN BLACK SHALE.
This shale variously called the Ohio, the Chattanooga and the New
Albany is one of the most widely known in Kentucky. Here is shown
an exposure 75 feet in thickness near Clay City, Powell County.

Black Shale.

One of the outstandingly plain stratigraphic horizon markers in
Kentucky is the black (Upper Devonian) shale, variously called the Ohio
shale in northeastern Kentucky, the New Albany shale in western Kentucky, and the Chattanooga shale in southern Kentucky.   It is well
known to every oil operator and driller, and is a convenient dividing line
for the stratigraphy of the State.   In southern Kentucky the Chattanooga
shale is a unit with the Sunbury (Mississippian) black shale of northeastern Kentucky and Ohio, and their line of demarkation is seldom
outlined.   These two shales, so alike in their lithologic characteristics,
may fortunately be regarded as one, where they so occur, in so far as the
driller is concerned, since neither one of them is productive of natural
petroleum, and only very occasionally of natural gas.   So infrequently
is gas secured in the Devonian black shale that it is not considered except
as a marker by the average driller.   And yet it is one of the most extensive bituminous sediments in Kentucky.

Meade County years ago produced considerable gas from a "sand" lens in the shale, and some Floyd County gas has been referred to a similar horizon. But outside of these two localities this formation is not known to carry any porous or reservoiring strata, and is unproductive of natural petroleum and natural gas in Kentucky. Lithologically it is of a finely laminated bituminous nature, and is the formation which it is proposed to retort for artificial oil through processes of oil shale destructive distillation. In thickness it ranges from about twenty feet in some central "Knobs" counties to several hundred feet in the extreme portions of eastern and western Kentucky.

Beaver Sand.

Intercalated between the blue green shale of (New Providence-Mississippian) age, the Beaver oil "sand" of Wayne, McCreary, Clinton, Barren, Allen and Warren counties is a producing horizon of much importance in the southern portion of Kentucky. It is not known to be productive elsewhere. The Beaver "sand" itself is a magnesian limestone ranging in thickness from 2 to 8 feet. It contains a considerable amount of chert, which occurs as more or less isolated nodules where examined on the outcrop. Drilling "sands" also reveal this characteristic. The Beaver "sand" has the qualities of fairly long life; and though none of the wells drilled into it are as large as many which have been drilled into the Corniferous, it is regarded as an important high grade oil producer. The Wayne County oil produced from the Beaver "sand" is the original Somerset grade of Kentucky, and is chiefly handled by the Cumberland Pipe Line Co.

Berea Sand.

Though long recognized as a petroliferous source, the Berea (Lower Mississippian) sandstone may be said to have come into its own as a producer of oil and gas in Kentucky only within the last few years. The chief areas of productivity in Kentucky are those within Lawrence County, though isolated wells occur in Martin, Johnson, Floyd, Elliott, and Boyd counties. The Berea ranges from 40 to 90 feet in thickness, is a true silica "sand," and exhibits one or two pay horizons ranging from 5 to 25 feet. The oil is high grade, and though the wells are for the most part small in oil production, they are long lived. The Busseyville, Fallsburg and Louisa, and adjoining pools are the most productive in Lawrence County.

Wier Sand.

Separated from the underlying Berea sandstone by the easily recognized black Sunbury (Mississippian) shale, the Wier (Lower Mississippian) sandstone of eastern Kentucky which has been correlated with the Cuyahoga (Lower Mississippian) group of sandstones and sandy shales, has become within the last year one of the most important oil and gas producing "sands" in Kentucky. Until 1917 it was unknown as an oil and gas producer in Kentucky. Its recognition came with the development of the several new and important gas pools of Johnson, Magoffin, Lawrence and Elliott counties. A true silica "sand," its character ranges from fine to medium. In its most productive localities it is found to be fairly soft, requiring frequently, however, a light shot to secure its best producing qualities. It is of a grayish white color when washed out, and in thickness ranges from 30 to 60 feet, with generally one or two pays

OUT CROP OF THE "BEAVER SAND."
This exposure of the well known "Beaver Sand" (New Providence Limestone) occurs on Beaver Creek in Wayne County. The cherty inclusions are well shown. The limestone which originally surrounded the exposed chert has been removed by weathering agencies.

of from 5 to 10 feet, producing a brownish green oil of high gravity. Individual wells show a productive variation of from five to thirty-five barrels. The Wier is regarded as one of the most important and undoubtedly long lived oil producing "sands" of Kentucky.

The Big Injun.

Of passing interest only is the Big Injun (Lower Mississippian) sand of eastern Kentucky. This formation which is irregular in its thickness, ranging from 5 to 25 feet in thickness, is a calcareous sandstone, productive of both oil and gas over a wide area, but undependable as to offset. The Big Injun oil wells are usually small, being under five barrels, but the oil is of an amber color and is of high gravity paraffin base. Scattered wells are recognized in Lawrence, Johnson, Martin, Floyd, Knox, and other adjoining eastern Kentucky counties. Its occurrence beneath the Big Lime (Mississippian) and the thick Pottsville (Pennsylvanian)

THE SUNBURY AND BEDFORD SHALES.

This exposure occurs at the Junction of siding to Bluestone Quarry one-half mile east of Rockville Station, Rowan County. The Sunbury is clifted above, and the Bedford forms the talus.—Photo by Chas. Butts.

Series in some parts of the southeastern section of this State, notably Pike County, places it at a considerable depth, so that it is drilled only infrequently. Were it not for this fact, more definite information concerning it might be available, including perhaps a better production record.

The Big Lime.

An oil and gas producing horizon of increasing importance in eastern Kentucky is gradually being revealed in the Big Lime (St. Genevieve-Gasper-Glen Dean Limestones) (Chester-Mississippian) by widespread "wild cat" tests. A recognized correlative of the Maxville limestone of southeastern Ohio, and known elsewhere variously as the Greenbrier, Mountain and Newman limestone, it may be seen in almost continual outcrop from the Ohio River near South Portsmouth southwestward along and slightly below the base of the Pottsville (Pennsylvanian) conglomerate. In Carter, Morgan, Wolfe, Powell, Lee, Whitley and Bell counties it may be seen with something of the same characteristic that is found by the driller in the counties to the east and southeast where, because of the normal dip, it occurs anywhere from a few to several hundred feet below the surface.

As a recognizable unit in eastern Kentucky the Big Lime consists of a heavy bed of hard gray-white limestones, ranging from 20 to 400 feet in thickness. In western Kentucky its correlative attains an even greater thickness and certain "sands" are known to be productive there of both oil and gas. It is generally separated from the Little Lime above it in eastern Kentucky by a thin shale, but frequently the Little Lime is absent or an integral part of the Big Lime below, and is therefore not recognized by the driller. The Big Lime is principally a gas producer, commercial production being secured in Martin, Knott, Knox, Whitley, and other counties of eastern Kentucky. It is probably productive of gas in many places yet undrilled and certainly contains oil, as recent drilling has shown, not only in Martin on the east and Whitley on the south, but also in Pulaski and

THE BEREA SANDSTONE.
This exposure is just south of Vanceburg, Ky., looking west. The sandstone is 22 feet thick.   —Photo by Charles Butts.

Rockcastle counties on the west. The oil produced is of high gravity, greenish brown in color, and similar in other characteristics to that produced from the Weir (Lower Mississippian) sand of Johnson, Magoffin and adjoining counties.

The Maxton.

Recent drilling campaigns of wide scope undertaken in the interest of natural gas production for public utility uses, have demonstrated that the Maxton (Mauch Chunk-Mississippian) sand of eastern Kentucky is one of the most important productive horizons in the State. Intercallated between "red rock" shales of varying thickness the Maxton occurs as a true silica sand, buff to white in color, and ranges from 5 to 30 feet in thickness. Occasionally it is split by a shale into two members, and not infrequently both sands are productive. The Maxton, foreshortened by many drillers to "Maxon," is now principally recognized as a gas sand in this State, and is productive in Johnson, Magoffin, Floyd, Martin, Pike, Knott, Knox, and Whitley counties. It has produced considerable oil in Floyd County, and is probably potentially productive elsewhere. The greatest negative factor experienced in drilling to "pay" in the Maxton is its notable irregularity. As an oil sand it exhibits the attractive feature of long life, and as a gas sand it is prolific, though the recent drilling up and extension of the Beaver Creek field in Floyd County has shown that the rock pressure, in some instances at least, has fallen off sooner than was anticipated for a true silica sand of competent thickness.

Penrod Sand.

A new western Kentucky petroliferous horizon, and a large producer of both oil and gas, the Penrod (Chester-Mississippian) sand of southeastern Muhlenberg County cannot fail to attract considerable attention. The sand occurs at the shallow depth of about 650 feet, and is composed of two members separated by a shale of about ten feet in thickness. The sands are reported to be true silica sands and are from seven to twenty feet in thickness. The upper sand is generally productive of gas, and the lower one shows oil. The upper sand has, however, been found to be barren, and the lower sand a large gas producer. The drilling up of this field will provide much important data.

Pottsville Sands.

Of wide-spread areal distribution and ample thickness the Pottsville (Lower Pennsylvanian) oil and gas sands have long attracted the attention of oil and gas producers. Recognized as three distinct producing "sands," the Beaver, Horton and Pike in Floyd and Knott counties, and the Wages, Jones and Epperson in Knox County, these petroliferous sands of the Coal Measures were among the first to give commercial oil production in Kentucky, and are today pointed to as possessed of the longest productive life of any Kentucky oil sand.* The Williamsburg (Whitley County, Ky.) oil and sand is also of Pottsville age, and shows qualities of long life.

The Caseyville (Pottsville) sands of western Kentucky are not now known to be large oil producers, but may be regarded as having excellent possibilities in certain localities. The Nolin River rock asphalt beds are of Pottsville age, and give concrete evidence of a great prehistoric oil pool in Edmonson, Grayson and Hart counties. The same conditions obtain, though on a lesser scale, in the rock asphalt regions of Carter, Rowan, Elliott, Morgan, and Johnson counties. Black tarry oil is now procurable from shallow Pottsville sands in Magoffin County and elsewhere in eastern Kentucky. The prolific petroleum "sand" of Crawford, Lawrence and Wabash counties, Ill., and Gibson County, Ind., located on the La Salle anticline directly north of Henderson and Union counties, Ky., are Pottsville age and suggest interesting oil producing possibilities for this large undrilled area just south of the Ohio River.

The oil and gas sands of the Pottsville, while varying somewhat in

---

*The Howard Purchase No. 1, drilled by Louis H. Gormley at the mouth of Salt Lick, Floyd County, in 1891, was the first well in the Beaver Creek field. It has produced oil continuously from that date to the present, and though now reduced in volume is still commercially important.

thickness, are ample, ranging generally from 50 to 200 feet. The Potts-ville conglomerate (basal Pennsylvanian) is much thicker and ranges from a little less than 50 feet in Carter County to about 1,000 feet at the Breaks of Sandy. The sands of Pottsville age are crystal white and angular. Frequently they are loosely and not uniformly cemented, giv-ing rise to irregularity of drilling, production and surface at the out-crop. The oil is of Somerset grade, and dark green in color. In western Kentucky the Pottsville formation reaches a maximum of about 600 feet.

The Sebree Sand.

Unrecognized as a producer of commercial oil and gas before the spring of 1922, the Sebree (Allegheny-Pennsylvanian) sandstone can truthfully be said to offer a large new field for exploration in western Kentucky. The Sebree "sand" is the lowest division of the Allegheny (composed of Carbondale and Mulford), and may be seen at the type outcrop in the range of hills north of the Steamport Ferry road running east of Sebree, Webster County. The Sebree sandstone is about 50 feet in thickness, somewhat massive in appearance, frequently crossbedded, and coarse grained. It is in places somewhat irregularly cemented.

Oil of good quality and in commercial quantity has been secured from the Sebree sandstone in the George Proctor well in Union County close to the Henderson County line just west of Corydon at a depth of 637 feet. Fifty-seven feet of sandstone was drilled, of which the bottom nine feet were "pay." The Sebree sand is known to contain adequate salt water, and favorable structure is assumed to exist with some degree of certainty. The coarse texture and thick shales surrounding the sand indicate for it a most interesting productive future.

Other Possible Sands.

Higher in the geologic time scale than the sediments of Pennsylvanian age, and consequently of more recent deposition, are the semi-consolidated and unconsolidated sands, gravels, shales, clays, marls, and chalky lime-stones of the Cretaceous and Quaternary Systems, which are found in the Purchase region of western Kentucky. Though some drilling has been done in this region during the last year, notably Fulton and Calloway counties, little has been found to indicate extensive oil and gas producing sands in this area and in these higher geologic divisions. This region has been the subject of a recent report presenting all the known data,* and while the conclusions thus reached are not too hopeful, it must be admitted that much is yet to be learned concerning the oil and gas pro-ducing possibilities of the loose or semiconsolidated sediments of this region. In the southwestern part of the Purchase Region these sediments attain a thickness of about 2,000 feet, and afford an interesting field for oil and gas exploration.

---

*Oil and Gas Possibilities of "the Jackson Purchase" Region. W. R. Jillson, Ky. Geol. Survey. Series Six, Vol. Six, pp. 191-220, 1921.

GEOLOGICAL SEQUENCE OF THE OIL AND GAS SANDS OF KENTUCKY.

(With General Lithology in Superimposed Order, and Thickness.)

(Paleozoic Sediments.)

System	Series or Formation	Sand	Lithology in Order	Thickness in Feet
Lower Pennsylvanian	Alleghany	"Sebree"	Massive sandstone	40—60
	Pottsville	"Beaver," "Horton," "Pike" in Floyd, Knott and Pike. "Wages," "Jones," "Epperson" in Knox "Williamsburg" in Whitley	Alternating sands and shales and coals with strong conglomerate base	60—1000
	MAJOR DISCONFORMITY			
	Mauch Chunk or Pennington (Eastern Kentucky)	"Maxton"	Red shale, Sandy shale, White sand, Shale, White sand, Calcareous-Shale, Bastard lime } E. Ky.	30—275
Upper Mississippian	Chester Group (Western Kentucky)	"Tar Springs" "Hardinsburg" "Cypress"	Sandstone, limestone and } W. Ky. thin shales	300—800 .
	Glen Dean, Gasper, and St. Genevieve	"Big Lime"	White limestone and some oolites	20—400 E. Ky.
			Tan sand lens	
	MINOR DISCONFORMITY			
	St. Louis	"Big Lime"	Fine gray white compact limestone, cherty	{ 300—650 W. Ky. 50—10 S. E. Ky.

DISCONFORMITY, EAST KENTUCKY

System	Series or Formation	Sand	Lithology in Order	Thickness in Feet
Lower Mississippian (Eastern Kentucky)	Waverly (Logan and Cuyhoga)	"Keener" "Big Injun" "Squaw" "Wier" "Berea"	Clastics—sandstones and shales in Eastern Kentucky	500 in N. E. 400—600 in E.
Lower Mississippian (Western Kentucky)	St. Louis-Warsaw	Warren County "Shallow"	Dark blue fine limestone	400—600 Warren
	New Providence	"Mt. Pisgah" "Beaver" "Otter" "Cooper" "Slickford" } Wayne	Blue-green shales and limestones in Western and Southern Kentucky	300—350 in S. 200 in S. E. 400 in W.
	Warsaw	"Amber Oil of Barren, Warren and Simpson		

DISCONFORMITY

Upper Devonian	Ohio or Chattanooga	"Black Shale—Strays"	Black, fissile Bituminous Fine shale	75—Southeast 240—Northeast 200—Southwest

DISCONFORMITY

Middle Devonian	Hamilton Onondaga	"Corniferous" "Irvine" "Ragland," or "Campton," etc.	Cement limestone West Kentucky only Cherty magnesian frequently porus limestone	9 0—24 0—45

System	Series or Formation	Sand	Lithology in Order	Thickness in Feet
MAJOR DISCONFORMITY				
Middle Silurian	Niagaran	"Niagaran" Warren County "Deep", "Olympia" and "Stanton"	Alternating thick shales and then sandy limestones	50—250 E. of Arch 50—200 W. of Arch
		"Clinton"	Light to dark blue to reddish sandy limestone	5—20
MINOR DISCONFORMITY				
Upper Ordovician	Cincinnatian	"Caney" "Upper Sunnybrook" Barren County "Deep", Cumberland "Shallow"	Limestone Blue shales Sandstone	450—700+or—
DISCONFORMITY				
Middle Ordovician	Champlainian	"Upper Trenton"— Lexington	Gray granular to Crystalline limestone	270
		"Lower Trenton"— High Bridge	Thick bedded and compact limestone	600+
MAJOR DISCONFORMITY				
Lower Ordovician	Canadian	"Magnesian", "Calciferous", and Lower Magnesian (Ind.)	Hard limestone Sandy limestone Magnesian limestone	500+
Upper Cambrian	Ozarkian	"Knox Dolomite"	Light and dark dolomitic limestones (all unexposed)	300 —2500

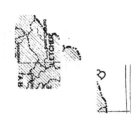

SKETCH MAP SHOWING THE AREAL GEOLOGY OF KENTUCKY

1. and 2. Ordovician
3. Silurian
4. Devonian
5. Mississippian
6. Pennsylvanian
7. Cretaceous
8. Quaternary
9. Recent

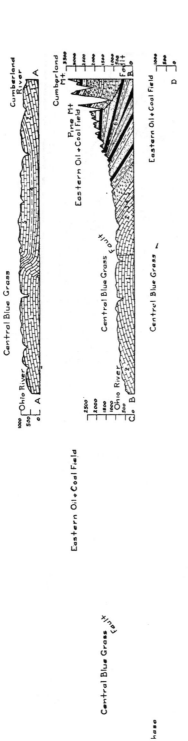

DIAGRAMATIC SECTIONS SHOWING THE STRUCTURAL GEOLOGY OF KENTUCKY.

The lettering of these sections corresponds to the lettering of the heavy lines on the opposite sketch map. The numbering of the formations in the sections corresponds to the numbering on the areal geologic map shown on the opposite page. These sections are all drawn to scale and are as accurate as the figures will allow.

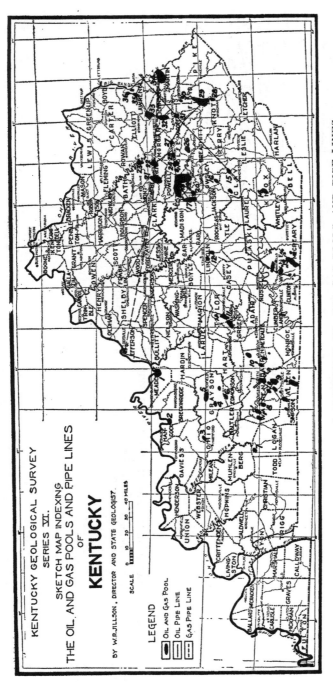

SKETCH MAP SHOWING KENTUCKY OIL AND GAS POOLS AND PIPE LINES.
(For list of pools see appendix.)

CHAPTER II.

ADAIR COUNTY.

Production: **Small oil and gas.** Producing Sands: **Unnamed, (Ordovician).**

Log. No. 1

J. S. Rector, No. 1, lessor. Roy Oil Co., lessee. Location: Darmon Creek Dome. Drilled 1920-21.

Strata.	Thickness	Depth
Mississippian & Devonian Systems.		
Limestone and shale (oil show 106, salt water 108)..	340	340
Ordovician System.		
Limestone, light blue, coarse	5	345
Limestone, fine, blue	35	380
Limestone, blue, coarse	20	400
Limestone, blue, very coarse	10	410
Limestone, blue, coarse	10	420
Limestone, blue, fine	10	430
Limestone, coarse, brown and blue...............	20	450
Limestone, blue, coarse, and black	20	470
Limestone, blue, coarse	10	480
Limestone, brown, coarse (oil show)	10	490
Limestone, blue, coarse	5	495
Limestone, blue, coarse	8	503
Limestone, brown, very hard	2	505
Total depth		505

NOTE—This well shows about the same relative distance between the oil sands as do the Creelsboro and Bakertown wells.

Log. No. 2

S. J. Roysè, No. 1, lessor. Palmer Oil and Gas Co., lessee. Begun about the base of the St. Louis.

Strata.	Thickness	Depth
Mississippian & Devonian Systems.		
Soil and clay	40	40
Limestone, dark, shaly (fresh water)	35	75
Limestone, darker, black shale	21	96
Silurian System.		
Limestone, black, fine (sulphur water)...........	19	115

Ordovician System.	Thickness	Depth
Limestone, black, coarse (salt water)	60	175
Limestone, light, coarse	35	210
Limestone, blue, very coarse	...	...
Incomplete depth		210

NOTE—Well not finished when samples of drilling supplied from which above record was made. Dec. 9, 1920.

Log. No. 3

Southern Oil and Refining Co., No. 1, lessee. Location: Dirigo P. O. Authority: L. Beckner.

Strata.	Thickness	Depth
Mississippian System.		
Limestone (oil show 102)	184	184
Shale, blue	100	284
Devonian System.		
Shale, black	40	324
Limestone (cap rock)	6	330
Limestone "sand" (oil)	15	345
Ordovician System.		
Limestone (oil show 560)	405	750
Total depth	...	750

ALLEN COUNTY.

Production: Oil and Gas. Producing Sands: Beaver (Mississippian); Corniferous (Devonian); and Niagaran (Silurian).

Log No. 4

Smith, No. 1. lessor. Location: Big Trammel Creek. Completed 1919.

Strata.	Thickness	Depth
Mississippian System.		
Limestone, black ....·............	196	196
Devonian System.		
Shale, brown ....·......	44	240
Limestone	12	252
Limestone, brown, first oil & gas	8	260
Limestone, white, broken up	28	288

Silurian and Ordovician Systems.

	Thickness	Depth
Shale, soft	132	420
Limestone, brown, "sand" oil	5	425
Shale, black, lime shells	45	470
Shale, blue	25	895
Shale, green	15	910
Limestone, hard & gray, oil & gas	26	936
Limestone, white, hard and sharp	21	957
Shale, blue	5	962
Limestone, red rock	16	978
Total depth		978

NOTE—The Silurian-Ordovician contact is about midway down in the 132 feet above 420 feet in depth.

Log No. 5

W. R. Cushenberry, No. 1, lessor. Commenced: June 4, 1919. Completed: June 14, 1919. Production: Dry. Authority: The Ohio Oil Company.

Strata.

Mississippian System.

	Thickness	Depth
Soil, red, soft	4	4
Limestone, blue, hard	89	93
Limestone, white, hard	16	109

Devonian System.

	Thickness	Depth
Shale, black, medium	34	143
Limestone, (cap rock), black, hard (salt water)	8	151
Limestone, "sand," dark, hard	15	166

Silurian System.

	Thickness	Depth
Limestone, dark, hard	22	188
Limestone, "sand," white, soft, (salt water)	20	208
Limestone, dark, hard	10	218
Limestone, white, soft (fine salt water)	18	236
Limestone, dark, hard, coarse	7	243
Limestone, white, hard	7	250
Total depth		250

Log No. 6

Deep test near Scottsville, Ky. Authority: Albert McGrain, Corydon, Indiana.

Strata.

Mississippian System.

	Thickness	Depth
Clay	3	3
Limestone, white	155	158

Devonian System.	Thickness	Depth
Shale, black (Chattanooga)	40	198
Limestone, rotten	798	. 996
Limestone, (salt water)	4	1,000
Shale, black	35	1,035
Shale, green	5	1,040
Limestone, rotten	40	1,080
Shale, green	2	1,082
Shale, brown	38	1,120
Shale, (Pencil Cave)	3	1,123
Limestone, brown	77	1,200
Limestone, white	80	1,280
Limestone, brown	224	1,504
Limestone, (salt water)	74	1,578
Limestone, brown	8	1,586
Limestone, black	42	1,628
Limestone, dark ......;......................	8	1,636
Limestone, brown (heavy salt water at 1,860..	154	1,790
Limestone, brown	82	1,872
Limestone, gray	8	1,880
Limestone, gray	30	1,910
Limestone, gray	5	1,915
Limestone, gray	15	1,930
Limestone, gray	10	1,940
Limestone, gray	5	1,945
Limestone, gray	5	1,950
Total depth		1,950

NOTE—The base of the Devonian and the top of the Silurian, the base of the Silurian and the top of the Ordovician are all contained within the 798 feet above 996 feet in depth. This record was not kept in detail.

Log No. 7.

Alfred Landers, No. 2, lessor. Commenced: June 25, 1920. Completed: July 15, 1920. Production: Dry. Authority: The Kenco Oil Company.

Strata.

Mississippian System.	Thickness	Depth
Clay, red	25	25
Limestone boulders ....(	15	40
Limestone, black (water)	50	90
Limestone, gray	10	100
Flint, blue, white	60	160
Limestone, white	60	220
Shale, green (New Providence)	40	260

Devonian System.	Thickness	Depth
Shale, black (Chattanooga)	48	308
Limestone, (cap rock)	2	310
Limestone, ''sand''	4	314
Limestone, ''sand'' (salt water)	2	316
Total depth		316

Log No. 8

E. Agee, No. 2, lessor. Location: near Allen Springs, Ky. Commenced: June 12, 1920. Completed: July 2, 1920. Authority: The Kenco Oil Company.

Strata.

Mississipian System.	Thickness	Depth
Soil, soft	40	40
Limestone	314	354
Devonian System.		
Shale, black (Chattanooga)	51	405
Limestone, gray, hard	4	409
Limestone, ''sand'' brown	4	413
Limestone, gray, soft	13	426
Limestone, brown, hard	25	451
Limestone, black, sandy, hard	10	461
Total depth		461

Log No. 9

Robert Mitchell, No. 1, lessor. Completed: June 26, 1919. Authority: The Kenco Oil Company.

Strata.

Mississippian System.	Thickness	Depth
Clay, red	29	29
Limestone, white	176	205
Limestone, sound	5	210
Limestone, blue	30	240
Shale, green	5	245
Devonian System.		
Shale, black (Chattanooga)	45	290
Limestone (cap rock)	6	296
Limestone, ''sand,'' (oil)	17	313
Silurian System.		
Shale, and limestone	35	348
Limestone, ''sand,'' (oil)	7	355
Shale	40	395
Limestone, ''sand,'' (oil)	10	405
Ordovician System.		
Limestone	32	437
Total depth		437

Log No. 10

Robert Mitchell, No. 2, lessor. Location: near Oak Hill, 4th Dist. Completed: in 1919. Authority: The Kenco Oil Company.

Strata.

Mississippian System.	Thickness	Depth
Clay, red	29	29
Limestone, gray	226	255
Devonian System.		
Shale, black (Chattanooga)	50	305
Limestone (cap rock)	5	310
Limestone and shale	127	437
Total depth		437

NOTE—The Devonian-Silurian-Ordovician contacts are included within the last 127 feet.

Log No. 11

Robert Mitchell, No. 3, lessor. Location: Oak Hill, 4th District. Commenced: October 4, 1919. Authority: The Kenco Oil Company.

Strata.

Mississippian System.	Thickness	Depth
Clay	5	5
Limestone	195	200
Shale, green	5	205
Devonian System.		
Shale, black	57	262
Limestone (cap rock), gas	4	266
Limestone, "sand" (oil)	2	268
Total depth		268

Log No. 12

Robert Mitchell, No. 4, lessor. Commenced: October 6, 1919. Completed: October 18, 1919. Authority: The Kenco Oil Company.

Strata.

Mississippian System.	Thickness	Depth
Clay	18	18
Limestone	42	60
Shale, hard	5	65
Limestone	130	195
Shale, green	5	200

Devonian System.   Thickness Depth

	Thickness	Depth
Shale, black (Chattanooga)	53	253
Limestone (cap rock)	4	257
Limestone, "sand,"	15	272
Total depth		272

Show of oil between 263 and 267 feet.
Salt water at 263 feet.

Log No. 13

Robert Mitchell, No. 5, lessor. Commenced: November 6, 1919.
Authority: The Kenco Oil Company.

Strata.

Mississippian System.   Thickness Depth

	Thickness	Depth
Clay	5	5
Limestone (water at 70 feet)	65	70
Shale, hard	30	100
Limestone	118	218

Devonian System.

	Thickness	Depth
Shale, black	52	270
Limestone (cap rock)	5	275
Limestone, "sand,"	5	280
Limestone (break)	4	284
Limestone, "sand,"	10	294
Total depth		294

Log No. 14

Robert Mitchell, No. 13, lessor. Production: Dry; casing pulled.
Abandoned: May 10, 1920. Authority: The Kenco Oil Company.

Strata.

Mississippian System.   Thickness Depth

	Thickness	Depth
Limestone	240	240

Devonian System.

	Thickness	Depth
Shale, black	69	309
Limestone (cap rock)	5	314
Limestone, "sand,"	4	318
Limestone, gray	35	353
Limestone, "sand," (salt water)	4	357
Total depth		357

Log No. 15

'Robert Mitchell, No. 16, lessor.   Completed: in 1920.   Authority: The Kenco Oil Company.

Strata.

Mississippian System.	Thickness	Depth
Clay, red	30	30
Limestone, gray	112	142
Limestone, white	65	207
Shale, green (New Providence)	56	263

Devonian System.		
Shale, black (Chattanooga)	54	317
Limestone (cap rock)	1½	318½
Limestone, "sand,"	4	322½
Shale, gray	29½	352
Limestone	4	356
Limestone, "sand" (salt water)	6	362
Total depth		362

NOTE—The second or deep "sand" in the Mitchell wells is probably Silurian.

Log No. 16

Fowler Mitchell, No. 1, lessor.   Commenced: July 15, 1919.   Completed: July 31, 1919.   Authority: The Kenco Oil Company.

Strata.

Mississippian System.	Thickness	Depth
Clay	13	13
Limestone	67	80
Shale, hard	5	85
Limestone	180	265

Devonian System.		
Shale	43	308
Limestone (cap rock)	2	310
Limestone and shale	30	340
Silurian System.		
Limestone "sand"	35	375
Total depth		375

Log No. 17

Simpson Long, No. 1, lessor. Completed: July 11, 1919. Authority: The Kenco Oil Company.

Strata.

Mississippian System.	Thickness	Depth
Clay, red	19	19
Limestone, gray	138	157
Shale and limestone	200	357

Devonian System.		
Shale, black (Chattanooga)	48	405
Limestone, (cap rock)	5	410
Limestone	15	425
Shale, hard, and limestone	114	539
Total depth		539

Log No. 18

Simpson Long, No. 2, lessor. Commenced: July 14, 1919. Completed: Aug. 8, 1919. Authority: The Kenco Oil Company.

Strata.

Mississippian System.	Thickness	Depth
Clay	19	19
Limestone and shale, hard	331	350

Devonian System.		
Shale, black	54	404
Limestone (cap rock)	2	406
Limestone, ''sand,''	5	411
Limestone and shale, hard	39	450
Limestone, ''sand,''	10	460
Limestone and shale, hard	81	541
Total depth		541

Fresh water at 80 feet.
Salt water at 420 feet.

Log No. 19

N. L. Hinton, No. 1, lessor. Commenced: Feb. 25, 1919. Completed: March 28, 1919. Authority: The Kenco Oil Company.

Strata.

Mississippian System.	Thickness	Depth
Clay	20	20
Limestone	10	30
Limestone, black	50	80
Limestone, blue	20	100
Limestone, white	70	170
Limestone and flint	20	190
Limestone, white	105	295
Shale, green	10	305
Devonian System.		
Shale, black	50	355
Limestone (oil sand)	10	365
Limestone and shale, hard	39	404
Total depth		404

Salt water at 404 feet.

Log No. 20.

N. L. Hinton, No. 2, lessor. Commenced: April 9, 1919. Completed: April 25, 1919. Authority: The Kenco Oil Company.

Strata.

Mississippian System.	Thickness	Depth
Limestone	214	214
Devonian System.		
Shale, black (Chattanooga)	38	252
Limestone (cap rock)	4	256
Limestone, "sand" (oil and gas from 260 to 264)	23	279
Shale, hard, and limestone	36	315
Sand	15	330
Limestone	20	350
Shale, hard	10	360
Limestone	30	390
Shale, hard	10	400
Limestone	1	401
Total depth		401

Fresh water at 40 feet.

Salt water at 320 feet.

NOTE—The base of the Devonian and the top of the Silurian is probably included within the 36 feet above 315 feet depth. These records show a difference of about 100 feet in the depths of the base of the Devonian shale in Hinton No. 1 and No. 2.

Log No. 21

W. H. Williams, No. 1, lessor. Commenced: July 18, 1919. Completed: July 30, 1919. Authority: The Kenco Oil Company.

Strata.

Mississippian System.	Thickness	Depth
Clay	10	1;0
Limestone	10	20
Shale, hard	2	22
Limestone	59	81
Limestone, gray	9	90
Limestone	95	185
Devonian System.		
Shale, black (Chattanooga)	49	234
Limestone (cap rock)	6	240
Limestone, oil "sand"	14	254
Shale, hard	106	360
Total depth		360

Fresh water at 22 feet.
Salt water at 265 feet.

Log No. 22

W. H. Williams, No. 2, lessor. Commenced: August 4, 1919. Completed: August 21, 1919. Authority: The Kenco Oil Company.

Strata.

Mississippian System.	Thickness	Depth
Clay	12	12
Limestone	158	170
Shale, hard	5	175
Limestone	80	255
Devonian System.		
Shale, black (Chattanooga)	50	305
Limestone (cap rock)	8	313
Limestone, soft	12	325
Limestone, sandy	30	355
Shale, hard	5	360
Limestone, sandy	5	365
Total depth		365

Fresh water at 87 feet.
Sulphur water 170 to 175 feet.

Log No. 23

J. R. Williams, No. 1, lessor. Commenced: July 18, 1919. Completed: August 30, 1919. Authority: The Kenco Oil Company.

Strata.

Mississippian System.	Thickness	Depth
Clay	12	12
Limestone and shale, hard	28	40
Limestone, sandy	140	180
Limestone (break)	2	182
Limestone	38	220
Limestone (break)	5	225
Limestone	61	286
Devonian System.		
Shale, black	55	341
Limestone, (cap rock)	1	342
Limestone, ''sand'' (oil at 342)	14	356
Limestone	5	361
Total depth		361

Fresh water at 60 to 65 feet.

Sulphur water from 180 to 182, and from 220 to 225 feet.

Salt water at 361 feet.

Log No. 24

Allen Lease, No. 20, lessor. Commenced: June 20, 1920. Completed: July 15, 1920. Production: 6 barrels of oil in the first 24 hours. Authority: The Kenco Oil Company.

Strata.

Mississippian System.	Thickness	Depth
Clay, red	20	20
Limestone, blue	90	110
Limestone, white	174	284
Shale, green	4	288
Devonian System.		
Shale, black (Chattanooga)	47	335
Limestone (cap rock), oil	3	338
Limestone, ''sand,'' brown	10	348
Total depth		348

Log No. 25

Effie Buchannon, No. 2, lessor. Drilled: in 1919. Authority: The Kenco Oil Company.

Strata.

Mississippian System.	Thickness	Depth
Clay	24	24
Limestone	251	275
Devonian System.		
Shale, black (Chattanooga)	50	325
Limestone (cap rock)	3	328
Limestone, "sand"	20	348
Limestone	22	370
Total depth		370

Log No. 26

G. D. Pruitt, No. 1, lessor. Authority: The Kenco Oil Company.

Strata.

Mississippian System.	Thickness	Depth
Mud, red	15	15
Limestone, gray	390	409
Shale, green	6	415
Devonian System.		
Shale, black (Chattanooga)	45	460
Limestone, brown	20	480
Limestone, blue	40	520
Limestone, white	110	630
Limestone, oil "sand,"	10	640
Shale and limestone	160	800
Total depth		800

NOTE—The base of the Devonian and the top of the Silurian is within the 40 feet above 520 in depth.

Log No. 27

Widow Lizzie Jewell, No. 4, lessor. Commenced: April, 1919. Completed: June 5, 1919. Authority: The Big Dipper Oil Company.

Strata.

Mississippian System.	Thickness	Depth
Limestone	175	175
Devonian System.		
Shale, black (Chattanooga)	60	235
Limestone, "sand," first	12	247
Limestone	42	289
Limestone, "sand," second	9	298
Limestone	11	309
Total depth		309

Log No. 28

Mrs. Lizzie Jewell, No. 5, lessor. Commenced: July 25, 1919. Completed: August 25, 1919. Authority: The Big Dipper Oil Company.

Strata.

Mississippian System.	Thickness	Depth
Limestone	195	195

Devonian System.		
Shale, black (Chattanooga)	48	243
Limestone	10	253
Limestone, "sand," first	9	262
Limestone	43	305
Limestone, "sand," second	9	314
Limestone	13	327
Total depth		327

Log No. 29

Mrs. Lizzie Jewell, No. 7, lessor. Commenced: October 23, 1919. Production: Dry. Authority: The Big Dipper Oil Company.

Strata.

Mississippian System.	Thickness	Depth
Limestone	232	232

Devonian System.		
Shale, black (Chattanooga)	56	288
Limestone	6	294
Limestone, "sand,"	9	303
Limestone	27	330
Total depth		330

Log No. 30

Gainesville Pool, No. 6. Drilled in 1920. Production: 12—15 bbls. daily. Authority: The Bowling Green Gas, Oil and Refining Co.

Strata.

Mississippian System.	Thickness	Depth
Clay, red	15	15
Limestone boulders	5	20
Limestone	30	50
Limestone, white	45	95
Limestone, gray	45	140
Flint, blue, soft	55	195
Limestone, blue	10	205

Devonian System.   Thickness Depth

    Shale, black (Chattanooga)   50   255
    Limestone (cap rock)   5   260
    Limestone, oil ''sand,''   15   275
    Limestone   20   295
        Total depth   295

Log No. 31

Gainesville Pool, No. 8. Drilled in 1920. Well abandoned, casing pulled. Authority: The Bowling Green Gas, Oil and Refining Co.

    Strata.

Mississippian System.   Thickness Depth

    Soil, limestone boulders, red, soft   20   20
    Limestone, blue, hard (little oil)   30   50
    Limestone, gray, hard   10   60
    Limestone, dark, soft   15   75
    Shale   5   80
    Limestone, white, hard   10   90
    Shale, blue, soft (cave from 103 to 105)   20   110
    Limestone, white, hard   40   150
    Limestone, dark blue, hard   5   155
    Limestone, white, hard   5   160
    Limestone and flint   15   175
    Limestone and flint, blue, white, hard   50   225
    Limestone (cap rock), blue, green, soft   15   240

Devonian System.

    Shale, black, soft (Chattanooga)   50   290
    Limestone (cap rock), gray, hard   3   293
        Total depth   293

Log No. 32

Gainesville Pool, No. 7. Drilled in 1920. Casing pulled, well abandoned. Authority: The Bowling Green Gas, Oil and Refining Co.

    Strata.

Mississippian System.   Thickness Depth

    Soil   3   3
    Limestone, dark (salt water at 55)   47   50
    Cotton Rock, dark blue, soft   10   60
    Limestone, white   120   180
    Limestone, green, soft   20   200
    Limestone, soft   5   205

Devonian System.	Thickness	Depth
Shale, black (Chattanooga)	50	255
Limestone (cap rock), very hard	4½	259½
Limestone, some oil	3½	263
Limestone, oil "sand,"	12	275
Limestone, dark	60	335
Limestone, oil "sand," light	15	350
Limestone, light	15	365
Limestone, dark	10	375
Limestone, white	20	395
Limestone, gray	5	400
Sandstone and limestone, light	2	402
Total depth		402

NOTE—The base of the Devonian and the top of the Silurian is within the 60 feet above 335 feet depth.

Log No. 33

Elizabeth Jewell, No. 1, lessor. Production: 15 bbls. oil.

Strata.

Mississippian System.	Thickness	Depth
Limestone	184	184

Devonian System.		
Shale, black (Chattanooga)	54	238
T. 1st pay	2	240
B. 1st pay	10	250
T. 2nd pay	..	...
B. 2nd pay		

Log No. 34

Elizabeth Jewell, No. 2, lessor. Production: oil.

Strata.

Mississippian System.	Thickness	Depth
Limestone	204	204

Devonian System.		
Shale, black (Chattanooga)	54	258
T. 1st pay	10	268
B. 1st pay	10	278
T. 2nd pay	..	...
B. 2nd pay	..	...

Log No. 35

Elizabeth Jewell, No. 3, lessor.   Production: oil.

Strata.

Mississippian System.	Thickness	Depth
Limestone	201	201

Devonian System.

	Thickness	Depth
Shale, black (Chattanooga)	63	264
T. 1st pay	3	267
B. 1st pay	10	277
T. 2nd pay	..	...
B. 2nd pay		

Log No. 36

Elizabeth Jewell, No. 4, lessor.   Production: oil.

Strata.

Mississippian System.	Thickness	Depth
Limestone	175	175

Devonian System.

	Thickness	Depth
Shale, black (Chattanooga)	60	235
T. 1st pay	2	237
B. 1st pay	10	247
T. 2nd pay	42	289
B. 2nd pay	9	298
Total depth		298

Log No. 37

George Jewell, No. 1, lessor.   Production: oil.

Strata.

Mississippian System.	Thickness	Depth
Limestone	187	187

Devonian System.

	Thickness	Depth
Shale, black (Chattanooga)	56	243
T. 1st pay	35	278
B. 1st pay	11	289
T. 2nd pay	16	305
B. 2nd pay	3	308
Total depth		308

Log No. 38

George Jewell, No. 2, lessor.   Production: Dry.

Strata.

Mississippian System.	Thickness	Depth
Limestone	187	187

Devonian System.

	Thickness	Depth
Shale, black (Chattanooga)	56	243
T. 1st pay	35	278
B. 1st pay	11	289
T. 2nd pay	16	305
B. 2nd pay	3	308
Total depth		308

Log No. 39

George Jewell, No. 3, lessor.   Production: oil.

Strata.

Mississippian System.	Thickness	Depth
Limestone	197	197

Devonian System.

	Thickness	Depth
Shale, black (Chattanooga)	57	254
T. 1st pay	20	274
B. 1st pay	10	284
T. 2nd pay	..	...
B. 2nd pay	..	...

Log No. 40

George Jewell, No. 4, lessor.   Production: oil.

Strata.

Mississippian System.	Thickness	Depth
Limestone	179	179

Devonian System.

	Thickness	Depth
Shale, black (Chattanooga)	54	233
T. 1st pay	37	270
B. 1st pay	3	273
T. 2nd pay	7	280
B. 2nd pay	10	290
Total depth		290

Log No. 41

Cook, No. 1, lessor. Location: Sledge Pool. Commenced: June 26, 1920. Completed: July 27, 1920.

Strata.

Mississippian System.	Thickness	Depth
Limestone	336	336
Devonian & Silurian Systems.		
Shale, black (Chattanooga)	47	383
Limestone, (salt water 548-58)	175	558
Total depth		558

Log No. 42

O. J. McDonald, lessor. Location: Sledge Pool. Commenced: Sept. 14, 1920. Completed: Sept. 25, 1920. Production: Dry.

Strata.

Mississippian System.	Thickness	Depth
Limestone	335	335
Devonian and Silurian Systems.		
Shale, black (Chattanooga)	50	385
Limestone	89½	474½
Total depth		474½

Log No. 43

Sol Williams, lessor. Location: Sledge Pool. Production: Dry; abandoned.

Strata.

Mississippian System.	Thickness	Depth
Limestone	357	357
Devonian & Silurian Systems.		
Shale, black (Chattanooga)	47	404
Limestone	201	605
Total depth		605

Log No. 44

Virgil Pruitt, No. 1, lessor. Location: Sledge Pool. Commenced: July 15, 1920. Completed: July 27, 1920.

Strata.

	Thickness	Depth
Mississippian System.		
Limestone	331	331
Devonian System.		
Shale, black (Chattanooga)	51½	382½
Limestone, (oil "sand")	4½	387
Total depth		387

Log No. 45

Virgil Pruitt, No. 2, lessor. Location: Sledge Pool. Commenced : July 31, 1920. Completed: Aug. 21, 1920, and shot at 403 with 20 qts.

Strata.

	Thickness	Depth
Mississippian System.		
Limestone	323	323
Shale, black (Chattanooga)	51	374
Limestone	3	377
Limestone, (1st "sand")	5	382
Limestone	16	398
Limestone, (2nd "sand")	5	403
Limestone	5½	408½
Total depth		408½

Log No. 46

Virgil Pruitt, No. 3, lessor. Location: Sledge Pool. Commenced: Aug. 7, 1920. Completed: Aug. 21, 1920.

Strata.

	Thickness	Depth
Mississippian System.		
Limestone	373	373
Devonian System.		
Shale, black (Chattanooga)	50	423
Limestone	2	425
Limestone, (1st "sand")	7	432
Limestone	13	445
Limestone, (2nd "sand")	5	450
Limestone	2	452
Total depth		452

A WARREN COUNTY GUSHER

This well is on the Briggs lease, Little Briggs pool,
Kerstetter, et al. operators, flowing from 6¼ inch casing.

Log No. 47

Virgil Pruitt, No. 4, lessor. Location: Sledge Pool. Commenced: Aug. 27, 1920. Completed: Sept. 20, 1920. Production: Dry.

Strata.

Mississippian System.	Thickness	Depth
Limestone	333	333
Devonian & Silurian Systems.		
Shale, black (Chattanooga)	47	380
Limestone	165	545
Total depth		545

Log No. 48

Virgil Pruitt, No. 5, lessor. Location: Sledge Pool. Commenced: Aug. 27, 1920. Completed: Sept. 25, 1920, and shot at 406 feet. Production: Dry.

Strata.

Mississippian System.	Thickness	Depth
Limestone	359	359
Devonian System.		
Shale, black (Chattanooga)	47	406
Limestone ''sand,''	42½	448½
Total depth		448½

Log No. 49

Virgil Pruitt, No. 6, lessor. Location: Sledge Pool. Commenced: Sept. 29, 1920. Completed: Oct. 23, 1920, and shot 428-438 feet. Production: 2d day, 4 bbls. oil.

Strata.

Mississippian System.	Thickness	Depth
Limestone	374½	374½
Devonian System.		
Shale, black (Chattanooga)	50	424½
Limestone, (oil 428—438)	13½	438
Total depth		438

Log No. 50

Virgil Pruitt, No. 7, lessor. Location: Sledge Pool. Commenced Oct. 2, 1920. Completed: Oct. 18, 1920. Production: produced first 24 hours, 22 bbls.; produced second 24 hours, 50 bbls.

Strata.

Mississippian System.	Thickness	Depth
Limestone	361	361
Devonian System.		
Shale, black (Chattanooga)	48	409
Limestone, (oil 413)	35	444
Total depth		444

Log No. 51

Virgil Pruitt, No. 8, lessor. Location: Sledge Pool. Commenced: Oct. 21, 1921. Completed: Nov. 10, 1921. Production: Dry.

Strata.

Mississippian System.	Thickness	Depth
Limestone	341	341
Devonian System.		
Shale, black (Chattanooga)	50	391
Limestone	41	432
Total depth		432

Log No. 52

Virgil Pruitt, No. 1, (Ten-Acre), lessor. Location: Sledge Pool. Drilled in 1920.

Strata.

Mississippian System.	Thickness	Depth
Limestone	440	440
Devonian System.		
Shale, black (Chattanooga)	50	490
Limestone ("cap")	5	495
Limestone (oil "sand,")	17	512
Limestone	9	521
Total depth		521

Log No. 53

Virgil Pruitt, No. 2, (Ten-Acre), lessor.   Location: Sledge Pool.
Drilled in 1920.
    Strata.
Mississippian System.   Thickness Depth
    Limestone   445   445
Devonian System.
    Shale, black (Chattanooga)   50   495
    Limestone ("cap")   5   500
    Limestone (oil "sand,")   17   517
    Limestone   13   530
        Total depth   530

Log No. 54

Bourbon Stamps, No. 2, lessor.   Elevation: 730.
    Strata.
Mississippian System.   Thickness Depth
    Soil and limestone   324   324
Devonian System.
    Shale, black (Chattanooga)   45   369
    Limestone ("cap")   5   374
    Limestone "sand"   5   379
    Limestone   5   384
    Limestone "sand"   12   396
    Limestone, blue   20   416
Silurian System.
    Limestone "sand,"   20   436
        Total depth   436

Log No. 55

W. M. Newman, No. 1, lessor.   Location: near Bourbon Stamps
lease.   Production: oil show only.
    Strata.
Mississippian System.   Thickness Depth
    Soil and limestone   311   311
Devonian System.
    Shale, black (Chattanooga)   41   352
    Limestone, blue   4   356
    Limestone (cap)   5   361
    Limestone "sand,"   18   379
        Total depth   379

Log No. 56

A. J. Wheat, No. 1, lessor. Location: adjoins W. M. Newman, No. 1. Production: Dry.

Record similar and sand in same place as in W. M. Newman, No. 1.

Log No. 57

Gerard, No. 1, lessor. Location: near Bourbon Stamps lease. Completed: July 31, 1907. Production: Dry.

Strata.

	Thickness	Depth
Mississippian System.		
Limestone	480	480
Devonian & Silurian Systems.		
Shale, black (Chattanooga)	45	525
Limestone	15	540
Limestone	270	810
Total depth		810

Log No. 58

Gerard, No. 2, lessor. Location: near Bourbon Stamps lease. Commenced: Dec. 3, 1907. Completed: Dec. 20, 1907. Production: gas at 310 and 440.

Strata.

	Thickness	Depth
Mississippian System.		
Limestone	515	515
Devonian System.		
Shale, black (Chattanooga)	45	560
Limestone	35	595
Total depth		595

Log No. 59

WHEAT POOL.

A. W. Stamp, No. 1, lessor. Production: Dry; abandoned Aug. 7, 1919.

Strata.

	Thickness	Depth
Mississippian and Devonian Systems.		
Soil	6	6
Limestone and shale	215	221
Total depth		221

Log No. 60

No. 2.

Commenced: March 29, 1919. Contractor: E. A. Dyer.

Strata.

Mississippian System.	Thickness	Depth
Limestone	55	55
Shale, green (New Providence)	2	57
Devonian System.		
Shale, black (Chattanooga)	48	105
Limestone (cap rock)	3	108
Limestone ''sand,'' (oil)	18	126
Limestone	$21\frac{1}{2}$	$128\frac{1}{2}$
Total depth		$128\frac{1}{2}$

Log No. 61

No. 3.

Strata.

Mississippian System.	Thickness	Depth
Soil	6	6
Limestone	50	56
Devonian System.		
Shale, black (Chattanooga)	63	119
Limestone, (cap rock)	1	120
Limestone ''sand,'' (oil)	14	134
Limestone	1	135
Total depth		135

Log No. 62

No. 4.

Location: at power house.

Strata.

Mississippian System.	Thickness	Depth
Soil	10	10
Limestone, gray	50	60
Devonian System.		
Shale, black (Chattanooga)	$48\frac{1}{2}$	$108\frac{1}{2}$
Limestone (cap rock)	$7\frac{1}{2}$	116
Limestone ''sand,'' (oil)	16	132
Limestone, gray	1	133
Total depth		133
Well shot April 1, 1920, 10 qts.		

Log No. 63

No. 5.

Location: On hill above powerhouse. Casing Record: 110 feet of casing.

Strata.

Mississippian System.	Thickness	Depth
Soil	8	8
Limestone	98	106
Devonian System.		
Shale, black (Chattanooga)	48	154
Limestone (cap rock)	7	161
Limestone "sand," (oil)	27	188
Limestone, gray	2	190
Total depth		190

Log No. 64

No. 6.

Location: Near Hinton lease.

Strata.

Mississippian System.	Thickness	Depth
Soil	6	6
Limestone, gray	98	104
Devonian System.		
Shale, black (Chattanooga)	55	159
Limestone "sand," (oil)	17	176
Limestone	2	178
Total depth		178

Log No. 65

No. 7.

Location: On hill towards house. Production: Strong flow of gas and oil.

Strata.

Mississippian System.	Thickness	Depth
Limestone, mixed	138	138
Limestone, white	10	148
Flint, blue, (New Providence)	12	160
Shale, green, (New Providence)	3	163
Devonian System.		
Shale, black (Chattanooga)	46	209
Limestone (cap rock)	4	213
Limestone "sand," (oil)	29	242
Limestone	1	243
Total depth		243

Log No. 66

A. Watkins, No. 1, lessor.   Location: Across branch from Stamp lease.

Strata.

Mississippian System.	Thickness	Depth
Limestone, mixed	45	45
Devonian System.		
Shale, black (Chattanooga)	51	96
Limestone (cap rock)	8	104
Limestone, ''sand,'' (oil)	15	119
Limestone, gray	2	121
Total depth		121

Log No. 67

Starks Well, No. 1, lessor.

Strata.

Mississippian System. ·	Thickness	Depth
Soil and subsoil	35	35
Limestone and flint	115	150
Devonian System.		
Shale, black (Chattanooga) ...:.............	48	198
Limestone (cap rock)	7	205
Limestone, gray, oil, ''sand''	10	215
Silurian System.		
Limestone, dark, rotten	30	245
Limestone, gray and sand	15	260
Limestone ''sand,'' (salt water)	5	265
Limestone ''sand,'' (oil)	3	268
Limestone, gray	2	270
Total depth		270

Log No. 68

Price, No. 1, lessor.   Location: Near mouth of Johns Creek. Drilled: in October, 1920.   Contractors: Brown Bros.   Elevation: 646 A. T.

Strata.

Mississippian System.	Thickness	Depth
Soil and limestone	242	242
Devonian System.		
Shale, black (Chattanooga)	52	294
Limestone, light, soft, brown	4	298
Limestone, harder, browner	4	302
Limestone, harder, lighter	8	310
Limestone, lighter, softer	4	314

Silurian System.   Thickness Depth
    Limestone, blue, harder 4   318
    Shale, blue, very soft 45   363
    Shale, blue, harder 4   367
    Limestone, brown, harder, crys. 4   371
    Limestone, blue, very soft, shaly 20   391
        Incomplete depth 391

Log No. 69

Payne, No. 1, lessor. Location: Between Harmony School and Mt. Aerial. Drilled in 1920. Operator: Stuart St. Clair, geologist. Elevation: 609 A. T. Authority: Stuart St. Clair, geologist.

    Strata.
Mississippian System.   Thickness Depth
    Soil and limestone shale 232   232

Devonian System.
    Shale, black (Chattanooga) 52   284
    Limestone 93   377
        Total depth 377

NOTE—Gas was found above the shale, which caught fire from a careless visitor and burned the rig. No other show. The lower part of the last 93 feet of limestone is Silurian.

Log No. 70

Bourbon Stamps, No. 1, lessor. Location: Near Harmony School. Production: Flowed oil. Elevation: 744 A. T.

    Strata.
Mississippian System.   Thickness Depth
    Soil and limestone 322   322

Devonian System.
    Shale, black (Chattanooga) 45   367
    Limestone (cap rock) 5   372
    Limestone "sand" 12   384

Silurian System.
    Limestone, blue 35   419
    Limestone "sand" 20   439
        Total depth 439

Log No. 71

Price Turner, lessor. Authority: Mr. De Caigny and J. H. Mc-Clurkin. Well No. 1. Elevation: 851.6 A. T.

Strata.

Mississippian System.	Thickness	Depth
Soil	32	32
Limestone, (water)	60	92
Limestone	180	272
Devonian System.		
Shale, black (Chattanooga)	44	316
Limestone (cap)	28	344
Limestone ''sand,'' (oil)	22	366
Limestone	9	375
Total depth		375

Log No. 72

Well No. 2. Elevation: 848.2 A. T.

Strata.

Mississippian System.	Thickness	Depth
Soil	30	30
Limestone, (water)	62	92
Limestone	176	268
Devonian System.		
Shale, black (Chattanooga)	48	316
Limestone (cap)	32	348
Limestone ''sand,'' (oil)	6	354
Limestone	14	368
Total depth		368

Log No. 73

Well No. 3. Elevation: 841.9 A. T.

Strata.

Mississippian System.	Thickness	Depth
Soil	19	19
Limestone, (water)	61	80
Limestone	160	240
Devonian System.		
Shale, black (Chattanooga)	48	288
Limestone (cap)	44	332
Limestone ''sand,'' (oil)	18	350
Limestone	9	359
Total depth		359

Log No. 74

Well No. 4.   Elevation: 809.4 A. T.

Strata.

Mississippian System.	Thickness	Depth
Soil	12	12
Limestone, (little water)	64	76
Limestone	142	218
Devonian System.		
Shale, black (Chattanooga)	43	261
Limestone (cap)	31	292
Limestone ''sand,'' (oil)	16	308
Limestone	57	365
Total depth		365

Log No. 75

Fed Shields, No. 1, lessor.   Seaboard Oil Co., lessee.   Completed and abandoned Feb. 19, 1921.   Authority: Seaboard Oil Co.

Strata.

Mississippian System.	Thickness	Depth
Soil and subsoils	10	10
Limestone, gray	40	50
Limestone, gray, and flint	120	170
Devonian System.		
Shale, black (Chattanooga)	45	215
Limestone (cap rock)	10	225
Limestone, blue	50	275
Shale or sugar lime or blue ''gumbo''	100	375
Shale, harder, blue	25	400
Limestone (salt water, heavy flow), white sands	10	410
Total depth		410

NOTE—From 50 to 170 feet all showed white flint, more or less gray and grayish lime mixed with green. Set casing 147, all pulled and well plugged with two plugs. Left 10 ft. 8¼ in. casing in hole; could not pull it.   The base of the Devonian and the top of the Silurian occurs midway in the 50 feet of limestone above 275 feet.

BARREN COUNTY.

Production: Oil and Gas. Producing Sands: Oil City (Amber Oil) (Mississippian); Corniferous (Devonian); "Second Sand" (Silurian); "Deep" (Ordovician).

Log No. 76

Lewis No. 1, lessor. Location: at Wathens Mills, 1 mile east of Haywood, near Oskamp. Production: 4 bbls. oil.

Strata.

	Thickness	Depth
Mississippian System.		
Limestone	78	78
Devonian System.		
Shale, black (Chattanooga)	49	127
Limestone (oil and gas, shallow)	21	148
Limestone (oil)	5	153
Silurian System.		
Limestone	264	417
Total depth		417

NOTE—The Silurian-Ordovician contact is included within the last 264 feet of the record.

Log No. 77

Lewis No. 2, lessor. Location at Wathens Mills, 1 mile east of Haywood, near Oskamp. Production: 4 bbls oil.

Strata.

	Thickness	Depth
Mississippian System.		
Limestone	105	105
Devonian System.		
Shale, black (Chattanooga)	45	150
Limestone (cap rock)	8	158
Limestone (oil)	7	165
Total depth		165

Log No. 78

Peden, No. 1, lessor. Location: one-half mile south of Temple Hill, on the crest of the Temple Hill Anticline. Initial Production: estimated at **1,000,000** cubic feet, rock pressure of **325** lbs. gauged.

Strata.

Mississippian System.	Thickness	Depth
Limestone	54	54

Devonian System.

	Thickness	Depth
Shale, black (Chattanooga)	35	89
Limestone, blue	271	360
Limestone (fissure gas, 366)	6	366
Limestone, gray, sandy	34	400
Limestone, blue (gas)	143	543
Total depth		543

NOTE—This was the first gas well drilled in at Temple Hill. The base of the Devonian top of the Silurian, base of the Silurian and top of the Ordovician are all included within the **271** feet of ''blue limestone'' above **360** feet in depth. This record lacks detail. The well finished in the Ordovician.

Log No. 79

Button, No. 1, lessor. Location: Junction of Skeggs and Beaver Creeks.

Strata.

Mississippian System.	Thickness	Depth
Limestone	102	102

Devonian System.

	Thickness	Depth
Shale, black (Chattanooga)	43	145
Limestone	12	157
Limestone, hard	1	158
Limestone (oil ''sand'')	4	162
Limestone	72	234
Limestone (oil ''sand'')	7	241
Limestone, hard	18	259
Total depth		259

NOTE—The Devonic-Siluric contact is within the **72** feet above **234** feet in depth.

Log No. 80

Button, No. 2, lessor.   Location: Junction of Skeggs and Beaver Creeks.   Production: Flush. 15 bbls. oil.

Strata.

Mississippian System.	Thickness	Depth
Limestone	152	152
Devonian System.		
Shale, black (Chattanooga)	43	195
Limestone	74	269
Limestone	5	274
Total depth		274

Casing head is 37', 8" higher than button No. 1.

NOTE—The Devonic-Siluric contact is within the 74 feet above 269 feet in depth.

Log No. 81

Robert Wayfield, No. 1, lessor.   Location: Junction of Upper Bowling Green and Stovall Roads.   Commenced: May 7, 1919.

Strata.

Mississippian System.	Thickness	Depth
Limestone ....(	250	250
Limestone	147	397
Limestone	8	405
Devonian System.		
Shale, black (Chattanooga)	65	470
Limestone (cap rock)	4	474
Limestone (gassy)	13	487
Limestone	113	600
Limestone	145	745
Fire clay	10	755
Limestone, gray and yellow	8	763
Shale	4	767
Total depth		767

NOTE—The Devonic-Siluric contact is within the 113 feet above 600 feet in depth.   The Siluric-Ordovicic contact is within the 145 feet above 745 feet in depth.

Log No. 82

Woodson, No. 1, lessor.   Location: on road north of Lucas, between Skeggs and Beaver Creeks.   Production: oil filled hole 242 feet.

Strata.

Mississippian System.	Thickness	Depth
Limestone	245	245

Devonian System.		
Shale, black (Chattanooga)	45	290
Limestone	62	352
Limestone, oil ''sand''	13	365
Total depth		365

NOTE—The oil in this well is found in the Silurian.   The Devonic-Siluric contact occurs in the 62 feet above 352 feet in depth.

Log No. 83

John Barrick, No. 1, lessor.   Location: 3 miles southwest of Beckton Station.   Production: 20 quarts.   Abandoned.

.Strata.

Mississippian System.	Thickness	Depth
Limestone	130	130
Limestone, (gassy)	5	135
Limestone	108	243
Limestone	102	345

Devonian System.		
Shale, black (Chattanooga)	65	410
Limestone (cap rock)	10	420
Limestone (oil ''sand'')	8	428

Silurian System.		
Limestone (oil show 522)	94	522
Limestone	34	556
Shale (oil)	8	564
Limestone	12	576
Total depth		576

Log No. 84

Dick Smith, No. 1, lessor.   Location: Dry Fork, southern Barren. Well abandoned.

Strata.

Mississippian System.	Thickness	Depth
Limestone	106	106

Devonian System.		
Shale, black (Chattanooga)	38	144
Limestone (gassy)	167	311
Limestone	341	652
Total depth		652

NOTE—The Devonic-Siluric contact occurs within and toward the top of the 167 feet of limestone above 311 feet in depth. The Siluric-Ordovicic contact occurs toward the top of the last 341 feet of the record.

Log No. 85

J. C. Cole, No. 1. lessor.   Authority: The Swiss Oil Corporation.

Strata.

Mississippian System.	Thickness	Depth
Soil	20	20
Limestone, brown, hard	70	90
Limestone, white, hard	61	151
Limestone, gray, hard	40	191

Devonian System.		
Shale, black, soft (Chattanooga)	52	243
Limestone, (cap rock), hard	4	247
Limestone, "sand" (first), hard	6	253
Shale	30	283
Limestone, (cap rock)	6	289
Limestone, "sand" (second)	20	309
Shale, blue	28	337
Limestone, "sand" (third)	3½	340½
Limestone and salt water	41	381½
Total depth		381½

NOTE—The Devonic-Siluric contact occurs within and toward the base of the 30 feet of "shale" above 283 feet in depth.

Log No. 86

J. C. Cole, No. 4, lessor. Commenced: January 28, 1920. Completed: February 10, 1920. Production: Dry. Authority: The Swiss Oil Corporation.

Strata.

Mississippian System.	Thickness	Depth
Clay, red, soft	15	15
Limestone, gray, hard	4	19
Clay, red	2	21
Limestone, gray, hard	2	23
Clay and limestone	4	27
Limestone, black, soft (water at 28)	37	64
Flint, white, hard	8	72
Limestone, black soft	30	102
Shale, hard, black, soft	5	107
Limestone, white, hard	12	119
Limestone, black, soft	110	229
Devonian System.		
Shale, black, soft	41	270
Shale, brown, hard	10	280
Limestone (cap rock)	2	282
Salt sand (heavy water)	3	285
Total depth		285

Log No. 87

H. M. Emmett, No. 1, lessor. Location: 1 mile west of Freedom P. O. Drilled by the Wabash Oil Co., May 31, 1920. Production: 1 bbl. amber oil.

Strata.

Mississippian System.	Thickness	Depth
Limestone	135	135
Devonian System.		
Shale, black (Chattanooga)	30	165
Limestone	42	207
Limestone (pay oil "sand")	3	210
Limestone	10	220
Total depth		220

NOTE—The oil horizon as given in this record is undoubtedly in the Niagaran (Silurian) limestone. The Devonian-Silurian contact is not recognized, but is within the 42 feet of limestone recorded just below the black shale. An amber oil in commercial quantities from this horizon is very unusual in Kentucky.

Log No. 88

Stephen Kinslow, No. 2, lessor. William Oskamp, lessee. Location: 4 miles south of Glasgow, on Boyd's Creek. Drilled in the summer of 1918. Production: in August, 1920, one barrel. Authority: Gordon Kinslow.

Strata.

Mississippian System.	Thickness	Depth
Soil	4	4
Limestone, variable	139	143
Devonian System.		
Shale, black (Chattanooga)	37	180
Limestone, brown	27½	207½
Limestone, ''sand'' (oil)	8	215½
Limestone	14½	230
Limestone, ''sand'' (oil)	4	234
Total depth		234

Log No. 89

Stephen Kinslow, No. 4, lessor. William Oskamp, lessee. Location: 4 miles south of Glasgow, on Boyd's Creek. Drilled in the summer of 1919. Production: summer of 1920, five barrels. Authority: Gordon Kinslow.

Strata.

Mississippian System.	Thickness	Depth
Limestone, variable	161	161
Devonian System.		
Shale, black (Chattanooga)	42	203
Limestone	27	230
Sand (oil)	2	232
Total depth		232

Log No. 90

William Oskamp, No. 3, lessor and lessee. Location: 4 miles south of Glasgow, on Boyd's Creek. Drilled in August, 1918. Production: August, 1920—3—4 barrels. Authority: Gordon Kinslow.

Strata.

Mississippian System.	Thickness	Depth
Limestone	97	97
Devonian System.		
Shale, black (Chattanooga)	40	137
Limestone, brown	30	167
Limestone, ''sand'' (oil)	7	174
Limestone	20	194
Total depth		194

Log No. 91

William Oskamp, No. 4, lessor and lessee. Location: 4 miles south of Glasgow, on Boyd's Creek. Drilled in the summer of 1918. Production: August, 1920, 2 barrels a day. Authority: Gordon Kinslow.

Strata.

Mississippian System.	Thickness	Depth
Limestone, variable	107	107
Devonian System.		
Shale, black (Chattanooga)	40	147
Limestone, brown	25	172
Limestone, "sand" (oil)	2	174
Total depth		174

Log No. 92

Wilson, No. 1, lessor. Location: 3 miles southeast of Cave City. Authority: E. T. Merry.

Strata.

Mississippian System.	Thickness	Depth
Soil	10	10
Clay	45	55
Shale, calcareous	60	115
Limestone	35	150
Limestone, (small gas)	50	200
Limestone	365	565
Devonian System.		
Shale, black (Chattanooga)	55	620
Limestone	30	650
Silurian System.		
Limestone, (oil show)	40	690
Limestone, sandy	35	725
Limestone, hard	75	800
Total depth		800

Casing, 8¼ to 75 feet.
" 6¼ to 275 feet.

Log No. 93

Richardson, No. 1, lessor. Elem Oil & Gas Co., lessee. Location: just north of Glasgow, Ky. Commenced:————— Completed: Sept. 20, 1921. Production: Oil, 5 bbls., gas small. Gravity of oil produced 44.7. Authority: W. H. Link.

Strata.

Mississippian System.	Thickness	Depth
Soil	36	36
Limestone, brown	10	46
Limestone, gray, soft	12	58
Limestone, dark gray	20	78
Limestone, brown	10	88
Limestone, light brown	15	103
Limestone, light gray	10	113
Limestone, gray, soft	7	120
Limestone, brown	5	125
Shale, gray	5	130
Limestone, gray	5	135
Limestone, white	10	145
Shale, gray	10	155
Limestone, gray	45	200
Limestone, light gray	5	205
Limestone, dark gray	5	210
Limestone, brown	8	218
Shale, gray	25	243
Shale, dark	20	263
Limestone, gray	5	268
Limestone, blue	15	283
Limestone, gray	15	298
Limestone, dark gray	10	308
Limestone, gray	12	320
Shale, gray	15	335
Limestone, gray	10	345
Shale, gray	10	355
Shale, gray	5	360
Limestone, gray	12	372
Devonian System.		
Shale, (Chattanooga), (cased below shale)	49	421
Cap rock, white	7	428
Limestone, gray	20	448
Silurian System.		
Limestone, brown	57	505
Sand, (oil show)	18	523

Silurian System.   Thickness Depth

    Limestone, blue 10   533

    Gas sand, brown, (small gas) 20   553

Ordovician System.

    Limestone, blue 34   587

    Cap rock, salt and pepper 7   594

    Oil sand, brown (showing) 20   614

    Limestone, dark 10   624

    Limestone, blue 66   690

    Cap rock, flint 6   696

    Oil & gas sand, white, (small oil & gas) 9   705

    Limestone, blue 44   749

    Flint, blue 10   759

    Limestone, blue 80   839

    Flint, blue 20   859

    Limestone, light blue 116   975

    Limestone, dark blue 38 1,013

    Limestone, dark gray 79 1,092

    Limestone, blue 5 1,097

    Cap rock (Trenton) 10 1,107

    Sand (Trenton), (commercial oil) 15 1,122

    Rock (Trenton) 89 1,211

       Total depth 1,211

BATH COUNTY.

Production: Oil. Producing Sand, Ragland (Corniferous) (Devonian), "Olympia" (Niagaran-Silurian).

Log No. 94

Ewing Heirs, No. 9, lessors. Location: Licking Union District. Completed: April 18, 1903. Authority: The New Domain Oil & Gas Company.

    Strata.

Mississippian System.   Thickness Depth

    Gravel, soft 18   18

    Limestone, hard 158   176

    Shale, hard, soft 274   450

Devonian System.

    Shale, black, soft (Chattanooga) 216   666

    Limestone, "sand," hard (oil at 670) 24   690

       Total depth 690

Log No. 95

Ewing Heirs, No. 10, lessors. Location: Licking Union District.
Completed: May 6, 1903. Authority: The New Domain Oil & Gas
Company.

Strata.

Mississippian System.	Thickness	Depth
Gravel	22	22
Limestone	153	175
Shale, white, hard	230	405
Sand	30	435
Devonian System.		
Shale, black (Chattanooga)	216	651
Limestone, ''sand,'' (oil at 655)	29	680
Total depth		680

Log No. 96

Ewing Heirs, No. 11, lessors. Location: Licking Union District.
Completed: May 28, 1903. Authority: The New Domain Oil & Gas
Company.

Strata.

Mississippian System.	Thickness	Depth
Gravel	13	13
Limestone	225	238
Shale, hard	273	511
Devonian System.		
Shale, black (Chattanooga)	215	726
Limestone, ''sand'' (oil pay at 729)	21	747
Total depth		747

Log No. 97

Ewing Heirs, No. 12, lessors. Location: Licking Union District.
Completed: August 1, 1903. Authority: The New Domain Oil & Gas
Company.

Strata.

Mississippian System.	Thickness	Depth
Limestone	50	50
Shale, white, hard	561	611
Devonian System.		
Shale, black (Chattanooga)	205	816
Shale (fire clay)	8	824
Shale, brown	15	839
Limestone, ''sand'' (oil at 842)	31	870
Total depth		870

Log No. 98

Ewing Heirs, No. 13, lessors. Location: Licking Union District.
Completed: August 14, 1903. Authority: The New Domain Oil & Gas
Company.

Strata.

Mississippian System.	Thickness	Depth
Limestone	50	50
Shale, blue	555	605
Devonian System.		
Shale, black (Chattanooga)	205	810
Shale (fire clay), white	5	815
Shale, brown	15	830
Limestone, "sand" (oil at 838)	25	855
Total depth		855

BOYD COUNTY.

**Production: Oil and Gas. Producing Sands: Pottsville (Pennsylvanian);
Big Injun, and Berea (Mississippian); "Tunnel Sand"
(Devonian).**

Log No. 99

W. I. Ross, No. 1, lessor. Good Losers Oil & Gas Co., (oil well
No. 3) lessee. Location: near Bolts Fork, in Boyd County. Commenced: Oct. 1, 1920. Completed: Dec. 14, 1920. Initial production:
25 bbls. oil. Authority: C. E. Bales.

Strata.

Pennsylvanian System.	Thickness	Depth
Soil and shale	40	40
Sandstone	30	70
Shale and shells	60	130
Coal	3	133
Limestone (shells)	7	140
Shale (Red Rock)	12	152
Shale, dark	100	252
Sandstone	75	327
Shale, light	40	367
Limestone, hard	13	380
Shale, dark	10	390
Sandstone, gray (water at 400)	40	430
Shale and shells	120	550
Limestone, gray	25	575
Shale, dark	125	700

Pennsylvanian System.	Thickness	Depth
Sandstone, gray	30	730
Shale and shells	90	820
"Salt sand"	70	890
Shale, dark	25	915
Sandstone, white	85	1,000

Mississippian System.		
Limestone, gray	40	1,040
Shale, dark	40	1,080
Limestone, hard	105	1,185
Shale, dark	35	1,220
Limestone, gritty	60	1,280
Sandstone, gray.........................	40	1,320
Limestone, hard	10	1,330
Waverly shale	250	1,580
Shale and shells	150	1,730
Shale, brown	27	1,757
Sandstone ("Berea Grit"), (oil at 20 ft. in sandstone, used 20 qts. nitro glycerine)	35	1,792
Shale, dark	32	1,824
Sandstone ("Berea Grit"), (all carried oil, used 20 qts. nitro glycerine)	6	1,830
Shale	36	1,866
Total depth		1,866

Log No. 100.

John Murphy, No. 1, lessor. Murphy Oil & Gas Co. (gas well No. 4), lessee. Location: east side of A. C. & I. Ry. Co. tunnel, just west of Ashland. Commenced: September 14, 1912. Completed: November 15, 1912. Initial production: 250,000 cu. ft. gas. Authority: C. E. Bales.

Strata.

Pennsylvanian System.	Thickness	Depth
Soil	11	11
Sandstone	70	81
Fire clay	9	90
Shale	50	140
Sandstone	10	150
Shale (fresh water set 8¼" casing at 180) ..	290	440
Salt sand	20	460
Shale, blue	30	490
Limestone, sandy	18	508
Salt sand (some gas)	60	568

DEEPEST WELL IN NORTHEASTERN KENTUCKY

This well drilled on the Martha Stewart farm is known as the Barrick Oil and Gas Co. No. 8. It is located about two miles East of Denton in Carter County on the A. C. & I. R. R. The total depth was 3920 feet. The drilling was started March 29, 1920, and was finished in the Ordovician limestone on Jan. 8, 1921.

Mississippian System.	Thickness	Depth
Limestone	37	605
Limestone, sandy	30	635
Shale, white	15	650
Sandstone	90	740
Limestone, sandy (salt water)	92	832
Shale, soft and muddy	8	840
Shale (set 6⅝" casing at 855)	25	865
Sandstone	45	910
Shale	347	1,257
Shale, "coffee"	15	1,272
Limestone	5	1,277
Sand ("Berea"), (show of oil)	15	1,292
Shale and sand	23	1,315
Sandstone, gray (gas)	33	1,348
Shale, blue	35	1,383
Shale (Red Rock)	11	1,394
Shale, blue	15	1,409
Shale, blue	176	1,585

Devonian System.		
Shale, black (Ohio)	115	1,700
Sandstone (fossil shells, gas)	5	1,705
Shale, black (Ohio)	200	1,905
Sandstone (fossil shells, gas)	5	1,910
Shale, black (Ohio)	20	1,930
Total depth		1,930

This well is still a good producer, the gas being pumped into the pipe line, serving Ashland, Boyd County, Kentucky.

Log No. 101.

Belle Ross, No. 1, lessor. Good Losers Oil & Gas Co., (oil well No. 2) lessee. Location: near Bolts Fork, in Boyd County, just north of Lawrance County line, and about 4½ miles east of Denton, Carter County. Commenced: June 25, 1920. Completed: Aug. 20, 1920, Initial production: 15 bbls. oil. Authority: C. E. Bales.

Strata.

Pennsylvanian System.	Thickness	Depth
Soil	20	20
Shale, black	65	85
Sandstone and limestone	170	255
Shale, white and brown	125	380
Sandstone (show of gas)	30	410

Pennsylvania System.

	Thickness	Depth
Sandstone and shale	170	580
Shale, black	200	780
Sandstone, white	175	955
Shale, black	15	970

Mississippian System.

Limestone (Big Lime)	105	1,075
Shale, white	108	1,183
Sandstone, white (salt water)	30	1,213
Waverly shale	424	1,637
Shale, black	20	1,657
Sandstone ("Berea Grit"), (oil)	44	1,701
Shale, black	8	1,709
Sandstone, dark	4	1,713
Shale, black	5	1,718
Sandstone ("Berea Grit"), (oil)	26	1,744
Shale, black	7	1,751
Total depth		1,751

Log No. 102.

Clara Williams, No. 1, lessor. Location: Lockwood, Ky. Authority: Associated Producers Company.

Strata.

Pennsylvanian System.

	Thickness	Depth
Shale and shell	400	400
Sandstone (cow run)	50	450
Sandstone (cow run) (second)	50	500
Sandstone (salt sand)	610	1,110
Shale	10	1,120

Mississippian System.

Sandstone (Maxon)	50	1,170
Limestone (Big Lime)	122	1,292
Shale, red and blue	40	1,332
Sandstone (Big Injun), (3 bbls. water in 24 hrs. 50 ft. in Big Injun sand)	200	1,532
Shale and shells	326	1,858
Shale, black	30	1,888
Sandstone (Berea sand) (top of Berea 1888)..	42	1,930
Total depth		1,930

Log No. 103

Drilled by the Huntington Gas & Development Co.   Location: on Big Sandy River, 13 miles north of Catlettsburg, Boyd County, Kentucky.

Strata.

Pennsylvanian System.	Thickness	Depth
Soil	46	46
Sandstone	39	85
Shale	10	95
Sand	10	105
Shale	20	125
Sand	65	190
Shale	10	200
Sand	20	220
Shale	20	240
Sand	60	300
Shale	227	527
Sand	15	542
Shale	274	816
Salt sand	224	1,040
Coal	2	1,042
Shale	18	1,060

Mississippian System.		
Sandstone (Maxon)	22	1,082
Shale	2	1,084
Limestone (Big Lime)	111	1,195
Shale	10	1,205
Sandstone (Big Injun)	175	1,380
Shale	21	1,401
Shale and shells	417	1,818
Sandstone (Berea)	44	1,862
Shale and shells	40	1,902

Devonian System.		
Shale (break)	658	2,560
Shale (Chattanooga)	186	2,746
Limestone and sand (Ragland)	34	2,780

Silurian System.		
Limestone	625	3,405
Shale	25	3,430
Limestone (Red Rock)	75	3,505
Shale, black	95	3,600
Limestone (shell)	5	3,605
Limestone (Red Rock)	25	3,630

Silurian System.

	Thickness	Depth
Shale, black	15	3,645
Limestone (shell), hard	4	3,649
Shale and sand (shell)	41	3,690
Shale	5	3,695
Limestone (Red Rock)	5	3,700
Shale, black	15	3,715
Limestone	20	3,735
Sand, broken	15	3,750
Shale, black	37	3,787
Total depth		3,787

NOTE—The base of the Silurian and the top of the Ordovician occurs near the top of the 625 feet of limestone above 3,405 feet in depth.

BRACKEN COUNTY.

Production: Small Oil and Gas. Producing Sands: Probably Trenton (Ordovician).

Log No. 104

Well A. Lessor and lessee unknown. Location: On the bank of the North Fork of the Licking River, near the Bracken-Robertson County Line. Drilling completed: 1903.

Strata.

Ordovician System.

	Thickness	Depth
Soil	15	15
Limestone, (oil show 85)	70	85
Limestone (oil ''sand'')	40	125
Limestone, gray	275	400
Total depth		400

Log No. 105

Well B. Lessor and lessee unknown. Location: About 800 feet east of Well A. Drilling completed: 1903. Production: Oil, flowed for months in the creek.

Strata.

Ordovican System.

	Thickness	Depth
Soil	15	15
Limestone, (oil show 70)	55	70
Limestone, (oil ''sand'')	40	110
Limestone, gray	290	400
Total depth		400

Log No. 106

Well C. Lessor and lessee unknown. Location: About 1,300 feet east of Well A. Drilling completed: 1903. Production: Gas flowed blowing white for months.

Strata.

Ordovician System.   Thickness Depth.

Limestone, (thickness unrecorded, record similar to A and B).

Log No. 107

Well D. Lessor and lessee unknown. Location: 500 feet west of Well A. Drilling completed: 1920. Production: Three barrels, est., some gas.

Strata.

Ordovician System.	Thickness	Depth
Limestone, (oil at 50)	50	50
Limestone	41	91
Limestone, (oil "sand")	40	131
Limestone	135	266
Limestone, gray	124	390
Total depth		390

NOTE—Wells A and D are standing about 90 feet in oil, and are estimated to be good for about three barrels.

BREATHITT COUNTY.

Production: Oil and Gas. Producing Sands: Pottsville (Pennsylvanian); Maxton, Big Injun, and Wier (Mississippian); Corniferous (Devonian); Niagaran (Silurian); and "Deep" or "Big Six" (Upper Ordovician).

Log No. 108

W. H. Pelfrey, No. 1, lessor. J. Fred Miles Oil & Gas Co. (oil well No. 1), lessee. Location: near Vancleve, about 200 yds. above iron bridge on the O. & K. R. R. Commenced: Aug. 2, 1919. Completed: Aug. 30, 1918. Initial production: 1 bbl. oil. Authority: C. E. Bales.

Strata.

Pennsylvanian System.	Thickness	Depth
Soil	12	12
Sandstone	38	50
Shale and shells	150	200
Sandstone	160	360
Shale	85	445

Mississippian System.

	Thickness	Depth
Limestone (Little Lime)	25	470
Shale	10	480
Limestone (Big Lime)	165	645
Shale (Waverly)	180	825
Shale, red	30	855
Shale	215	1,070
Sandstone ("Berea Grit")	25	1,095

Devonian System.

	Thickness	Depth
Shale, brown	249	1,344
Limestone (oil at 1,347 to 1,352)	155	1,499
Limestone, brown	36	1,535
Limestone, blue	40	1,575
Shale	19	1,594
Shale, red	7	1,601
Total depth		1,601

NOTE—The base of the Devonian and the top of the Silurian occurs within the 155 feet above 1,499 feet in depth. The oil occurring for 1347 to 1352 feet is in the Corniferous.

Log No. 109

Terrell, No. 1, lessor. Big Six Oil-Co. (well No. 3), lessee. Location: Terrell farm, ¾ mile up Sulphur Fork, Frozen Creek. Casing head elevation: 820 ft. A. T. Production: Gas, 5,820,000 cu. ft.

Strata.

Pennsylvanian System.

	Thickness	Depth
Soil	17	17
Shale, black	83	100
Sand, white	60	160
Shale, black	30	190
Sand, white	210	400
Shale, black	100	500
Sand, white	145	645
Shale, black	20	665

Mississippian System.

	Thickness	Depth
Limestone (Little Lime)	10	675
Shale, black	16	681
Limestone (Big Lime)	139	820
Shale (Waverly)	600	1,420

Devonian System.	Thickness	Depth
Shale, black (Chattanooga)	170	1,590
Shale, white	14	1,604
Limestone	174	1,778
Gas sand (sample of gas sand was a coarse-grained, pebbly sand)	25	1,803
Total depth		1,803

NOTE—The Devonian limestone and the Silurian limestone are both included within the 174 feet above 1,778 feet in depth. As in the Taulbee wells, it is quite probable that the gas production occurring in this field about 175 feet below the base of the Devonian shale comes from the uppermost Ordovician beds.

Log No. 110

J. S. Taulbee, No. 1, lessor. Location: ⅜ miles up Sulphur Fork of Frozen Creek. Casing head elevation: 805. A. T. Production: 5,000,000 cu. ft. gas, 175 ft. in sand.

Strata.

Pennsylvanian System.	Thickness	Depth
Soil	16	16
Shale, black	54	70
Sand, white	90	160
Sand, black	30	190
Sand, white	210	400
Sand, black	100	500
Sand, white	145	645
Shale, black	20	665
Mississippian System.		
Limestone (Little Lime)	10	675
Shale, black	5	680
Limestone (Big Lime)	150	830
Shale (Waverly)	600	1,430
Devonian System.		
Shale, black (Chattanooga)	170	1,600
Shale, white	14	1,614
Limestone (gas 1,790)	176	1,790
Limestone	25	1,815
Total depth		1,815

NOTE—The gas production in this well coming from a depth of 1,790 feet is either basal Silurian or uppermost Ordovician, and probably the latter. The Devonian limestone (Corniferous) and the Silurian limestone (Niagaran) are included within the 176 feet above 1,790.

Log No. 111.

J. S. Taulbee, No. 2, lessor. Big Six Oil Co. (well No. 2), lessee.
Location: J. S. Taulbee farm, ½ mile up Sulphur Fork, Frozen Creek.
Casing head elevation: 795. A. T.

Strata.

Pennsylvanian System.	Thickness	Depth
Soil	10	10
Shale, black	190	200
Sand, white	50	250
Shale, black	150	400
Shale, sandy	200	600
Sand, white	40	640
Shale, white	25	665
Mississippian System.		
Limestone (Little Lime)	5	670
Shale, black	20	690
Sand, dark	10	700
Limestone (Big Lime)	100	800
Shale (Waverly)	520	1,320
Devonian System.		
Shale, black (Chattanooga)	277	1,597
Shale, white	30	1,627
Limestone (gas show 1,803)	176	1,803
Limestone (finished in Red Rock)	78	1,881
Total depth		1,881

NOTE—Base of Devonian and top of Silurian indefinite, but included in 176 feet of limestone above 1,803 feet. The gas is either Silurian or Ordovician, and not unlikely the latter.

BRECKINRIDGE COUNTY.

Production: **Small Oil and Gas.** Producing Sands: **Cloverport Gas Sand (Lower Mississippian) Corniferous (Devonian).**

Log No. 111A.

John Gibson, No. 1, lessor. Location: 1½ miles southwest of Sample Station. Completed: Spring, 1922. Authority: C. F. Dunbar, driller.

Strata.

Mississippian System.	Thickness	Depth.
Shales, sandstones and limestones	1,280	1,280
Devonian System.		
Shale, black	100	1,380
Limestone	57	1,437
Limestone "sand," (oil show)	10	1,447
Total depth		1,447

BUTLER COUNTY.

**Production: Oil and Gas. Producing Sands: Unnamed (Mississippian);
"Deep" (Devonian-Silurian).**

Log No. 112

M. D. Duncan, No. 1, lessor. The Arkansas Natural Gas Co., lessee. Location: near Flat Rock P. O. Commenced: Jan. 15, 1921. Drillers: O. L. Drake and L. C. Jones. Casing head elevation by aneroid: 625. Authority: W. C. Eyl.

Strata.

Mississippian System.	Thickness	Depth
Clay	15	15
Limestone	25	40
Shale	25	65
Limestone	15	80
Shale	56	136
Limestone	24	160
Shale	5	165
Limestone, gray	16	181
Shale, blue	24	205
Shale, brown	15	220
Shale, white	5	225
Limestone, white	35	260
Shale, blue, soft	25	285
Limestone (pay sand)	25	310
Limestone (pay sand), soft	10	320
Shale, blue	10	330
Limestone, white	35	365
Shale, blue	5	370
Limestone, white	80	450
Limestone, gray	130	580
Limestone, white, (salt water 700, 775 8¼ in. casing set)	195	775
Limestone, yellow	50	825
Limestone, gray	45	870
Limestone, white	8	878
Limestone, yellow	4	882
Limestone, black	10	892
Limestone, white	10	902
Limestone, black	20	922
Limestone, brown	58	980
Shale, brown	10	990
Limestone, brown	15	1,005
Limestone, black	10	1,015
Limestone, brown	73	1,088
Limestone, gray	187	1,275
Limestone, black	265	1,540

Devonian System.	Thickness	Depth
Shale, black (Chattanooga)	122	1,662
Shale, black, and limestone	13	1,675
Limestone, white	42	1,717
Silurian System.		
Limestone, gray	5	1,722
Limestone, brown, (oil show)	26	1,748
Limestone, grayish brown, soft	15½	1,763½
Limestone, light gray	39½	1,803
Limestone, dark gray	49	1,852
Limestone, (salt water)	19	1,871
Total depth		1,871

CALDWELL COUNTY.

Production: Neither Oil or Gas to Date. Producing Sands: Tar Springs and Cypress occur but are not known to be productive in the county.

Log No. 113

Mrs. W. F. O'Hara (widow), No. 1, lessor. Climax Oil Corporation, lessee. Location: 3 miles east of Princeton, Ky., near Cedar Hill P. O. Drillers: Ray Brown and Scott Dalton. Tool dressers: Sid Hunter and Oscar Boyd. Contractors: Brown & Dalton.

Strata.

Mississippian System.	Thickness	Depth
Soil	15	15
Limestone	20	35
Shale, (water	5	40
Limestone, white	20	60
Granite	5	65
Shale, white	5	70
Limestone, blue	110	180
Limestone, sandy, (water)	20	200
Limestone, gray	45	245
Limestone "sand" (oil)	5	250
Limestone, white	25	275
Limestone, sandy	35	310
Limestone, gray	15	325
Limestone, white	20	345
Limestone, sandy	5	350
Limestone, white, sandy	25	375
Limestone, dark	25	400
Limestone, light gray	20	420
Limestone, gray, sandy	55	475

	Thickness	Depth
Mississippian System.		
Limestone, gray	25	500
Limestone, black	10	510
Limestone, dark gray	20	530
Limestone, light gray	20	550
Limestone, broken	10	560
Limestone, dark	25	585
Limestone, white	55	640
Limestone, dark gray	35	675
Limestone, light gray	5	680
Limestone, light, sandy	70	750
Limestone, gray	50	800
Limestone, shelly	15	815
Limestone, gray	15	830
Limestone, sandy	10	840
Limestone, dark	20	860
Limestone, gray	15	875
Limestone, dark	15	890
Limestone, shelly	20	910
Limestone, dark, sandy	5	915
Limestone, white	15	930
Limestone, light gray	10	940
Limestone, white	15	955
Limestone, gray	10	965
Limestone, gray	10	975
Liimestone, dark	36	1,011
Shale	4	1,015
Limestone, dark	585	1,600
Shale and limestone	20	1,620
Limestone, white	10	1,630
Devonian System.		
Shale, black (Chattanooga)	125	1,755
Limestone, white	10	1,765
Shale and limestone	5	1,770
Total depth		1,770

CARTER COUNTY.

Production: **Oil and Gas.** Producing Sands: Pottsville (Pennsylvanian); **Big Injun, Wier and Berea (Mississippian).**

Log No. 114

Levi Porter, No. 1, lessor. Elcaro Oil & Gas Co. (well No. 1), lessee. Location: Near Lawton, Tygarts Creek above C. & O. R. R. Commenced: September 28, 1920. Completed: November 12, 1920. Initial production: 3 bbls oil. Authority: C. E. Bales.

Strata.

Pennsylvanian System	Thickness	Depth
Soil	15	15
Shale and blue slate	65	80
Shale, sandy	10	90
Shale, black, and fire clay	25	115
Sandstone	55	170
Mississippian System.		
Shale, green, and red rock	30	200
Limestone ("Big Lime")	60	260
Sandstone, limey, (little show oil & gas)	12	272
Shale, blue	306	578
Sandstone, blue (salt water at 612)	34	612
Sandstone, white	24	636
Shale, blue	16	652
Sandstone, white	24	676
Shale, blue, soft	54	730
Sandstone, gray (show of oil)	23	753
Shale, black	17	770
Sandstone, blue (good showing of oil)	18	788
Total depth		788

Log No. 115

William Offill, No. 1, lessor. Lawton Oil & Gas Co. (well No. 2), lessee. Location: Near Lawton, about 1 mile due south from well No. 3, on waters of Tygart Creek. Commenced: January, 1920. Completed: July, 1920. Initial Production: ——bbls. oil. Authority: C. E. Bales.

Strata.

Mississippian System.	Thickness	Depth
Soil	6	6
Rock, blue, hard	80	86
Sandstone, (little show of oil)	7	93
Shale, and blue slate	212	305
Sandstone, gray	125	430
Shale, blue, and slate	47	477
Sandstone (Wier), (show of oil)	23	500
Shale, black (Sunbury)	17	517
Sandstone (Berea), (show of oil and gas)	36	553
Total depth		553

Salt water at 316 feet.
This well was never shot.

Log No. 116

J. W. Jacobs, No. 1, lessor. Lawton Oil & Gas Co., (well No. 1), lessee. Location: Near Tygart; near Lawton P. O., on waters of Tygart Creek. Commenced: October, 1919. Completed: May, 1920. Initial production: 2 bbls. oil. Authority: C. E. Bales.

Strata.

Mississippian System.	Thickness	Depth
Soil	12	12
Rock, blue, hard	58	70
Sandstone, gray (show of oil & gas) (non-productive)	7	77
Shale, blue, and slate	228	305
Sandstone, light gray	95	400
Shale, sandy	20	420
Shale, blue, soft	37	457
Sandstone (Wier), (show of oil)	21	478
Shale, black (Sunbury)	16	494
Sandstone (Berea), (show of oil & gas)	50	544
Total depth		544

Small amount of salt water at 73 feet.
Salt water in bottom sandstone.

Log No. 117

J. W. Jacobs, No. 2, lessor. Lawton Oil & Gas Co. (well No. 3), lessee. Location: Near Lawton, about 800 ft. from well No. 1. Commenced: March, 1920. Completed, May, 1920. Initial production: —— bbls. oil. Authority: C. E. Bales.

Strata.

Mississippian System.	Thickness	Depth
Soil	15	15
Rock, blue, hard	20	35
Sandstone (good show of oil) (salt water)	11	46
Shale, blue, hard	253	299
Sandstone, gray (salt water at 325)	126	425
Sandstone (Wier), (little show of oil)	22	447
Shale, black (Sunbury)	16	463
Sandstone (Berea), (show of oil)	48	511
Total depth		511

This well was lost, due to collapse of casing when well was shot.

Log No. 118

L. C. Glancy, No. 1, lessor. Location: Near Grayson. Completed: March 3, 1904. Production: Dry. Authority: New Domain Oil & Gas Company.

Strata.

Pennsylvanian System.	Thickness	Depth
Sand, white, hard	80	80
Limestone, white, hard	10	90
Fire clay, white, soft	20	110
Shale, black, hard, soft	45	155
Shale, dark, hard	10	165
Shale, white, hard, soft	30	195
Fire clay, white, soft	30	225
Shale, dark, hard	10	235
Sand, white, soft	10	245
Coal, black, soft	4	249
Shale, white, hard, soft	30	279
Sandstone, white, hard	10	289
Shale, white, hard, soft	20	309
Sand, white, hard	10	319
Shale, black, hard, soft	60	379
Sandstone, white, hard	15	394
Shale, dark, hard, soft	60	454
Sandstone, white, hard	40	494
Sand, white, open	70	564
Fire clay, white, soft	10	574
Sand, white, open	51	625
Shale, black, hard, soft	45	670

Mississippian System.		
Limestone (St. Louis), very hard	100	770
Shale (Waverly), white, very hard	330	1,100
Shale, white, hard, soft	110	1,210
Sandstone shells, white, soft and hard	165	1,375

Devonian System.		
Shale, brown, hard	510	1,885
Limestone, white, hard, soft	110	1,995
Limestone, white, sandy, hard	89	2,084
Total depth		2,084

NOTE—The base of the Devonian and the top of the Silurian occurs within the 110 feet of limestone above 1995 feet in depth. The top of the Ordovician may also be included within the last 89 feet of the record.

Log No. 119

Murphy and Burdette, No. 1, lessors. Barrick-Kentucky Oil & Gas
Co. (well No. 3), lessees. Location: Near Denton, about 1½ miles
east, on the A. C. & I. R. R. Commenced: September 8, 1919. Completed: October 18, 1919. Initial production: 500,000 cu. ft. gas per
day. Authority: C. E. Bales.
Strata.

Pennsylvanian System.	Thickness	Depth
Soil	12	12
Shale, black (fresh water at 30 ft.)	28	40
Sandstone	30	70
Shale, black	46	116
Sandstone	15	131
Shale, black	184	215
Limestone	10	225
Shale, black	10	235
Sandstone, gray	4	239
Shale, black	61	300
Coal	3	303
Shale, black	29	332
Shale, white	7	339
Sandstone, gray	15	354
Coal	3	357
Sandstone, gray	20	377
Shale, black	28	405
Sandstone, gray (gas from 405 to 409)	20½	425½
Total depth		425½

NOTE—This record is entirely within the Pottsville.

Log No. 120

Murphy and Burdette, No. 2, lessors. Barrick-Kentucky Oil &
Gas Co. (well No. 4), lessees. Location: near Denton, about 1½ miles
east, on the A. C. & I. R. R. Commenced: October 21, 1919. Completed: November 28, 1919. Initial production: 1,000,000 cu. ft. gas.
Authority: C. E. Bales.
Strata.

Pennsylvanian System.	Thickness	Depth
Soil	14	14
Sandstone	61	74
Coal	4	78
Shale, black	12	90
Sandstone	20	110
Shale, blue	20	130

Pennsylvanian System.

	Thickness	Depth
Coal	1	131
Shale, black	54	185
Sandstone	30	215
Shale, black	70	285
Sandstone	10	295
Shale, black	20	315
Sandstone	5	320
Shale, black	85	405
Sandstone	17	422
Shale, black	29	451
Sandstone (gas)	7	458
Shale, black	3	461
Sandstone, (gas at 465)	13	474
Total depth		474

NOTE—This well is entirely within the Pottsville.

Log No. 121

Richard Fraley, No. 1, lessor. Barrick-Kentucky Oil & Gas Co. (well No. 5), lessee. Location: Near Denton, about 2 miles east of Denton, on the A. C. & I. R. R. Commenced: November 10, 1919. Completed: December 12, 1919. Initial production: 900,000 cu. ft. gas. Authority: C. E. Bales.

Strata.

Pennsylvanian System.

	Thickness	Depth
Soil, ...	21	21
Sandstone	9	30
Shale, black	25	55
Fire clay	18	73
Shale, black	69	142
Sandstone	6	148
Shale, white	147	295
Sandstone	5	300
Shale, black	50	350
Sandstone	20	370
Sandstone, (gas)	14	384
Total depth		384

NOTE—This well is entirely within the Pottsville.

Log No. 122

Richard Fraley, No. 2, lessor. Barrick-Kentucky Oil & Gas Co. (well No. 7), lessee. Location: near Denton, about 2 miles east of Denton, on the A. C. & I. R. R. Commenced: January 6, 1920. Completed: March 27, 1920. Initial production: 1,000,000 cu. ft. gas & salt water.

Strata.

Pennsylvanian System.	Thickness	Depth
Soil	12	12
Sandstone	7	19
Shale, black	11	30
Sandstone, gray	35	65
Shale, black	25	90
Fire clay	10	100
Shale, gray	40	140
Shale, black	10	150
Shale, white	15	165
Sandstone	15	180
Shale, black, (gas at 225)	80	260
Sandstone	25	285
Shale, black	67	352
Coal	2	354
Shale, black	11	365
Shale, blue	31	396
Sandstone	7	403
Shale, black	37	440
Sandstone, (salt water and gas)	16	456
Total depth		456

NOTE—The record is entirely within the Pottsville.

Log No. 123

Oil well ½ mile east of Denton, north of C. & O. R. R. right-of-way. Commenced: December 29, 1916. Completed: April 6, 1917. Production: 1 bbl. oil and some gas. Authority: C. E. Bales.

Strata.

Pennsylvanian System.	Thickness	Depth
Soil	38	38
Shale, hard	287	325
Sandstone	20	345
Shale	15	360
Sandstone, (gas at 380)	90	450
Shale	25	475
Sandstone	69	544
Mississippian System.		
Limestone (Big Lime)	66	610
Shale (Waverly)	508	1,118
Shale, black (Sunbury)	18	1,136
Sandstone (Berea)	40	1,176
Shale,	12	1,188
Sandstone (Berea), (oil show at 1,223)	67	1,255
Shale,	14	1,269

Devonian System. Thickness Depth
  Shale, brown-black 471 1,740
  Shale, white 91 1,831
  Limestone (Ragland Sand) 29 1,860
Silurian System.
  Limestone 60 1,920
    Total depth 1,920

Log No. 124

J. C. Riffe, No. 1, lessor. Good Losers Oil Co. (well No. 1), lessee. Location: on Bolt's Fork, 4 miles east of Denton. Commenced: March 24, 1920. Completed: May 7, 1920. Production: 15.20 bbls. oil per day. Drilled by Patton and Foreman.
  Strata.

Pennsylvanian System. Thickness Depth
  Soil 35 35
  Shale 65 100
  Sandstone and shale 175 275
  Shale, white and brown 125 400
  Sandstone, white, (show of gas) 30 430
  Sand, broken, and shale 170 600
  Shale, black 200 800
  Sandstone, white, (show of gas) 175 975
  Shale, black 15 990
Mississippian System.
  Limestone (Big Lime), (set 6⅝ casing at 990) 32 1,022
  Shale, white 208 1,230
  Sandstone (Big Injun), (salt water) 30 1,260
  Shale (Waverly) 406 1,666
  Shale, black (Sunbury) 20 1,686
  Sandstone (Berea), (about 2 bbls. oil per day) 47 1,733
  Shale, black 9 1,742
  Sandstone, dark 5 1,747
  Shale, dark 6 1,753
  Sand (Berea) 30 1,783
    Total depth 1,783

Log No. 125

Martha Stewart, No. 1, lessor. Barrick-Kentucky Oil & Gas Co. (well No. 8), lessee. Location: near Denton, about 2 miles east of Denton, on the A. C. & I. R. R. Commenced: March 29, 1920. Completed: January 8, 1921. Authority: C. E. Bales.
  Strata.

Pennsylvanian System. Thickness Depth
  Soil and shale 90 90
  Coal 2 92

	Thickness	Depth
Pennsylvania System.		
Shale	13	105
Sandstone	75	180
Shale, black	10	190
Sandstone (show of oil at 200)	20	210
Shale	100	310
Sandstone	10	320
Shale, black	145	465
Sandstone	10	475
Coal	2	477
Shale	50	527
Sandstone	23	550
Shale	10	560
Sandstone	5	565
Shale	5	570
Sandstone	100	670
Shale	30	700
Mississippian System.		
Sandstone (Maxon)	90	790
Limestone	40	830
Shale, (Pencil Cave), (caved somewhat)	10	840
Limestone (Big Lime), not very hard	265	1,105
Shale, black	5	1,110
Limestone, blue	50	1,160
Shale, black	10	1,170
Limestone	60	1,230
Shale, white	20	1,250
Limestone	45	1,295
Shale and shells	35	1,330
Shale, black (Sunbury)	22	1,352
Sandstone (Berea Grit)	48	1,400
Shale	5	1,405
Sandstone (Berea Grit)	34	1,439
Shale	3	1,442
Sandstone (Berea Grit), hard	21	1,463
Devonian System.		
Shale, black	97	1,560
Shale, white	20	1,580
Shale, black (Chattanooga)	460	2,040
Shale, white	26	2,066
Limestone (Corniferous), (show of oil)	64	2,130
Silurian and Ordovician Systems.		
Shale, (show of oil)	1	2,131
Limestone, (show of oil and water)	329	2,460
Shale, black	15	2,475
Red rock	135	2,610
Shale, light	25	2,635

Devonian System.	Thickness	Depth
Red rock	43	2,678
Limestone and shells	4	2,682
Shale, blue	8	2,690
Limestone and shells	5	2,695
Shale, blue	10	2,705
Red rock	10	2,715
Shale, blue, sandstone, shells	90	2,805
Red rock	15	2,820
Shale, blue, and shells	180	3,000
Shale and shells	115	3,115
Limestone, sandy	100	3,215
Shale, sandstone and shells (Calciferous)	230	3,445
Limestone, gray	10	3,455
Shale, limestone and shells	10	3,465
Shale, blue, soft	10	3,475
Shale, white, limestone and shells	155	3,630
Limestone	290	3,920
Shale, black	5	3,925
Total depth		3,925

NOTE—The base of the Silurian and the top of the Ordovician undoubtedly occurs in the 329 feet of limestone above 2,450 feet in depth. This well finished in the Knox Dolomite (Cambrian) and is the deepest record in northeastern Kentucky to date.

Log No. 126

Martha Stewart, No. 2, lessor. Barrick-Kentucky Oil & Gas Co. (well No. 6), lessee. Location: near Denton, about 1½ miles east of Denton, on the A. C. & I. R. R. Commenced: January 13, 1920. Completed: February 21, 1920. Initial production: 1,600,000 cu. ft. gas. Authority: C. E. Bales.

Strata. Pennsylvanian System.	Thickness	Depth
Soil	16	16
Shale	49	65
Sandstone	10	75
Shale, white	15	90
Black mud (clay?)	85	175
Sandstone	10	185
Shale, black, soft	13	198
Sandstone	7	205
Shale, black, soft	27	232
Sandstone	18	250
Shale, black, soft	5	255
Sandstone	3	258

Pennsylvania System.	Thickness	Depth
Coal	2	260
Shale, blue	30	290
Sandstone	20	310
Shale	20	330
Sandstone, (gas from 334 to 355)	28	358
Total depth		358

NOTE—This record is entirely within the Pottsville.

CHRISTIAN COUNTY.

Production: Small Oil and Gas. Producing Sands: Unnamed (Mississippian).

Log No. 127

W. E. Denton, No. 1, lessor. Location: 1¼ miles east of Crofton. Commenced: August, 1919. Completed: March, 1920. Authority: J. M. Huggins, driller.

Strata.

Mississippian System.	Thickness	Depth
Broken limestone	6	6
Limestone, hard	12	18
Limestone, hard, flinty	20	38
Limestone, hard	23	61
Shale, hard	8	69
Limestone	57	126
Shale, black, hard	9	135
Shale (red rock)	5	140
Shale, dark blue, hard	75	215
Shale, hard	15	230
Limestone, "niggerhead," hard	1	231
Shale, white, hard	33	264
Limestone, hard	9	273
Shale, black, hard	4	277
Limestone, hard, gritty, sandy	3	280
Shale (red rock)	3	283
Shale, blue	16	299
Shale, black, hard	11	310
Shell	2	312
Shale, blue, hard	12	324
Shell, hard, coarse, sandy	8	332
Sand, (little oil)	3	335
Shale, black, sandy	2	337
Shale, blue, hard	23	360
Limestone, hard	3	363
Shale, blue, hard	10	373

A SLEDGE POOL DRILLING.

One of the most productive shallow oil pools of the Allen-Warren County field was the Sledge Pool on Bays Creek. Portable rigs similar to that shown above were used and the pool was rapidly drilled up.

Mississippian System.

	Thickness	Depth
Shale, pink	2	375
Limestone, shell	4	379
Rock, pink	10	389
Shale, brown, hard	4	393
Shale, gray, hard	11	404
Limestone, variable	66	470
Shale, hard, rotten	6	476
Shale	5	481
Limestone, hard	5	486
Shale, gray, hard	4	490
Limestone	10	500
Limestone, (little salt water)	6	506
Shale, black, hard	10	516
Shale, black, hard	9	525
Sand, gray, coarse, (gas)	10	535
Sand, gray, coarse	20	555
Limestone, white, hard	5	560
Limestone, hard	20	580
Shale, hard	5	585
Limestone	40	625
Limestone	7	632
Shale (break), hard	2	634
Limestone	6	640
Sand, (show of oil) (shot)	8	648
Limestone, dark brown	10	658
Limestone, light brown, (water)	119	777
Limestone	43	820
Limestone, gray, gritty, (water)	2	822
Limestone, white, (water)	16	838
Limestone, gray, dark	3	841
Limestone, brown, fine	26	867
Total depth		867

NOTE—This record is entirely within the Mississippian.

Log No. 128

Croft, No. 1, lessor. Location: 1 mile northeast of Crofton. Completed: May 18, 1920. Authority: J. M. Huggins, driller.

Strata.

Mississippian System.

	Thickness	Depth
Soil, red, and clay	4	4
Limestone, loose	6	10
Shale, hard, and limestone	10	20

Mississippian System.	Thickness	Depth
Shale, hard, and shells	22	42
Limestone	3	45
Shale, hard	2	47
Sandstone, gray	15	62
Shale, hard	6	68
Shale, sandy	4	72
Sand, white, (water)	4	76
Shale, black	10	86
Shale, dark, hard	7	93
Limestone, gray, brown	17	110
Shale (break)	1	111
Limestone, dark and brown	36	147
Shale, and shale, hard	65	212
Sand, coarse and gray	6	218
Shale, hard	8	226
Sand, gray, coarse	6	232
Sand, (water)	9	241
Shale, dark, hard	8	249
Shell	1	250
Shale, black, hard	8	258
Limestone, hard	4	262
Shale, hard	4	266
Shale, white, hard	18	284
Limestone, brown, hard	12	296
Shale, black, hard	4	300
Limestone, gray, coarse, gritty	4	304
Shale, black, hard	4	308
Shale, red	4	312
Shale, blue	28	340
Limestone, hard	3	343
Shale, black, hard	10	353
Sand, (show of oil)	7	360
Shale, black, hard	20	380
Limestone, dark, hard	4	384
Shale, gray, hard	14	398
Shell	2	400
Shale, black, hard	3	403
Shale, red	4	407
Limestone, blue, and red rock	3	410
Shale, brown, hard	3	423
Limestone, brown	51	474
Shale, hard, rotten	4	478
Limestone, brown	4	482
Shale, black, hard	4	486
Limestone, brown	4	490
Shale, black, hard	3	493

Mississippian System.	Thickness	Depth
Shale (break)	10	535
Sandstone, black	5	540
Limestone, dark, hard	60	600
Sandstone, black, (oil and gas show)	6	606
Limestone, black	4	610
Limestone, white	20	630
Limestone, dark	21	651
Limestone, black, (sulphur water)	5	656
Limestone, white	20	676
Limestone, black	10	686
Limestone, black	14	700
Limestone, dark	20	720
Limestone, black	20	740
Limestone, white, fine	6	746
Limestone, light	10	756
Limestone, dark	30	786
Limestone, light	14	800
Limestone, black, sandy	30	830
Limestone, dark	40	870
Limestone, dark	40	910
Limestone, light	10	920
Limestone, black	30	950
Limestone, gray	45	995
Shale (break), (4⅞ casing)	9	1,004
Limestone, brown	20	1,024
Limestone, sandy	10	1,034
Limestone, sandy, hard	16	1,050
Limestone, brown	10	1,060
Limestone, light	15	1,075
Limestone, light brown	25	1,100
Limestone, brown, (top of black limestone)..	20	1,120
Limestone, black	35	1,155
Shale, limy	45	1,200
Shale, limy	28	1,228
Limestone, black, hard	7	1,235
Shale, limy	50	1,285
Shale, hard, limy	11	1,296
Shale, dark, limy	64	1,360
Limestone, light brown	40	1,400
Limestone light gray	80	1,480
Limestone, light	25	1,505
Limestone, white	5	1,510
Limestone (cap rock), hard	2	1,512
Limestone, light, sandy, (oil show)	16	1,528
Limestone (cap rock), white, hard	5	1,533
Sand, white, (oil show)	3½	1,536½

Mississippian System.	Thickness	Depth
Sand, light, limy	4½	1,541
Sand, hard	9	1,550
Sand, (oil show)	10	1,560
Sand, shaly	5	1,565
Sand, light, coarse	5	1,570
Sand, (oil)	10	1,580
Total depth		1,580

Well shot at from 1518 to 1538, 40 qts.
Well shot at from 1518 to 1580, 165 qts.

Log No. 129

Earnest Lowther, No. 1, lessor. Huggins and Son, Drillers. Location: near Crofton.

Strata.

Mississippian System.	Thickness	Depth
Clay	6	6
Limestone	10	16
Shale, black	18	34
Limestone, (8¼" casing, 41')	51	85
Shale, sandy	30	115
Sand, white, limy, hard	15	130
Shale, black	35	165
Shale, light	40	205
Limestone, dark	40	245
Shale, black, sandy	15	260
Limestone	32	292
Limestone and shale, hard	3	295
Sandstone	5	300
Shale, black	20	320
Limestone, dark, hard	10	330
Shale, sandy, red	10	340
Limestone, hard	5	345
Shale, light	5	350
Limestone, sandy, (oil show)	15	365
Limestone, white	21	386
Limestone, dark	4	390
Limestone, sandy, (oil show)	6	396
Limestone, dark brown, (6½" casing) ..	24	420
Limestone, hard	40	460
Shale, black	40	500
Limestone, sandy, (gas and oil show) .	6	506
Limestone	19	525

Mississippian System.	Thickness	Depth
Limestone shell	1	494
Shale, gray, hard	10	504
Shale, hard	2	506
Limestone	28	534
Shale, black, hard	36	570
Incomplete depth		570

The tools became lodged in the well, and the drilling was stopped temporarily at 570 feet. Remainder of record not secured. The part given is entirely in the Mississippian.

CHAPTER III.

CLAY COUNTY.

Production: Gas.   Producing Sand: Corniferous (Devonian).

Log No. 130

Peabody Coal Co., No. 1, Unit No. 1.   Location: Heeter Creek, 4½ miles east of Manchester.   Commenced: April 7, 1919.   Completed: June 3, 1919.   Production: Dry.

Strata.

Pennsylvanian System.	Thickness	Depth
Soil, clay	16	16
Shale	14	30
Sandstone, hard	70	100
Sandstone, hard	20	120
Shale, black, soft	8	128
Sandstone, white, hard	22	150
Shale	10	160
Coal	3	163
Sandstone, hard	132	305
Shale, dark	25	330
Sandstone, hard, fine	10	340
Shale	60	400
Sandstone, white, hard	45	445
Shale, black	50	535
Sandstone, hard, fine	50	585
Sandstone, hard	5	590
Shale	40	630
Sandstone	35	665
Sandstone, white, hard, (salt)	369	1,034
(small gas flow 740, water 950)		
Shale	10	1,044
Sandstone, white, hard	26	1,070
Shale	45	1,115
Mississippian System.		
Sandstone, black, hard	6	1,121
Sandstone, gray, hard	39	1,160
Shale, red rock and shells	20	1,180
Shale, white	80	1,260
Limestone (Little Lime), dark, hard	20	1,280
Shale (pencil cave), blue, soft	5	1,285
Limestone (Big Lime)	238	1,523
Sandstone, limy	57	1,580
Shale, red, soft	40	1,620
Limestone, red, hard	20	1,640
Shale	100	1,740

Strata.

Mississippian System.	Thickness	Depth
Limestone, hard	30	1,770
Shale	30	1,800
Sand, (little gas in top, 1,800)	20	1,820
Shale and limestone shells	30	1,850
Devonian System.		
Shale, black (Chattanooga)	171	2,021
Limestone (Irvine Sand)	94	2,115
Shale	30	2,145
Total depth		2,145

Log No. 131

Peabody Coal Co. Well No. 2. Location: Sutton Branch of Goose Creek, 5 miles northeast of Manchester, Clay County. Production: Dry.

Strata.

Pennsylvanian System.	Thickness	Depth
Clay	16	16
Sandstone, hard	24	40
Shale, black	15	55
Sandstone, hard	65	120
Shale, light	45	165
Sandstone, (salt water)	55	220
Sandstone, (small gas at 362)	256	476
Coal	2	478
Shale, black	82	560
Sandstone	145	705
Shale, black	10	715
Limestone	5	720
Shale, red	95	815
Shale black	55	870
Sandstone, hard, fine	50	920
Mississippian System.		
Shale	35	955
Limestone (Little Lime)	25	980
Shale (pencil cave)	4	984
Limestone (Big Lime)	264	1,230
Shale, red, soft	60	1,290
Limestone, red	30	1,320
Limestone, black	40	1,360
Shale, black	40	1,400
Limestone	60	1,460
Shale, white	50	1,510

Devonian System.

	Thickness	Depth
Shale, black	132	1,642
Sand	13	1,655
Shale, brown	35	1,690
Limestone (Irvine Sand)	95	1,785
Shale, black	35	1,820
Sandstone, (small gas)	25	1,845
Shale, red	65	1,910
Shale, red	25	1,935
Limestone shell	25	1,960
Shale, red	20	1,980
Shale	11	1,991
Total depth		1,991

Dry and plugged, with all casing pulled. The Irvine ''sand'' was principally all 'limestone, with 20 feet in the center which was nearly all ''sand'' and very hard, and no show for oil or gas.

NOTE—The Devonian-Silurian contact and the Silurian-Ordovician contact is not defined. The well pierced the top of the Ordovician rocks.

Log No. 132

Peabody, No. 3, lessor. Location: On Long Fork of Hector's Creek of Red River, in Clay County. Commenced: August 18, 1919. Completed: November 19, 1919. Authority: E. H. Mould, Pineville, Ky.

Strata.

Pennsylvanian System.

	Thickness	Depth
Clay	15	15
Sand	65	80
Shale	40	120
Sand	35	155
Shale, black	235	390
Limestone, white	50	440
Sand, white	50	490
Shale and limestone shells	125	615
Salt sand, white, hard	390	1,005
Shale, black	18	1,023

Mississippian System.

	Thickness	Depth
Limestone, red	17	1,040
Shale, sandy, red	30	1,070
Limestone, white	20	1,090
Shale, sandy, red	50	1,140
Shale, black	35	1,175
Limestone, black	5	1,180

Mississippian System.	Thickness	Depth
Shale	20	1,200
Limestone	20	1,220
Shale (pencil cave), soft, cave	6	1,226
Limestone (Big Lime)	274	1,500
Shale, black	10	1,510
Shale, sandy, red	70	1,580
Shale, white	100	1,680
Shale and limestone shells	115	1,795
Devonian System.		
Shale, brown	162	1,957
Limestone (Irvine "sand"), dark, hard	74	2,031
Shale and limestone shells (principally lime- stone, with neither oil or gas)	59	2,090
Total depth		2,090

Water hole, 160.

Dry and plugged, with all casing pulled.

Conductor, 16 feet pulled.

10" casing, 64 feet pulled.

3¼" casing, 1020 feet pulled.

6⅝" casing, 1270 feet pulled.

Log No. 133

Peabody No. 4, lessor. Location: 12 miles southeast of Manchester, on Otter Creek of Goose Creek. Commenced: April 24, 1920. Completed July 30, 1920. Production: Dry, Casing pulled, hole plugged. Authority: P. Kennedy, Barbourville, Ky.

Strata.

Pennsylvanian System.	Thickness	Depth
Soil	31	31
Shale, blue, hard	49	80
Coal	2	82
Shale, blue, hard	53	135
Sand	10	145
Shale, black, hard	20	165
Sand, white	128	293
Shale, black, hard	12	305
Shale, and limestone shells, black	15	320
Sand, white	100	420
Shale, blue, hard	65	485
Shale, hard, and limestone shells	45	530
Shale, black, hard	35	565
Sand, broken, (gas at 590)	30	595
Shale, blue, hard	10	605

Pennsylvanian System.	Thickness	Depth
Salt sand, (water at 800)	440	1,045
Shale, blue, hard	10	1,055
Sand, white	100	1,155
Shale, blue, hard	55	1,210
Limestone, black	20	1,230
Shale, blue, hard	10	1,240
Shale, red	40	1,280
Sand, blue	25	1,305
Sand, white	135	1,440
Shale, blue, hard	10	1,450
Mississippian System.		
Limestone (Little Lime)	20	1,470
Sandstone (pencil cave)	10	1,480
Limestone (Big Lime)	225	1,705
Sand, white	20	1,725
Limestone, red	23	1,748
Red rock	27	1,775
Sand, red	55	1,830
Limestone, blue	5	1,835
Sand, blue	210	2,045
Devonian System.		
Shale, hard	25	2,070
Shale, black	129	2,199
Sand, Irvine	68	2,267
Shale, white, hard	17	2,284
Total depth		2,284

Log No. 134

Oneida Institute, No. 1, lessor. C. P. Kennedy, et al., No. 1, lessees. Location: just north of South Fork of Kentucky River, near Oneida, Clay County, Kentucky. Commenced: 1917. Completed: 1918. Production: Dry; some little gas. Altitude: 735 feet. Authority: D. C. Moffett, contractor.

Strata.

Pennsylvanian System.	Thickness	Depth
Soil	23	23
Sandstone, very hard	73	96
Shale, brown	7	103
Sandstone	17	120
Shale, brown	43	163
Sandstone, very hard	10	173
Shale, white	3	176
Sandstone, hard	103	278
Shale, brown	14	292

	Thickness	Depth
Pennsylvanian System.		
Sandstone	23	315
Sand, dark and shale	3	318
Sandstone, white, hard	32	350
Shale, gray	15	365
Sand, limy	4	369
Shale, brown	29	398
Sand, limy	7	405
Shale, brown	10	415
Sandstone, white, hard	137	552
Shale, brown	5	557
Limestone (?)	7	564
Shale, brown	31	595
Red rock	15	610
Shale, white	10	620
Limestone	6	626
Shale, gray	60	686
Limestone	2	688
Shale, white	8	696
Limestone	13	709
Sandstone	10	719
Shale, brown	2	721
Shale, white	11	732
Mississippian System.		
Big Lime (St. Louis), (gas at 1,025)	259	991
Shale, gray	43	1,034
Shale, red	53	1,087
Shale, white	56	1,143
Shale, brown, and limestone shells	67	1,210
Limestone and shale	7	1,217
Shale, gray	60	1,277
Devonian System.		
Shale, brown, (gas at 1,300)	145	1,422
Shale, gray	15	1,437
Shale, black	17	1,454
Limestone (Irvine Sand), cap very hard	10	1,464
Limestone (Irvine Sand)	92	1,556
Shale, blue	44	1,600
Shale, gray	90	1,690
Shale, red	42	1,732
Shale, green	31	1,763
Shale, white	42	1,805
Shale, red	10	1,815
Limestone	5	1,820
Shale, gray, and limestone shells	6	1,826
Limestone	10	1,836
Sand, white, limy	20	1,856

Devonian System.   Thickness Depth

    Limestone and shale, sandy   14   1,870
    Limestone   15   1,885
    Shale, white   10   1,895
    Limestone, very hard   126   2,021
        Total depth   2,021

        8 inch casing, 23 feet.
        6¼ inch casing, 742 feet.
        Gas at 235, 1,025 and 1,300 feet.
        Ist water 73 feet, 1st salt water 285 feet and again at
           350, 440, 490, and 665 feet.

   NOTE—This well finished in the Ordovician. Devonian-Silurian and Silurian-Ordovician contacts are not defined. The record is not very accurate.

Log No. 135

   Beverly Burns, No. 1, lessor. Oneida Oil & Gas Co. (formerly C. T. Cherry), No. 3, lessee. Location: Bullskin Creek, 2 miles southeast of Oneida, and near Seth post office, Clay County. Commenced: in 1918. Completed: in 1918. Production: 780,000 cu. ft. gas. Rock pressure: 270 lbs. Altitude: 795 feet.

     Strata.

Pennsylvanian System.   Thickness Depth

    Soil   15   15
    Shale, gray   40   55
    Coal   3   58
    Sandstone   175   233
    Shale   5   238
    Sandstone   222   460
    Sandstone, hard and fine   28   488
    Sandstone   65   553
    Sandstone, hard and fine   50   603
    Shale   20   623
    Sandstone, hard   46   669
    Shale, white   145   814
    Sandstone   54   868

Mississippian System.

    Limestone (Big Lime)   225   1,093
    Shale (Red Rock)   15   1,108
    Shale, sandy   70   1,178
    Shale, brown   190   1,368

Devonian System.	Thickness	Depth
Shale, black (Chattanooga)	135	1,503
Shale, gray	50	1,553
Limestone (Irvine "sand")	68	1,621
Shale, black, hard	2	1,623
Total depth		1,623

Log No. 136

Irven Hensley, No. 1, lessor. Oneida Oil & Gas Co., No. 1, lessee. Location: on Red Bird Creek, 2 miles above Oneida. Commenced: Jan. 1, 1920. Completed: Feb. 3, 1920. Production: 1,350,000 cu. ft. gas. Rock pressure: 310 lbs. Altitude: 780 feet.

Strata.

Pennsylvanian System.	Thickness	Depth
Soil	30	30
Shale, hard, and sand	120	150
Sandstone, hard	490	640
Shale and red rock	60	700
Shale and limestone	145	845
Mississippian System.		
Limestone, (Big Lime)	260	1,105
Shale	45	1,150
Shale (red rock)	60	1,210
Shale, hard, gritty	170	1,380
Devonian System.		
Shale, brown	160	1,540
Shale, black	30	1,570
Limestone (Irvine "sand"), (gas)	10	1,580
Total depth		1,580

Log No. 137

H. M. Burns, No. 1, lessor. Oneida Oil & Gas Co., No. 4, lessee. Location: on Bullskin Creek, 2½ miles southeast of Oneida. near Seth P. O., Clay County. Commenced: Aug. 18, 1920. Completed: Nov. 18, 1920. Production: 474,000 cu. ft. gas. Rock pressure: 300 lbs. (Apr. 29, 1921.) Altitude: 800 feet.

Strata.

Pennsylvanian System.	Thickness	Depth
Alluvium, yellow, sandy clay	20	20
Sandstone, yellow, hard	80	100
Shale, blue, soft	20	120
Sandstone, yellow, hard	500	620
Shale, blue, soft	245	865

Mississippian System.	Thickness	Depth
Limestone (Big Lime), white, hard	335	1,200
Shale, blue, soft	230	1,430
Devonian System.		
Shale, black, soft (Chattanooga)	205	1,635
Shale, blue, soft, fire clay	13	1,648
Limestone (Irvine "sand"), brown, hard (gas)	37	1,685
Total depth		1,685

8¼ inch casing, 20 feet.
6⅝ inch casing, 650 feet.

CLINTON COUNTY.

Production: Oil and Gas. Producing Sands: Beaver (Mississippian); Sunnybrook and Trenton (Ordovician); and Beech Bottom (Knox Dolomite age?) (Cambro-Ordovician).

Log No. 138

G. W. Ward, No. 1, lessor. Completed: October 3, 1907. Production: Dry. Abandoned. Authority: The New Domain Oil & Gas Co.

Strata.

Mississippian System.	Thickness	Depth
Clay, red, soft	11	11
Limestone, blue, hard	219	230
Shale, blue, hard, soft (New Providence)	40	270
Devonian System.		
Shale, black, soft (Chattanooga)	20	290
Ordovician System.		
Limestone, light gray (Saluda)	325	615
Limestone, dark gray, hard	200	815
Shale (pencil cave), dark blue, soft	2	817
Limestone, dark gray, hard	18	835
Total depth		835

Log No. 139

G. W. Boles, No. 1, lessor. Completed: June 11, 1907. Production: Dry. Authority: The New Domain Oil & Gas Company. Location: Fannis Creek.

Strata.

Mississippian System.	Thickness	Depth
Soil	5	5
Limestone, blue, hard	323	328
Shale, blue, hard, soft	55	383

Devonian System.	Thickness	Depth
Shale, black, soft (Chattanooga)	20	403
Ordovician System.		
Limestone, gray, hard	413	816
Shale (pencil cave), blue, soft	2	818
Limestone, light gray, hard	97	915
Total depth		915

Log No. 140.

Jacob Speck, No. 1, lessor. Completed: March 1, 1907. Production: Dry. Authority: The New Domain Oil & Gas Co. Location: 2 miles south of Albany.

Strata.

Mississippian System.	Thickness	Depth
Limestone, black, hard, (sulphur water at 40)	55	55
Limestone, blue, hard	10	65
Shale, black, hard, soft	10	75
Limestone, variable	250	325
Shale, blue, hard, soft (New Providence)	20	345
Devonian System.		
Shale, black, soft (Chattanooga)	20	365
Ordovician System.		
Limestone, white, (gas at 580 to 885)	565	930
Shale (pencil cave), soft	5	935
Limestone, hard, variable	995	1,930
Limestone, blue, soft	20	1,950
Total depth		1,950

Log No. 141

J. T. Tompkins, No. 1, lessor. Completed: December 27, 1906. Production: first day 40 bbls. Authority: The New Domain Oil & Gas Company. Location: Fannis Creek of Illwill Creek.

Strata.

Devonian System.	Thickness	Depth
Clay, soft	5	5
Shale, black, soft (Chattanooga)	18	23
Ordovician System.		
Limestone, white, hard (Saluda)	27	50
Limestone, hard, variable, (oil)	226	276
Total depth		276

Log No. 142

C. L. Holsapple, No. 1, lessor. Completed: November 24, 1904. Production: Dry. Abandoned. Authority: New Domain Oil & Gas Co. Location: near Forrest Co. Hage P. O., headwaters of Willis Creek.

Strata.

Mississippian System.	Thickness	Depth
Limestone, blue, hard	30	30
Limestone, gray, soft	200	230
Limestone, blue, hard	110	340
Devonian System.		
Shale, black, soft (Chattanooga)	25	365
Ordovician System.		
Limestone, light, hard	435	800
Shale, blue, hard, soft	15	815
Limestone, light, hard	707	1,522
Total depth		1,522

Show of oil at 750.
Vein of gas at 238 and 1,135 feet.

Log No. 143

J. F. Brentz, No. 1, lessor. Completed: October 17, 1904. Production: Dry. Authority: The New Domain Oil & Gas Company. Location: Near Ida Post Office.

Strata.

Mississippian System.	Thickness	Depth
Limestone, blue, hard	180	180
Limestone, white, hard	170	350
Limestone, white, hard	180	530
Limestone, gray, hard	150	680
Devonian System.		
Shale, black, medium (Chattanooga)	20	700
Ordovician System.		
Limestone, dark blue, (small gas at 904)	270	970
Shale (pencil cave)	7	977
Limestone, brown, hard	110	1,087
Limestone, dark blue, medium	258	1,345
Limestone, dark blue, medium	15	1,360
Total depth		1,360

Log No. 144

John Johnson, No. 1, lessor. Completed: November 6, 1906. Authority: The New Domain Oil & Gas Company. Location: Wolfe River, near Tenn. Line.

Strata.

Mississippian System.	Thickness	Depth
Soil and shells	14	14
Shale, black, hard, medium (water 32)	18	32
Devonian System.		
Shale, black, soft (Chattanooga)	22	54
Ordovician System.		
Limestone, black, variable (gas 330)	546	600
Shale (pencil cave), blue, soft	2	602
Limestone, gray, variable (gas 745)	738	1,340
Limestone, gray, hard (oil 1,342)	130	1,470
Limestone, gray, hard	180	1,650
Limestone, white, hard (salt water 1,655) ...	150	1,800
Limestone, gray, hard, gritty	200	2,000
Total depth		2,000

Log No. 145

L. D. Bow, No. 1, lessor. Completed: November 8, 1907. Production: Dry. Authority: The New Domain Oil & Gas Company. Location: Fannis Creek.

Strata.

Mississippian System.	Thickness	Depth
Clay, red, soft	14	14
Shale, blue, hard soft	290	304
Devonian System.		
Shale, black, soft (Chattanooga)	24	328
*Ordovician System.		
Limestone, gray, medium	502	830
Shale (pencil cave), soft	6	836
Limestone, dark gray, medium	37	873
Total depth		873

Log No. 146

E. Luttrell, No. 1, lessor. Completed: January 7, 1905. Production: Dry. Authority: The New Domain Oil & Gas Company. Location: near Cumberland River at Ida P. O.

Strata.

Mississippian System.	Thickness	Depth
Limestone, blue, hard	20	20
Limestone, gray, hard ,	40	60
Limestone, white, soft	200	260
Devonian System.		
Shale, black, soft (Chattanooga)	25	285
Ordovician System.		
Limestone, white, hard	615	900
Shale (pencil cave), brown, soft	3	903
Limestone, brown, hard	197	1,100
Limestone, white, hard	172	1,272
Shale (pencil cave), brown, soft	15	1,287
Limestone, white, hard	213	1,500
Total depth		1,500

Log No. 147

G. A. Thurman, No. 1, lessor. Completed: August 14, 1907. Production: Dry. Well abandoned. Authority: The New Domain Oil &
Gas Company. Location: Fannis Creek of Illwill Creek just above
forks.

Strata.

Mississippian System.	Thickness	Depth
Soil	12	12
Limestone, blue, soft	6	18
Clay	34	52
Limestone, dark gray, soft	252	304
Shale, blue, hard, soft (New Providence)	60	364
Devonian System.		
Shale, black, soft (Chattanooga)	25	389
Ordovician System.		
Limestone, gray, medium (Saluda)	171	560
Shale (pencil cave); soft	3	563
Limestone, gray, hard	327	890
Total depth		890

NOTE—Only the upper part of the 171 feet above 560 feet in
depth is Saluda.

Log No. 148

W. F. Braswell, No. 1, lessor. Completed: July 26, 1907. Production: Dry, abandoned. Authority: The New Domain Oil & Gas Company. Location: 2 miles east of Beech Bo Hon.

Strata.

Mississippian System.	Thickness	Depth
Clay, red, soft	45	45
Gravel	10	55
Clay, blue, soft	10	65
Limestone, gray, hard	40	105
Limestone, dark, hard	60	165
Limestone, dark, hard	60	225
Limestone, white, medium	240	465
Devonian System.		
Shale, black, soft (Chattanooga)	20	485
Ordovician System.		
Limestone, white, soft (Saluda in part)	315	800
Limestone, black, white, hard	150	950
Limestone, white, hard	325	1,275
Total depth		1,275

NOTE—Only the upper portion of the 315 feet above 800 feet in depth is referable to the Saluda.

Log No. 149

J. T. Tompkins, No. 3, lessor. Completed: October 16, 1907. Production: 10 bbls. oil. Authority: The New Domain Oil & Gas Company. Location: Fannis Creek.

Strata.

Mississippian and Devonian Systems.	Thickness	Depth
Limestone and black shale	280	280
Ordovician System.		
Limestone, gray, medium	270	550
Shale (pencil cave), dark blue, soft	3	553
Limestone, gray, dark, hard	29	582
Total depth		582

Log No. 150

George Smith, No. 1, lessor. Beech Bottom Oil & Gas Co., No. 3, lessee. Location: On Kogar Creek, Clinton County, Ky. Commenced: Feb. **6, 1922.** Completed: March **28, 1922.** Drillers: Otha Dalton, Geo. Davison. Field Manager: Less Combest. Production: 5 bbls. estimated. Edge well.

Strata.

	Thickness	Depth.
Mississippian System.		
Limestone, white and gravel	43	43
Limestone (fresh water)	37	80
Limestone, brown, sticky, sandy	10	90
Limestone, black (black sulphur water, gas) ..	10	100
Limestone, white and gray (set. 185 feet 8¼)	85	185
Limestone, white and gray	115	300
Limestone, black	35	335
Limestone, black, mixed with shale	5	340
Shale	10	350
Limestone rock, hard, gray	2	352
Limestone (Beaver "sand"), (oil)	8	360
Devonian System.		
Shale, black (Chattanooga)	35	395
Ordovician System.		
Limestone, blue and gray	280	675
Limestone, soft (gas)	15	690
Limestone, gray	60	750
Limestone, brown, sandy, sticky	15	765
Limestone, gray, mixed	75	840
Limestone, soft, mixed with shale, (Sunnybrook formations)	85	925
Shale, 1st, (pencil cave)	8	933
Limestone, soft (soapstone)	27	960
Shale, 2nd, (pencil cave), very soft, (set. 40 feet 6-5/8)	5	965
Limestone, grey	30	995
Limestone, brown, sandy, (good looking sand, no oil show) (Calciferous)	105	1,100
Limestone, blue, mixed	150	1,250
Limestone, blue	445	1,695
Cambro-Ordovician System.		
Sandstone, brown, mixed with limestone	33	1,728
Limestone, (oil show)	30	1,758
Sandstone, hard, close	7	1,765
Sandstone, soft, brown	5	1,770

Cambro-Ordovician System.	Thickness	Depth
Sandstone, (oil, high gravity, green)	10	1,780
Sandstone, dry, brown	5	1,785
Limestone	44	1,829
Limestone, dry, brown to gray	20	1,849
Limestone, sandy, dry, brown	41	1,890
Limestone, blue	10	1,900
Limestone, blue, sandy, hard	25	1,925
Limestone, brown, hard, sandy, (salt?)	9½	1,934½
Total depth		1,934½

Only pay of importance, 1770-1780.

Small showing in Beaver formation, 352.

Small showings in Trenton formation or Knox Dolomite, 1728-1758.

Well was completed without any fishing jobs, water troubles or caves.

Set 8¼ to 185 feet. Set 40 feet 6-5/8 at 965.

Beveled at both ends to cut off caves.

NOTE—The stratigraphic position of the 10 feet of oil ''sand'' above 1,780 is in dispute. By some it is claimed that the Trenton overlies the 1st Pencil Cave at 925 feet, which is undoubtedly the Bentonite of Pickett County, Tenn., wells. The record from 1,728 to 1,934½ is then Knox Dolomite (Cambro-Ordovician), showing 52 feet of oil ''sand'' with two pays. This well compares favorably with the record of the Cinda Sells, No. 1, Holbert Creek, near Wolfe River, Pickett County, near Fentress County line. Sfr. Tenn. Geol. Surv., Bull. No. 25, p. 57, 1921. Other authorities reject all of the above and claim this oil ''sand'' is lower Ordovician.

CRITTENDEN COUNTY.

Production: Neither oil or gas to date. Producing Sands: None recognized to date.

Log No. 151

O. C. & G. G. Cook, lessees. Location: ⅓ mile east of Marion P. O. Commenced: April 25, 1921. Driller: J. R. Butts. Casing: 290 feet of 6¼ in. Stratigraphic determinations by Stuart Weller, Ass't Geologist.

Strata.

Mississippian System.	Thickness	Depth
Clay, red, Cypress	7	7
Sandstone, white, Cypress	45	52
Mud, red, Paint Creek	10	62
Shale, blue, (1st water 70), Paint Creek	30	92
Limestone, dark, Paint Creek	4	96
Shale, gray, (2nd water 169), Paint Creek ..	73	169

Mississippian System.   Thickness Depth

	Thickness	Depth
Limestone, dark, Paint Creek	2	171
Shale, gray, Paint Creek	70	241
Sand, white, (Bethel)	47	288
Shale, gray, Renault	20	308
Limestone, black, Renault	4	312
Limestone, hard, sandy, Renault	38	350
Shale, blue, Renault	2	352
Limestone, gray, Renault	15	367
Limestone, gray, and shale, mixed, Renault ..	30	397
Limestone, blue, Renault	2	399
Limestone, light brown, (oil show 400), St. Genevieve	50	449
Sand, dark, St. Genevieve	2	451
Shale, blue, St. Genevieve	6	457
Limestone, gray, St. Genevieve	70	527
Limestone, dark, St. Genevieve	10	537
Limestone, gray, St. Genevieve	75	612
Limestone, gray, oolite specks, St. Genevieve	30	642
Flint, hard, (sea level), St. Louis	15	657
Limestone, gray, St. Louis	40	697
Flint, blue, St. Louis	30	727
Limestone, light brown, St. Louis	12	739
Limestone, gray, St. Louis	30	769
Chert, white & blue, very hard, St. Louis	33	802
Limestone, light brown, Spergen	15	817
Limestone, gray, (oolite), Spergen	40	857
Limestone, blue, Spergen	20	877
Limestone, dark, Warsaw passing down into Keokuk and possibly Burlington	8	885
Limestone, brown, Warsaw passing down into Keokuk and possibly Burlington	40	925
Limestone, gray, Warsaw passing down into Keokuk and possibly Burlington	8	933
Limestone, dark, Warsaw passing down into Keokuk and possibly Burlington	25	958
Limestone or shale, dark, Warsaw passing down into Keokuk and possibly Burlington	3	961
Sand, Warsaw passing down into Keokuk and possibly Burlington	5	966
Limestone, dark, Warsaw passing down into Keokuk and possibly Burlington	52	1,018
Limestone, very dark, Warsaw passing down into Keokuk and possibly Burlington	5	1,023

Mississippian System.   Thickness Depth
   Limestone, little lighter, Warsaw passing down
      into Keokuk and possibly Burlington   10   1,033
   Limestone, still lighter, Warsaw passing down
      into Keokuk and possibly Burlington   1,040
      Incomplete depth   1,040

NOTE—It is not possible in this record to determine the Renault-St. Genevieve contact. The Renault should be from 75 to 100 feet thick. The black shale (Devonian) should be expected beneath the lowest recorded limestones, at some depth.

CUMBERLAND COUNTY.

Production: **Oil and Gas.** Producing Sands: **Sunnybrook and Trenton (Ordovician).**

Log No. 152

A. M. Fudge, No. 1, lessor. Location: near Burkesville. Completed: in 1903. Authority: The New Domain Oil & Gas Co.
     Strata.

Ordovician System.   Thickness Depth

	Thickness	Depth
Limestone, blue, black, hard, close	200	200
Limestone, blue, gray, soft (oil show 452) ..	255	455
Limestone, blue, black, soft, open	115	570
Limestone, blue, gray, soft, open	430	1,000
Total depth		1,000

Gas at 150, 285 and 340 feet.

Log No. 153

W. M. Bryant, No. 1, lessor. Location: Eighth Precinct. Completed: September 2, 1903. Authority: The New Domain Oil & Gas Company.
     Strata.

Ordovician System.   Thickness Depth

	Thickness	Depth
Limestone, white, hard, close	50	50
Limestone, blue, soft, loose	200	250
Limestone, gray, soft, loose	50	300
Limestone, blue, soft, open	75	375
Limestone, gray, soft, close	50	425
Limestone, gray, hard, close	125	550
Limestone, dark gray, soft, loose	50	600
Limestone, white, hard, close	100	700
Limestone, gray, hard, close	30	730
Limestone, dark gray, soft, loose	100	830
Limestone, dark gray, hard, loose	50	880
Limestone, dark, soft, loose	75	955
Limestone, dark, hard, open	46	1,001
Total depth		1,001

A little gas at 225 feet.

THE KENTUCKY "TRENTON."
The view shows the Kentucky River cliffs of Garrard County just
above the Mouth of the Dix River. The section from the cave (left) up,
is the Trenton which is productive in Southern Kentucky.

Log No. 154

W. R. Neeley, No. 2, lessor. Completed: October 6, 1904. Production: 15 bbls. oil per day. Authority: The New Domain Oil & Gas Company.

Strata.

Ordovician System.	Thickness	Depth
Limestone, dark gray, hard	121	121
Limestone, gray, hard	60	181
Shale, blue, hard, soft	10	191
Limestone, brown, hard	10	201
Shale, blue, hard, soft	40	241
Limestone, gray, loose	100	341
Shale, blue, hard, soft	15	356
Limestone, brown, soft	269	625
Limestone, dark gray, hard	105	730
Limestone, gray, loose	20	750
Limestone, dark gray, loose	33	783
Total depth		783

Log No. 155

Cloyd Heirs, No. 2, lessors. Completed: May 5, 1903. Production: Dry, following shot. Authority: The New Domain Oil & Gas Company.

Strata.

Ordovician System.	Thickness	Depth
Limestone	250	250
Limestone, blue, medium	100	350
Sand, gray, hard	125	475
Limestone, soft	33	508
Shale, white, hard, soft	2	510
Limestone, white, soft, (oil show 522)	35	545
Limestone, hard, soft	100	645
Limestone, white, soft	'55	700
Limestone ''sand,'' hard	150	850
Shale, white, hard	30	880
Limestone, gray, soft	10	890
Limestone, dark, hard	35	925
Limestone, white, hard	25	950
Total depth		950

Log No. 156

Radford, No. 1, lessor. Location: Brush Creek Pool. Casing head elevation: 550 feet, approx. Drilled about 1867. Structural Location: Tip of pronounced dome on which are also located the Glidewell, Melton and Parrish wells. Authority: L. Beckner.

Strata.

Ordovician System.

	Thickness	Depth
Soil, ...	15	15
Limestone, (salt water & gas 190)	175	190
Limestone, (uncontrollable gas 290)	100	290
Total depth		290

NOTE—Large and uncontrollable gas was struck at "about 290 feet, which blew Mr. Classon, the driller, off his stool and 30 feet away into a gulley." The well was allowed to blow open for a week or more, when it was finally abandoned with the tools in the hole. Statement of Jacob Radford, an eyewitness, July, 1920.

Log No. 157.

Glidewell, No. 1, lessor. Location: Across the Cumberland River from Bakerton P. O., in Brush Creek Pool. Drilled: about 1867. Production: a good oil show.

Log No. 158

Glidewell, No. 2, lessor. Location: Just across the branch from Glidewell, No. 1. Drilled: about 1892. Production: oil at 390 feet depth.

NOTE—Fragmentary information upon the further development of this tract is as follows: Glidewell, No. 3, was drilled about 1906, complete log and depth unknown. Glidewell, Nos. 4 and 5, were drilled subsequently, and the record is said to have been the same as Glidewell No. 3. The Wes Melton Nos. 1 and 2 had a similar record to Glidewell Nos. 3, 4 and 5. The Parrish wells Nos. 1, 2, 3, and 4 were also similar to the Glidewell records it is said, but the records have not been secured. All of these wells started in the Maysville (Ordovician) and struck oil at 380 to 420 feet. Casing head elevation from about 540 feet A. T., and all in Brush Creek Pool.

DAVIESS COUNTY.

Production: Small oil and gas. Producing Sands: Unnamed of Alleghany and Pottsville age (Pennsylvanian).

Log No. 159

England, No. 1, lessor. Location: Across the road, east of the Eiglehard wells about 800 feet, between Calhoun and Owensboro. Operators: Henry O'Hara, St. Louis; B. A. Kinney, Penn.; Luckett and Boggett, St. Louis. Authority: J. G. Stuart.

Strata.

Pennsylvanian System.	Thickness	Depth
Clay, green, and chert	25	25
Sandstone, soft	5	30
Limestone and shale	10	40
Shale, black, and coal	5	45
Shale, blue, hard	10	55
Shale, gray, hard	35	90
Shale, black, coal	4	94
Fire clay	1	95
Broken limestone and shale	16	111
Flint rock, gray	5	116
Limestone, broken	10	126
Shale, blue	20	146
Shale, black, coal	5	151
Fire clay	1	152
Limestone, blue	10	162
Shale, limy, (water)	5	167
Shale, blue, carbonaceous	17	184
Shale, black	5	189
Shale, limy	15	204
Sandstone	26	230
Total depth		230

Two sands, or rather, sand with parting.
Top sand good show: 2nd sand much better.

Log No. 160

Roy Haggerman, No. 1, lessor. Location: 3 miles southwest of Panther. Operator: Elmer Little, Gunther Petrie, and others. Authority: C. Shadwick, driller.

Strata.

Pennsylvanian System.	Thickness	Depth
Clay, yellow	4	4
Sandstone, brown	15	19

Pennsylvanian System

	Thickness	Depth
Shale, soft	1	20
Sandstone, blue	10	30
Shale, blue, sandy	30	60
Shale, blue, soft	10	70
Shale, black, coal	5	75
Fire clay	3	78
Limestone, blue	2	80
Limestone, gray, and sand	20	100
Shale, blue	10	110
Shale, blue	45	155
Shale, black	5	160
Fire, clay	5	165
Shale, brown	5	170
Shale, blue, sandy	15	185
Shale, black, sandy	10	195
Shale, black	6	201
Fire clay	3	204
	1	205
Sandstone, (oil)	3	208
Shale, blue	7	215
Shale, gray	10	225
Sand, white, (oil)	20	245
Shale, blue, hard	10	255
Shale, black, and coal	14	269
Fire clay	1	270
Shale, gray	20	290
Shale, soft, dry	5	295
Total depth		295

Log No. 161

School House Well, 3 miles northwest of Panther.

Strata.

Pennsylvanian System.

	Thickness	Depth
Clay, yellow	20	20
Sandstone	6	26
Shale, black	5	31
Fire clay	3	34
Limestone, not hard	4	38
Sand and limestone	25	63
Shale, dark blue	24	87
Shale, blue, soft	30	117
Shale	5	122
Fire clay	6	128
Total depth		128

Log No. 162

R. A. Alvey, No. 1, lessor. Location: 1½ Miles southeast of Panther, on Bushy Fork. S. L. elevation 415' (about). Well No. 1 is located about 300 ft. east of Well No. 2. Authority: Turner Burns, Mgr. Panther Creek Oil Co.

Strata.

Pennsylvanian System.	Thickness	Depth
Sandstone	6	6
Shale, sandy	10	16
Shale, brown and blue	65	81
Shale	2	83
Fire clay	2	85
Limestone and shale	20	105
Sandstone, (oil) (show gas)	12	117
Total depth		117

Log No. 163

R. A. Alvey, No 2, lessor. Panther Creek Oil Co., Owensboro, Ky., lessee. Location: 300 ft. from No. 1. Log by driller, C. Shadwick.

Strata.

Pennsylvanian System.	Thickness	Depth
Clay	21	21
Sandstone	4	25
Shale	5	30
Shale, brown and blue (called by driller soap stone)	50	80
Shale, black	3	83
Fire clay	6	89
Limestone, broken, and shale, 113	18	107
Oil sand 117	13	120
Break, parting not identified by driller	2	122
Sandstone (oil), (gas at 127)	9	131
Shale, gray	18	149
Shale, soft	9	158
Shale, black	6	164
Fire clay	4	168
Shale, gray, sandy	24	192
Limestone, gray	3	195
Fire clay	3	198
Shale, blue, sandy	5	203
Sandstone and limestone shale	20	223
Shale, black, sandy	35	258
Incomplete depth		258

An incomplete log. This well was drilled somewhat deeper. Well left in condition to be shot. Authority: J. G. Stuart.

Log No. 164

Eiglehardt, No. 1, lessor. Location: between Owensboro and Calhoun, 16 miles from Owensboro, 8 miles from Calhoun. Operators: B. A. Kinney, Oil & Gas Inspector for State of Indiana, Henry O'Hara. Luckett & Baggett, of St. Louis, Mo.

Strata.

Pennsylvanian System.	Thickness	Depth
Clay, yellow	20	20
Shale, black	4	24
Fire clay	6	30
Shale, hard	20	50
Shale, blue	10	60
Shale, black	5	65
Shale, blue	5	70
Shale, blue, sandy	25	95
Shale, black, coal	4	99
Fire clay	1	100
Broken limestone shale	15	115
Limestone, blue, cherty	5	120
Broken limestone (gravel?)	5	125
Shale, blue	15	140
Shale, black	3	143
Fire clay	2	145
Limestone, blue	5	150
Shale, gray, sandy	25	175
Shale, blue, limy	15	190
Shale, broken, limy	7	197
Sand	8	205
Sand (oil)	7	212
Total depth		212

These wells had from 16 to 26 feet good sand according to the operators and the driller. Three wells on this farm. All logs run alike. All promise pay oil. Authority: J. G. Stuart.

EDMONSON COUNTY.

Production: Oil and Gas. Producing Sands: Pottsville (Pennsylvanian); "Shallow" of Warren County (Mississippian); Corniferous (Devonian); "Deep" (Silurian).

Log No. 164-A

Location: Branch of Dismal Creek. Production: Dry. (Oil shows only.) Authority: J. Owen Bryant.

Strata.

Pennsylvanian System.	Thickness	Depth
Clay	8	8
Sand, black	25	33

Mississippian System.   Thickness Depth
  Shale   25   58
  Limestone   9   67
  Shale   15   82
  Limestone   34   116
  Shale   8   124
  Limestone   42   166
  Shale   17   183
  Limestone   20   203
  Sand, black   15   218
  Sand and shale   42   260
  Shale   20   280
  Limestone, gray   524   804
     Total depth   804

NOTE—This well started just below the lowest coal.

ELLIOTT COUNTY.*

Production: **Oil and Gas.**   Producing Sands: **Wier and Berea**
(Mississippian).

Log No. 165

Ad. Johnson, No. 1, lessor.   Elcaro Oil & Gas Co. (No. 2), lessee.
Location: southeast of Lawton, near the head of Big Sinking Creek.
Commenced: December 4, 1920.   Completed: January 15, 1921.   In-
itial production: .... bbls. oil.   Authority: C. E. Bales.

       Strata.
Pennsylvanian System.   Thickness Depth
  Soil   14   14
  Sandstone   45   59
Mississippian System.
  Shale (red rock) and fire clay   12   71
  Limestone (Big Lime)   91   162
  Limestone, sandy   20   182
  Shale, blue   73   255
  Sandstone, gray (little show of oil) ..   18   273
  Shale, blue   202   475
  Sandstone, dark blue (strong gas pressure) .. 10   485
  Shale, blue   140   625
  Sandstone (Wier), (show of oil)   27   652
  Shale, brown (Sunbury)   15   667
  Sandstone (Berea), (little show of oil) ... 31   698

*For additional records Elliott County, see "Economic Papers on
Kentucky Geology"—W. R. Jillson, Ky. Geological Survey, Series VI,
Vol. II, 1921.

Mississippian System.	Thickness	Depth
Shale, gray	5	703
Sandstone, gray	14	717
Shale, green	4	721
Sandstone	8	729
Shale, green, sandy	54	783
Devonian System.		
Shale, black (Chattanooga)	28	811
Total depth		811

Log No. 166

Dr. Wallace Brown, No. 1, lessor. Washington Oil Company, lessee. Location: One-half mile southeast of Ordinary P. O. Elevation: 922 feet. Authority: C. T. Dabney, Winchester, Ky.

Strata.		
Pennsylvanian System.	Thickness	Depth
Soil	4	4
Sand	158	162
Mississippian System.		
Limestone	2	164
Shale, white, (fire clay, muck, water on top)	3	167
Limestone	86	253
Shale, white	2	255
Limestone	30	285
Shale, black	73	358
Shale (Waverly), white	238	596
Shale, dark	9	605
Shale, white, and shells	12	617
Shale, black	45	662
Sandy shells	37	699
Sand and shale	20	719
Shale, black (Sunbury)	19	738
Sandstone (Berea grit)	92	830
Shale, white	24	854
Devonian System.		
Shale, brown, Ohio shale	60	914
Shale, white (gray), Ohio shale	16	930
Shale, brown, Ohio shale	176	1,106
Fire clay	19	1,125
Shale, white	2	1,127
Limestone, brown (Corniferous)	43	1,170
Silurian System.		
Shale, white, and red rock	30	1,200
Limestone, brown, dolomitic	9	1,209

Silurian System.	Thickness	Depth
Limestone	267	1,476
Shale, white	38	1,514
Shale, limy, red	80	1,594
Shale, white and gray	87	1,681
Limestone, red (Clinton)	19	1,700
Ordovician System.		
Shale, gray, and shells (very dark, almost black)	35	1,735
Shale, blue	40	1,775
Shale and shells	48	1,823
Shale, white	74	1,897
Limestone shells	20	1,917
Shale, white	13	1,930
Limestone, black	12	1,942
Shale and shells	66	2,008
Shale, white, and shells	16	2,024
Shale, white	30	2,054
Incomplete depth		2,054

Steel tape used here. Cannot locate error. 2,000 to 2,463 Trenton lime. Bottom of hole puffs of gas toward bottom of hole.

ESTILL COUNTY.

Production: **Oil and Gas.**   Producing Sand: **Corniferous (Devonian).**

Log No. 167

Isom Ballard, No. 12, lessor. Commenced: August 27, 1919. Completed: October 13, 1919. Authority: The Superior Oil Corporation.

Strata.

Mississippian System.	Thickness	Depth
Shale, blue, soft	218	218
Devonian System.		
Shale, black, hard (Chattanooga)	96	314
Shale, red, hard	12	326
Fire clay, soft	7	333
Limestone ''sand,'' hard (Corniferous)	8	341
Total depth		341

Log No. 168

Isom Ballard, No. 13, lessor.   Authority: The Superior Oil Corporation.

Strata.

Mississippian System.	Thickness	Depth
Soil, black, soft	20	20
Shale, blue, soft	178	198
Clay, blue, soft	77	275

Devonian System.		
Shale, black, hard (Chattanooga)	96	371
Shale, red, hard	12	383
Fire clay, blue, soft	7	390
Limestone "sand," gray, hard (Corniferous)	7½	397½
Total depth		397½

Log No. 169

Isom Ballard, No. 14, lessor.   Commenced: November 17, 1919. Completed: December 8, 1919.   Authority: The Superior Oil Corporation.

Strata.

Mississippian System.	Thickness	Depth
Soil yellow, black, soft	40	40
Shale, blue, soft	208	248
Clay, blue, soft	78	326

Devonian System.		
Shale, black, soft (Chattanooga)	96	422
Shale, red, soft	12	434
Fire clay, white, yellow, soft	7	441
Limestone "sand," brown, hard (Corniferous) (oil)		448
Total depth		448

AN UNCONFORMITY AT IRVINE, KY.

This outcrop of the Onondaga Limestone and subjacent greenish gray shale below the River New Hotel, Irvine, Ky., shows cross bedding and suggests a local unconformity.

Log No. 170

Isom Ballard, No. 16, lessor. Completed: March 19, 1920. Production: Dry. Authority: The Superior Oil Corporation.

Strata.

Mississippian System.	Thickness	Depth
Soil, yellow, soft	20	20
Limestone, white, hard	40	60
Shale, blue, soft	383	443
Devonian System.		
Shale, black, hard (Chattanooga)	96	539
Shale, red, soft	12	551
Fire clay, white, soft	10	561
Limestone ''sand,'' white, hard (Corniferous)	2	563
Limestone ''sand,'' brown, soft (Corniferous)	6	569
Limestone ''sand,'' white, hard	2½	571½
Total depth		571½

Log No. 171

Thomas Henderson, No. 19, lessor. Commenced: May 27, 1920. Completed: June 9, 1920. Authority: The Superior Oil Corporation.

Strata.

Mississippian System.	Thickness	Depth
Soil black, soft	11	11
Shale, blue, soft	250	261
Devonian System.		
Shale, black, hard	125	386
Fire clay, yellow, soft	10½	396½
Limestone (cap rock), black, hard	1	397½
Limestone, oil ''sand,'' brown, hard (Corniferous)	8	405½
Total depth		405½

Log No. 172

Thomas Tipton, No. 30, lessor. Commenced: September 25, 1919. Completed: November 25, 1919. Producing oil December 9, 1919. Authority: The Superior Oil Corporation.

Strata.

Mississippian System.	Thickness	Depth
Clay, yellow, sandy, soft	55	55
Limestone, white, hard	65	120
Shale, blue, soft	130	250
Shale, blue, soft, and mud	350	600
Shale (Red Rock), soft	10	610
Clay, blue, soft	18	628

Devonian System.

	Thickness	Depth
Shale, black, soft (Chattanooga)	110	738
Fire clay, white, yellow, soft	11	749
Limestone "sand," soft, (Corniferous)	3½	752½
Limestone "sand," hard, (Corniferous)	3½	756
Limestone "sand," soft, (Corniferous)	3½	759½
Limestone "sand" broken, (Corniferous) ..	2½	762
Total depth		762

Log No. 173

Grant Shoemaker, No. 2, lessor. Commenced: September 21, 1919. Completed: January 25, 1920. Production: Dry; casing pulled, well plugged. Authority: The Ohio Oil Co.

Strata.

Mississippian System.

	Thickness	Depth
Soil red, soft	20	20
Sandstone, red, medium	180	200
Limestone, white, hard	140	340
Shale, blue, hard, medium	450	790
Shale, hard, pink, soft	15	805
Shale, hard, white, soft	25	830

Devonian System.

	Thickness	Depth
Shale, brown, medium (Chattanooga)	116	946
Fire clay, white, soft	21	967
Limestone "sand," hard, dark, coarse (little oil)	4	971
Limestone "sand," hard, light	6	977
Total depth		977

NOTE—The last 10 feet of the record is in the Onondaga limestone (Corniferous "sand").

Log No. 174

G. R. Srac, No. 2, lessor. Commenced: October 20, 1919. Completed: January 31, 1920. Production: Dry. Authority: The Ohio Oil Co.

Strata.

Mississippian System.

	Thickness	Depth
Soil, red, soft	20	20
Sandstone, red, soft	80	100
Limestone, white, hard	93	193

Mississippian System.	Thickness	Depth

	Thickness	Depth
Shale, blue, soft	460	653
Fire clay, white, soft	25	678
Shale (Red Rock), soft	15	693

Devonian System.

	Thickness	Depth
Shale, black, soft (Chattanooga)	147	840
Fire clay, white, soft	14	854
Limestone "sand," hard, dark, fine	2	856
Limestone "sand," gray, soft, coarser, (little oil)	2	858
Limestone "sand," hard, white, fine, (salt water)	2	860
Total depth		860

Log No. 175

William McIntosh, No. 1, lessor. Commenced: October 1, 1915. Completed: October 5, 1915. Production: 10 to 15 bbls. oil. Authority: The Wood Oil Company.

Strata.

Mississippian & Devonian Systems.

	Thickness	Depth
To top of Irvine Sand	218	218
Limestone (Irvine "sand")	19	237
Total depth		237

A lead plug was put in this well on April 18, 1917.

Log No. 176

William McIntosh, No. 2, lessor. Commenced: January 3, 1916. Completed: January 7, 1916. Production: 4 bbls. oil. Authority: The Wood Oil Company.

Strata.

Mississippian & Devonian Systems.

	Thickness	Depth
Limestone and shale	306	306
Limestone (Irvine "sand")	12	318
Total depth		318

Best pay oil from 310 to 314 feet. No gas.

Log No. 177

Dan McCoy, No. 5, lessor. Completed: June 13, 1917. Production: 5 bbls. oil. Authority: The Wood Oil Company.

Strata.

Mississippian & Devonian Systems.	Thickness	Depth
Sand and shale	615	615
Limestone ''sand,'' blue, hard	1	616
Limestone ''sand,'' blue, (slight show of oil)	12	628
Limestone ''sand,'' blue, muddy (no pay) ..	16	644
Limestone ''sand,'' brown and white (no pay)	7	651
Shale, soft	3	654
Total depth		654

Log No. 178

Dan McCoy, No. 4, lessor. Commenced: May 30, 1917. Completed: June 13, 1917. Production: Dry. Authority: The Wood Oil Company.

Strata.

Mississippian & Devonian Systems.	Thickness	Depth
Limestone and shale	730	730
Limestone ''sand,'' fine, dark, (salt water 200)	2	732
Limestone ''sand,'' dark	23	755
Limestone ''sand,'' coarse, dark	5	760
Limestone ''sand,'' lighter	4	764
Silurian System.		
Limestone ''sand,'' very coarse, gray-brown and soft	9	773
Limestone ''sand,'' blue and gray mixed	8	781
Limestone ''sand,'' light brown (smell of oil)	27	808
Shale, very soft	3	811
Total depth		811

Log No. 179

George Lile, No. 2, lessor. Commenced: July 31, 1917. Completed: August 17, 1917. Production: Dry. Authority: The Wood Oil Company.

Strata.

Mississippian & Devonian Systems.	Thickness	Depth
Limestone and shale	797	797
Limestone (Irvine ''sand'')	38	835
Total depth		835

Log No. 180

George Lile, No. 1, lessor. Commenced: November 13, 1916. Completed: December 5, 1916. Abandoned: December 7, 1916. Authority: The Wood Oil Company.
Strata.

Mississippian & Devonian Systems.	Thickness	Depth
Limestone and shale	759	759
Limestone (Irvine "sand"), hard, light brown	1	760
Limestone, (oil show 900)	511	1,271
Total depth		1,271

Stopped drilling in blue limestone.

Log No. 181

Elizabeth Gibson, No. 1, lessor. Commenced: July 17, 1916. Completed: August 2, 1916. Production: 30 bbls. oil. Authority: The Wood Oil Company.
Strata.

Mississippian & Devonian Systems.	Thickness	Depth
Limestone and shale	729	729
Limestone (Irvine "sand")	20	749
Limestone, blue, hard	3	752
Total depth		752

Log No. 182

Elizabeth Gibson, No. 2, lessor. Commenced: August 5, 1916. Completed: August 16, 1916. Production: 20 bbls. oil. Authority: The Wood Oil Company.
Strata.

Mississippian & Devonian Systems.	Thickness	Depth
Limestone and shale	730	730
Limestone (Irvine "sand")	23	753
Total depth		753

Log No. 183

Elizabeth Gibson, No. 3, lessor. Commenced: August 19, 1916. Completed: September 9, 1916. Production: 25 bbls. Authority: The Wood Oil Company.
Strata.

Mississippian & Devonian Systems.	Thickness	Depth
Limestone and shale	785	785
Limestone (Irvine "sand")	19	804
Total depth		804

Log No. 184

Elizabeth Gibson, No. 4, lessor. Commenced: September 13, 1916. Completed: September 26, 1916. Production: 25 bbls. oil. Authority: The Wood Oil Company.

Strata.

Mississippian & Devonian Systems.	Thickness	Depth
Limestone and shale	732	732
Limestone (Irvine "sand")	16	748
Total depth		748

Remarks: The sand was all fairly good.

Log No. 185

E. Gibson, No. 5, lessor. Commenced: September 29, 1916. Completed: October 6, 1916. Production: 10 bbls. oil.

Strata.

Mississippian & Devonian Systems.	Thickness	Depth
Limestone and shale	739	739
Limestone (Irvine "sand")	14	753
Total depth		753

Log No. 186

Widow Garrett, No. 1, lessor. Commenced: April 20, 1916. Completed: May 5, 1916. Production: 25 bbls. oil. Authority: The Wood Oil Company.

Strata.

Mississippian & Devonian Systems.	Thickness	Depth
Limestone and shale	782	782
Limestone (Irvine "sand")	20	802
Total depth		802

There was a showing of oil from 782 to 786 feet.

Log No. 187

Widow Garrett, No. 2, lessor. Commenced: May 6, 1916. Completed: May 17, 1916. Production: 10 bbls. oil after shot. Authority: The Wood Oil Company.

Strata.

Mississippian & Devonian Systems.	Thickness	Depth
Soil, limestone and black shale	750	750
Limestone (Irvine "sand")	25	775
Total depth		775

Remarks: The sixth screw showed salt water.

Log No. 188

Widow J. M. Garrett, No. 4, lessor. Commenced: June 7, 1916. Completed: June 21, 1916. Production: 20 bbls. oil. Authority: The Wood Oil Company.

Strata.

Mississippian & Devonian Systems.	Thickness	Depth
Limestone and shale	800	800
Limestone (Irvine "sand")	19	819
Total depth		819

Remarks: Show of oil at 802 feet. The sand was dark gray. 810 to 815 feet change in sand to light gray.

Log No. 189

Mrs. J. M. Garrett, No. 3, lessor. Commenced: May 25, 1916. Completed: June 5, 1916. Authority: The Wood Oil Company.

Strata.

Mississippian & Devonian Systems.	Thickness	Depth
Soil, sandy shale and black shale	747	747
Limestone (Irvine "sand")	9	756
Total depth		756

Remarks: Only a light show of oil in this well.

Log No. 190

Mrs. J. M. Garrett, No. 5, lessor. Commenced: June 24, 1916. Completed: July 4, 1916. Production: 2 bbls. natural. Authority: The Wood Oil Company.

Strata.

Mississippian & Devonian Systems.	Thickness	Depth
Sandstone and shale	803	803
Limestone (Irvine "sand")	34	837
Total depth		837

Remarks: Stopped drilling in gritty limestone formation.

Log No. 191

Mrs. J. M. Garrett, No. 6, lessor. Commenced: July 7, 1916. Completed: July 18, 1916. Production: 25 bbls. Authority: The Wood Oil Company.

Strata.

Mississippian & Devonian Systems.	Thickness	Depth
Sandstone and shale	772	772
Limestone (Irvine "sand"), lower part blue	27	799
Total depth		799

Log No. 192

Mrs. J. M. Garrett, No. 7, lessor. Commenced: July 20, 1916. Completed: July 29, 1916. Production: 25 bbls. Authority: The Wood Oil Company.

Strata.

Mississippian & Devonian Systems.	Thickness	Depth
Sandstone and shale	728	728
Limestone (Irvine "sand")	29	757
Total depth		757

Log No. 193

Mrs. J. M. Garrett, No. 8, lessor. Commenced: August 1, 1916. Completed: August 18, 1916. Production: 20 bbls. Authority: The Wood Oil Company.

Strata.

Mississippian & Devonian Systems.	Thickness	Depth
Sandstone and shale	695	695
Limestone (Irvine "sand")	46	741
Total depth		741

Remarks: Stopped drilling in hard, bluish-gray sand, with no pay.

Log No. 194

Mrs. J. A. Garrett, No. 9, lessor. Completed: August 28, 1916. Production: 20 bbls. Authority: The Wood Oil Company.

Strata.

Mississippian & Devonian Systems.	Thickness	Depth
Sandstone and shale	715	715
Limestone (Irvine "sand")	29	744
Total depth		744

Remarks: The tenth and eleventh screws showed hard shale and mud.

Log No. 195

Mrs. J. A. Garrett, No. 10, lessor. Commenced: September 12, 1916. Completed: September 21, 1916. Production: 5 bbls. Authority: The Wood Oil Company.

Strata.

Mississippian & Devonian Systems.	Thickness	Depth
Sandstone and shale	598	598
Limestone (Irvine "sand")	48	646
Total depth		646

Log No. 196

Mrs. J. A. Garrett, No. 11, lessor. Commenced: September 11, 1916. Completed: September 21, 1916. Production: 20 bbls. Authority: The Wood Oil Company.

Strata.

Mississippian & Devonian Systems.	Thickness	Depth
Sandstone and shale	715	715
Limestone (Irvine "sand")	43	758
Total depth		758

Log No. 197

Mrs. J. A. Garrett, No. 12, lessor. Commenced: September 25, 1916. Completed: October 3, 1916. Authority: The Wood Oil Company.

Strata.

Mississippian & Devonian Systems.	Thickness	Depth
Sandstone and shale	594	594
Limestone (Irvine "sand"), (gas in top) ...	37	631
Total depth		631

Log No. 198

Mrs. J. A. Garrett, No. 13, lessor. Commenced: September 25, 1916. Completed: October 3, 1916. Production: 15 bbls. Authority: The Wood Oil Company.

Strata.

Mississippian & Devonian Systems.	Thickness	Depth
Sandstone and shale	766	766
Limestone (Irvine "sand")	33	799
Total depth		799

Log No. 199

Mrs. J. A. Garrett, No. 14, lessor. Commenced: October 15, 1916. Completed: October 20, 1916. Production: 15 bbls. Authority: The Wood Oil Company

Strata.

Mississippian & Devonian Systems.	Thickness	Depth
Sandstone and shale	493	493
Limestone (Irvine "sand")	42	535
Total depth		535

Log No. 200

Mrs. J. A. Garrett, No. 15, lessor. Commenced: October 4, 1916. Completed: October 18, 1916. Production: 25 bbls. Authority: The Wood Oil Company.

Strata.

Mississippian & Devonian Systems.	Thickness	Depth
Limestone, sandstone and shale	739	739
Limestone (Irvine ''sand'')	38	777
Total depth		777

Best pay from 742 to 746 feet and from 754 to 758 feet.

Log No. 201

Joseph Fox, No. 1, lessor. Commenced: December 2, 1916. Completed: December 7, 1916. Authority: The Wood Oil Company.

Strata.

Mississippian & Devonian Systems.	Thickness	Depth
Soil sandy shale and black shale	635	635
Limestone (Irvine "sand"), white, hard (salt water)	29	664
Total depth		664

The hole filled up 150 feet with salt water.

Log No. 202

B. Brinegar, No. 8, lessor. Commenced: June 17, 1916. Completed: June 22, 1916. Production: 15 bbls. oil. Authority: The Wood Oil Company.

Strata.

Mississippian & Devonian Systems.	Thickness	Depth
Soil, sandy shale and black shale	564	564
Limestone (Irvine "sand"), fine oil	9	573
Total depth		573

The first screw showed a little oil. The sand was dark. The second screw showed no increase in oil. The sand was gray. The third screw showed a little more oil. The sand was fine.

Log No. 203

B. Brinegar, No. 9, lessor. Commenced: June 24, 1916. Completed: June 30, 1916. Production: 15 bbls. Authority: The Wood Oil Company.

Strata.

Mississippian & Devonian Systems.	Thickness	Depth
Soil, sandy shale and black shale	558	558
Limestone (Irvine ''sand''), gray	12	570
Total depth		570

Log No. 204

B. Brinegar, No. 1, lessor. Location: Irvine District. Commenced:
March 1, 1916. Completed: March 13, 1916. Production: 15 bbls.
Authority: The Wood Oil Company.

Strata.

Devonian System.	Thickness	Depth
Soil and black shale (Chattanooga)	186	186
Limestone (Irvine ''sand''), quite soft	12½	198½
Total depth		198½

Log No. 205

B. Brinegar, No. 2, lessor. Commenced: March 16, 1916. Completed: March 22, 1916. Authority: The Wood Oil Company.

Strata.

Devonian System.	Thickness	Depth
Soil and black shale (Chattanooga)	211	211
Limestone (Irvine ''sand''), ...	21½	232½
Total depth		232½

Remarks: Showings of each screw were as follows:
   (1) Shelly, with a very light show in the bottom.
   (2) A slight increase in oil.
   (3) Blue and shelly, no increase in the oil.
   (4) Filled 5 feet over the tools.
   (5) Showed good looking sand, with fairly strong gas in the
       top, filled 30 feet of oil over the tools.
   (6) Filled 75 feet of oil over the tools.
   (7) Filled 90 feet of oil over the tools.
The best pay was between 225 and 232 feet.

Log No. 206

B. Brinegar, No. 3, lessor. Commenced: May 5, 1916. Completed: May 9, 1916. Production: 3 bbls. Authority: The Wood Oil
Company.

Strata.

Mississippian & Devonian Systems.	Thickness	Depth
Soil and black shale	313	313
Limestone (Irvine ''sand'')	15½	328½
Total depth		328½

Gas at 323 feet.
Best pay from 313 to 323 feet.

Log No. 207

B. Brinegar, No. 4, lessor. Commenced: May 9, 1916. Completed: May 15, 1916. Production: 15 bbls. Authority: The Wood Oil Company.

Strata.

Mississippian & Devonian Systems.	Thickness	Depth
Soil and black shale	372	372
Limestone (Irvine "sand"), (oil & gas)	20½	392½
Total depth		392½

Log No. 208

B. Brinegar, No. 5, lessor. Commenced: May 18, 1916. Completed: May 24, 1916. Production: 15 bbls. Well abandoned and plugged Nov. 4, 1917. Authority: The Wood Oil Company.

Strata.

Mississippian & Devonian Systems.	Thickness	Depth
Soil, sandy shale and black shale	568	568
Limestone (Irvine "sand")	20	588
Total depth		588

Log No. 209

B. Brinegar, No. 6, lessor. Commenced: May 26, 1916. Completed: May 31, 1916. Production: 20 bbls. Authority: The Wood Oil Company.

Strata.

Mississippian & Devonian Systems.	Thickness	Depth
Soil, sandy shale and black shale	564	564
Limestone (Irvine "sand"), (oil)	16	580
Total depth		580

Log No. 210

B. Brinegar, No. 7, lessor. Commenced: June 2, 1916. Completed: June 16, 1916. Production: 15 bbls. Authority: The Wood Oil Company.

Strata.

Mississippian & Devonian Systems.	Thickness	Depth
Soil, sandy shale and black shale	561	561
Limestone (Irvine "sand," (oil and gas)	14	575
Total depth		575

Log No. 211

Prewitt, Miller and Goff, No. 106, lessors.   Completed: April 23, 1918.   Authority: The Petroleum Exploration Company.

Strata.

Pennsylvanian System.	Thickness	Depth
Shale and sandstone	75	75
Sandstone (Pottsville)	50	125
Limestone	10	135
Fire clay	15	150

Mississippian System.		
Limestone (Big Lime)	100	250
Sandstone and shale	475	725

Devonian System.		
Shale, black (Chattanooga)	142	867
Fire clay	15	882
Limestone "sand," (oil at 924, 936½ to 960)	96½	978½
Total depth		978½

Log No. 212

Prewitt, Miller and Goff, No. 108, lessors.   Commenced: April 8, 1918.   Completed: April 23, 1918.   Authority: The Petroleum Exploration Company.

Strata.

Mississippian System.	Thickness	Depth
Soil	15	15
Limestone (Big Lime)	87	102
Sandstone and shale	508	610

Devonian System.		
Shale, brown	130	740
Fire clay	15	755
Limestone (cap rock) at 755	42	797
Limestone, 1st "sand" oil	13	810
Limestone, 2nd "sand" oil	22	832
Limestone, 3rd "sand" oil	10½	842½
Limestone	2½	845
Total depth		845

Log No. 213

Prewitt, Miller and Goff, No. 110, lessors. Authority: The Petroleum Exploration Company.

Strata.

Mississippian System.	Thickness	Depth
Soil	10	10
Limestone (Big Lime)	55	65
Limestone, sandstone and shale	507	572
Devonian System.		
Shale, brown (Chattanooga)	135	707
Fire clay	15	722
Limestone (Cap rock), oil "sand"	88	810
Total depth		810

NOTE—The lower part of the last 88 feet of this record is undoubtedly Silurian.

Log No. 214

Prewitt, Miller and Goff, No. 111, lessors. Commenced: June 11, 1918. Completed: June 28, 1918. Authority: The Petroleum Exploration Company.

Strata.

Pennsylvanian & Mississippian Systems.	Thickness	Depth
Sandstone and shale	225	225
Limestone (Big Lime)	90	315
Clay, blue	504	819
Devonian System.		
Shale, brown (Chattanooga)	135	954
Fire clay	15	969
Limestone (cap rock) and oil "sand" (oil at 1,019)	88	1,057
Total depth		1,057

Remarks: Bottom of oil pay, 1,039. The lower part of the last 88 feet of this record is undoubtedly Silurian.

Log No. 215

Prewitt, Miller and Goff, No. 116, lessors. Commenced: September 18, 1918. Completed: October 11, 1918. Production: 25 to 30 bbls. Authority: The Petroleum Exploration Company.

Strata.

Mississippian System.	Thickness	Depth
Soil	4	4
Limestone (Big Lime)	81	85
Shale, brown	605	690

Devoniän System.

	Thickness	Depth
Shale, black (Chattanooga)	50	740
Fire clay	9	749
Limestone (cap rock) (Steel Line Measurement)	42	791
Limestone, 1st oil "sand"	2	793
Shale, hard, brown	4	797

Silurian System.

Limestone, 2nd oil "sand"	3	800
Shale, hard	2	802
Limestone, 3rd oil "sand," (oil 400 feet high)	9	811
Limestone and shale, hard	22	833
Limestone, 4th oil "sand"	3	836
Limestone	2	838
Limestone	3	841
Total depth		841

Log No. 216

Prewitt, Miller and Goff, No. 119, lessors. Commenced: February 26, 1919. Completed: March 26, 1919. Production: 10 bbls. oil. Authority: The Petroleum Exploration Company.

Strata.

Pennsylvanian System.

	Thickness	Depth
Soil	6	6
Sandstone (Pottsville)	39	45
Shale and sandstone	80	125

Mississippian System.

Limestone (Little Lime)	15	140
Limestone (Big Lime)	115	255
Clay, blue	15	270
Shale, hard	215	485
Shale, hard, and shells	240	725

Devonian System.

Shale, brown	135	860
Shale, hard	15	875
Limestone (cap rock) and "sand"	98	973
Total depth		973

Remarks: Oil at 915 to 938. The lower part of the last 98 feet of "sand" is Silurian limestone.

Log No. 217

Prewitt, Miller and Goff, No. 120, lessors. Commenced: December 19, 1918. Completed: January 18, 1919. Production: 3 to 4 bbls. oil. Authority: The Petroleum Exploration Company.

Strata.

	Thickness	Depth
Pennsylvanian System.		
Sandstone (Pottsville)	50	50
Shale, blue	45	95
Mississippian System.		
Limestone	125	220
Shale, blue	490	710
Devonian System.		
Shale, brown	150	860
Fire clay	12	872
Limestone ''sand''	96	968
Total depth		968

Remarks Salt water at 885 and 935. Oil pay from 921 to 934· The lower part of the last 96 feet of this record is Silurian.

Log No. 218

Prewitt, Miller and Goff, No. 121, lessors. Commenced: March 29, 1919. Completed: April 16, 1919. Authority: The Petroleum Exploration Company.

Strata.

	Thickness	Depth
Pennsylvanian System.		
Sandstone and shale (Pottsville)	120	120
Mississippian System.		
Limestone (Little Lime), (water at 125)	30	150
Shale	15	165
Limestone (Big Lime)	100	265
Shale, hard, and shells	505	770
Devonian System.		
Shale, brown, hard	140	910
Limestone (cap rock) and ''sand''	110	1,020
Total depth		1,020

Remarks: Salt water at 922. Oil pay, light from 958 to 978. The lower part of the last 110 feet of this record is Silurian.

Log No. 219

Prewitt, Miller and Goff, No. 125, lessors. Commenced: August 8, 1919. Completed: September 20, 1919. Production: 12½ bbls. per day. Authority: The Petroleum Exploration Company.

Strata.

Pennsylvanian System.	Thickness	Depth
Soil	8	8
Shale, hard	72	80
Mississippian System.		
Limestone (Little Lime)	20	100
Shale, hard	54	154
Limestone (Big Lime)	85	239
Shale, hard	189	428
Shale, brown	250	678
Shale, hard, light	37	715
Shale, red	15	730
Devonian System.		
Shale (Chattanooga)	145	875
Fire clay	17	892
Limestone (cap rock), (oil and gas 889)	7	899
Limestone "sand," (water 900, oil 935-950)	92	991
Total depth		991

NOTE—The lower part of the last 92 feet of this record is Silurian.

Log No. 220

Prewitt, Miller and Goff, No. 123, lessors. Commenced: March 14, 1919. Completed: April 3, 1919. Production: 10 bbls. after shot; 4 bbls. natural. Authority: The Petroleum Exploration Company.

Strata.

Pennsylvanian System.	Thickness	Depth
Soil	10	10
Sandstone and shale	210	220
Mississippian System.		
Limestone (Big Lime), (water at 210)	45	265
Shale, green	185	450
Shale, hard	350	800
Devonian System.		
Shale, brown (Chattanooga)	150	950
Fire clay	15	965
Limestone "sand," (water 988, oil 1,024-1,036)	114	1,079
Total depth		1,079

NOTE—The lower part of the last 114 feet of this record is Silurian.

Log No. 221

Prewitt, Miller and Goff, No. 127, lessors. Commenced: May 10, 1919. Completed: May 29, 1919. Production: 20 bbls. after first day. Authority: The Petroleum Exploration Company.

Strata.

Pennsylvanian System.	Thickness	Depth
Soil	4	4
Sandstone and shale	81	85
Sand (Pottsville)	40	125
Fire clay	10	135

Mississippian System.		
Limestone	80	215
Shale, brown	45	260
Limestone	10	270
Shale, brown	390	660
Shale, red	25	685
Shale, gray	20	705

Devonian System.		
Shale, brown	140	845
Shale, gray	23	868
Limestone (cap rock) and "sand"	99	967
Total depth		967

Remarks: Oil as gas, light show, at 886. Salt water, hole full, at 887. Oil pay from 924 to 949. The lower part of the last 99 feet of limestone in this well is Silurian.

Log No. 222

Prewitt, Miller and Goff, No. 128, lessors. Commenced: May 20, 1919. Completed: June 21, 1919. Production: 5 bbls. after first day. Authority: The Petroleum Exploration Company.

Strata.

Pennsylvanian System.	Thickness	Depth
Sandstone and shale	80	80

Mississippian System.		
Limestone (Big Lime)	86	166
Limestone, green	15	181
Shale, green, hard	67	248
Shale, brown	62	310

Mississippian System.

	Thickness	Depth
Limestone and shells	5.	315
Shale, black, hard	315	630
Pink rock	20	650
Shale, white, hard	15	665

Devonian System.

	Thickness	Depth
Shale, brown (Chattanooga)	145.	810
Fire clay	18	828
Limestone (cap rock) and "sand" 878 to 898 ..	107	935
Total depth		935

NOTE—The lower part of the last 107 feet of this record is in the Silurian.

Log No. 223

J. F. West, No. 1, lessor. Location: Rock House Fork, 1½ miles N. E. Pitts P. O. Completed: Feb. 27, 1903. Authority: New Domain Oil & Gas Co

Strata.

Devonian System.

	Thickness	Depth
Clay	21	21
Shale, hard, black (Chattanooga)	43	64
Limestone, gray, hard (Corniferous)	30	94
Total depth		94

Log No. 224.

J. F. West, No. 2, lessor. Location: Rock House Fork 1½ miles N. E. Pitts P. O. Completed: May 15, 1903. Production: Dry. Authority: New Domain Oil & Gas Co.

Strata.

Devonian System.

	Thickness	Depth
Clay	45	45
Shale, black (Chattanooga)	24	69
Limestone, gray, hard (Corniferous)	25	94

Silurian System.

	Thickness	Depth
Limestone, light gray, hard, Niagaran	36	130
Sandstone, light gray, soft, Niagaran	145	275
Limestone, gray, hard, Niagaran	30	305
Sandstone light, soft, Niagaran	10	315
Limestone, gray, hard, Niagaran	8	323
Limestone, red, hard, Niagaran	10	333

Ordovician System.

	Thickness	Depth
Limestone bastard gray, hard	17	350
Limestone, bastard brown, hard	40	390
Limestone, bastard gray, hard	839	1,229
Total depth		1,229

Log No. 225

J. F. West, No. 3, lessor. Location: Rock House Fork, 1½ miles
N. E. Pitts P. O. Completed: spring of 1903. Production: first day,
estimated at 4 bbls. Authority: New Domain Oil & Gas Co.

Strata.

Devonian System.	Thickness	Depth
Clay	14	14
Shale, black (Chattanooga)	49	63
Limestone, gray, hard (Corniferous), (salt water in last 2 feet)	20	83
Total depth		83

Log No. 226

J. F. West, No. 4, lessor. Location: Rock House Fork, 1½ miles
N. E. Pitts P. O. Completed: May 23, 1903. Authority: New Domain
Oil & Gas Co.

Strata.

Devonian System.	Thickness	Depth
Clay	3	3
Shale, black (Chattanooga)	69	72
Limestone or Estill "sand," gray, hard	20	92
Total depth		92

Log No. 227

J. F. West, No. 5, lessor. Location: Rock House Fork, 1½ miles
N. E. Pitts P. O. Completed: May 30, 1903. Very light show of oil,
good and dry; salt water in the last foot of the sand. Authority: New
Domain Oil & Gas Co.

Strata.

Devonian System.	Thickness	Depth
Clay	25	25
Shale, black (Chattanooga)	50	75
Limestone, gray, hard (Corniferous)	18	93
Total depth		93

Log No. 228

J. F. West, No. 6, lessor. Location: Rock House Fork, 1½ miles N. E. Pitts P. O. Completed: Nov. 24, 1903. Estimated production: 1 bbl. the first day Authority: New Domain Oil & Gas Co.
Strata.

Devonian System.	Thickness	Depth
Clay, yellow	45	45
Shale, black, (oil 62)	17	62
Limestone "sand"	13	75
Total depth		75

Log No. 229

J. F. West, No. 7, lessor. Location: Rock House Fork, 1½ miles N. E. Pitts P. O. Completed: Nov. 25, 1903. Estimated production: ½ bbl. the first day. Authority: New Domain Oil & Gas Co.
Strata.

Devonian System.	Thickness	Depth
Clay, yellow	19	19
Shale, black, (oil)	. 45	64
Limestone "sand" (Corniferous)	12	76
Total depth		76

Log No. 230

J. F. West, No. 8 lessor. Location: Rock House Fork, 1½ miles N. E. Pitts P. O. Completed: Nov. 27, 1903. Estimated production: 1 bbl. the first day. Authority: New Domain Oil & Gas Co.
Strata.

Devonian System.	Thickness	Depth
Clay, yellow	8	8
Shale, black (Chattanooga)	63½	71½
Limestone "sand" (Corniferous)	15	86½
Total depth		86½

Log No. 231

J. F. West, No. 9, lessor. Location: Rock House Fork 1½ miles N. E. Pitts P. O. Completed: Nov. 30, 1903. Production: much water. Authority: New Domain Oil & Gas Co.
Strata.

Devonian System.	Thickness	Depth
Clay, yellow	22	22
Shale, black (Chattanooga)	52	74
Limestone "sand" (Corniferous)	13	87
Total depth		87

Log No. 232

J. F. West, No. 10, lessor. Location: Rock House Fork, 1½ miles N. E. Pitts P. O. Completed: Spring of 1903. Estimated production: 1 bbl. the first day. Authority: New Domain Oil & Gas Co.

Strata.

Devonian System.	Thickness	Depth
Clay, yellow	32	32
Shale black (Chattanooga)	13	45
Limestone "sand" (Corniferous), (oil 54) ...	16	61
Total depth		61

Log No. 233

C. P. Rogers, No. 1, lessor. Completed: Sept. 10, 1904. Production: The well was dry. Authority: New Domain Oil & Gas Co.

Strata.

Mississippian System.	Thickness	Depth
Clay, yellow, soft	9	9
Sandstone, blue, soft	36	45
Shale, blue, soft	36	81
Devonian System.		
Shale, black, hard (Chattanooga)	113	194
Shale, white, soft	2	196
Limestone "sand" (Irvine), gray, hard	101	297
Shale, blue, soft	33	330
Silurian System		
Shale, pink, soft	60	390
Shale, blue, soft	50	440
Limestone, blue, hard	7	447
Shale, blue, soft	6	453
Shale, pink, soft	7	460
Shale, blue, soft	5	465
Limestone, blue, hard	5	470
Shale (red rock), hard	15	485
Shale, blue, soft	5	490
Limestone, blue, hard	40	530
Shale, blue, hard	14	544
Ordovician System.		
Limestone, blue, hard	63	607
Total depth		607

Log No. 234

Burnside Tipton, No. 1, lessor. Completed: Aug. 23, 1904. Production: Dry. Authority: New Domain Oil & Gas Co.

Strata.	Thickness	Depth
Mississippian System.		
Clay, yellow, soft	15	15
Soapstone, blue, soft	45	60
Shale, blue, soft	56	116
Devonian System.		
Shale, black, hard (Chattanooga)	101	217
Shale, blue, soft	8	225
Limestone, gray, hard	8	233
Limestone, blue, hard	8	241
Silurian System.		
Limestone, gray, hard	67	308
Shale, blue, soft	20	328
Shale, pink, soft	28	356
Shale, blue, soft	78	434
Limestone, blue, hard	4	438
Shale, blue, soft	16	454
Ordovician System.		
Limestone, blue, hard	4	458
Shale, blue, hard	7	465
Shale (red rock), hard	3	468
Shale, blue, hard	6	474
Shale (red rock), blue, hard	3	477
Shale, blue, hard	11	488
Limestone, gray, hard	2	490
Shale, blue, soft	3	493
Shale (red rock), hard	6	499
Limestone, gray, hard	8	507
Limestone, blue, hard	19	526
Shale, blue, soft	14	540
Limestone, blue, hard	3	543
Shale, blue, soft	5	548
Limestone, blue, hard	137	685
Shale, blue soft	3	688
Limestone, blue, hard	23	711
Total depth		711

FLOYD COUNTY.

Production: **Oil and Gas.** Producing Sands: **Pottsville (Pennsylvanian);
Maxton, Bradley, Big Injun, and Berea (Mississippian).**

Log No. 235

Frank D. Hopkins, No. 1, lessor. A. Fleming, et al., lessees. Location: Near mouth of Bull Creek, on the Big Sandy River, below Dwale P. O. Completed: June 8, 1920. Production: Gas from Maxton sand, 500,000 cu. feet with over 500 lbs. rock pressure. Authority: A. Fleming. King Drilling Co., by A. P. Brookover, Driller.

Strata.

	Thickness	Depth
Pennsylvanian System.		
Soil	12	12
Sand	40	52
Shale, blue	70	122
Coal	5	127
Sand	45	172
Shale	75	247
Sand, white	60	307
Shale, black	155	462
Salt sand (Beaver)	225	687
Shale, black	82	769
Bottom salt sand	20	789
Shale, black	29	818
Mississippian System.		
Sand (Maxton), 30 ft. oil and gas pay near top	90	908
Sandstone, red, show of oil	15	923
Limestone shells	50	973
Limestone (Little Lime)	15	988
Limestone (Big Lime)	140	1,128
Sandstone (Big Indian), red	12	1,140
Shale, white	110	1,250
Sand and limestone shells	203	1,453
Shale, brown	47	1,500
Sand (Wier), oil show in 82 ft.	90	1,590
Shale, brown	135	1,725
Sandstone (Berea), Rainbow	40	1,765
Devonian System.		
Shale, brown	155	1,920
Shale, white	30	1,950
Shale, brown	180	2,130

Devonian System.

	Thickness	Depth
Shale, white	165	2,295
Shale, brown	23	2,318
Limestone (Corniferous "sand") 5½ ft. streak of oil show 7 ft. from top, 6 ft. of bottom show of oil and much gas	62	2,380
Sandy shale (black and white)	3	2,383
Total depth		2,383

Log No. 236

Isaac Bradley, No. 1, lessor. Yolanda Oil Company, lessee. Location: 1½ miles from Wayland, on Right Beaver Creek. Completed: November 27, 1916. Casing head: 961.5 A. T. Production: 50,000 cubic feet gas. Well abandoned. Authority: The Eastern Gulf Oil Company.

Strata.

Pennsylvanian System.

	Thickness	Depth
Soil and alluvium	20	20
Sandstone, white, hard	20	40
Shale, hard	5	45
Sandstone ,...........................	7	52
Coal	5	57
Sandstone, (fresh water)	20	77
Shale, hard	5	82
Sandstone, white, hard	18	100
Shale, hard	10	110
Sandstone, white, hard	25	135
Shale, hard	19	154
Sandstone, (large fresh water)	18	172
Shale, hard	18	190
Sandstone	10	200
Shale, hard	15	215
Sandstone	20	235
Shale, hard	5	240
Sandstone	40	280
Shale, hard	30	310
Limestone, shaly, black	15	325
Shale, hard	40	365
Sandstone, (fresh water 370 to 380)	15	380
Shale, hard	15	395
Shale, calcareous, hard, black	25	420
Shale, hard, and shells	16	436
Sandstone	17	453
Shale hard	3	456
Sandstone	15	471

Pennsylvanian System.   Thickness Depth
    Shale, hard   5   476
    Sandstone   20   496
    Shale, hard   5   501
    Sandstone   39   540
    Shale, hard   5   545
    Shale, dark, hard   10   555
    Shale, hard, light colored   10   565
    Sandstone   10   575
    Shale, hard   62   637
    Sandstone   15   652
    Shale, hard   10   662
    Sandstone   13   675
    Shale, hard   5   680
    Sandstone, white, hard   35   715
    Shale, hard   10   725
    Sandstone, light colored   10   735
    Shale, hard   5   740
    Sandstone, (salt water at 900) 195   935
    Shale, black, hard   10   945
    Shale, hard   10   955
    Shale, hard, dark   15   970
    Shale, hard, light   40   1,010
    Sandstone   55   1,065
    Shale, hard, black   2   1,067
    Sandstone, white   7   1,074
    Shale, dark, hard   12   1,086
    Sandstone (Berea Sand), (gas 1086 to 1090
        estimated 50,000 cu. ft. per 24 hours.
        Salt water flooded hole at 1172) 171   1,257
    Shale, hard   14   1,271
    Sandstone   30   1,301
    Shale, hard   18   1,319

Mississippian System.
    Sandstone (Maxon), (salt water at 1463) ... 161   1,480
    Shale, hard, black   14   1,494
    Sand shells and shale, hard   12   1,506
    Limestone, hard, black   9   1,515
    Shale, hard, black   10   1,525
    Limestone, gray   15   1,540
    Shale, hard, black   14   1,554
    Sandstone (Bradley), (oil and gas 1554 to
        1559)   29   1,583
    Limestone, dark gray 4½   1,587½
    Limestone, white (Big Lime) 166½   1,754
    Shale, sandy and red (Big Injun)   2   1,756

Mississippian System.   Thickness Depth
    Shale and sandstone (Big Injun) 249   2,005
    Sandstone 40   2,045
    Shale, brown 100   2,145
Devonian System.
    Shale, hard, black (Chattanooga) 155   2,300
    Shale, and sand shells (Chattanooga) 3   2,303
    Shale, hard, black (Chattanooga) 102½ 2,405½
        Total depth 2,405½

NOTE—The sandy phase is the middle of the Devonian (Chattanooga) black shale. In one well on Aker Branch of Left Beaver Creek in Floyd County this sandy shale produced gas, but it never has produced oil. The Corniferous was not reached by this well.

Log No. 237

    Station Well, lessor. Pennagrade Oil & Gas Co., lessee. Location: Maytown.
        Strata.
Pennsylvanian System.   Thickness Depth
    Soil 55   55
    Shale 75   130
    Sandstone 80   210
    Shale, black 215   425
    Sandstone 60   485
    Shale, black, (gas 525, gas and water 635.... 265   750
    Shale, gray 38   788
    Shale, blue 10   798
    Sandstone 42   840
Mississippian System.
    Shale, red, sandy 54   894
    Sandstone (Maxon) 55   949
    Sandstone, (Water and gas 987) 41   990
        Total depth   990

Log No. 238

    S. May, lessor. Pennagrade Oil & Gas Co., lessee. Location: Mouth of Wilson Creek. Completed: October 31, 1920.
        Strata.
Pennsylvanian System.   Thickness Depth
    Soil 27   27
    Shale 75   102
    Sandstone 58   160
    Shale, black 200   360

Pennsylvanian System.   Thickness   Depth

	Thickness	Depth
Sandstone	40	400
Sandstone, shaly	118	518
Sandstone, (540,400,000 cu. ft. gas)	232	750
Shale, black (salt water 600)	20	770
Shell, black (gas show 630)	38	808
Sandstone, (250,000 cu. ft. gas, 810)	44	852
Shale, blue, (gas pay)	15	867
Sandstone, (salt)	41	908
Total depth		908

This record is all in the Pottsville.

Log No. 239

K. Moore, No. 1, lessor. Pennagrade Oil & Gas Co., lessee. Completed:. October 16, 1920.

Strata.

Pennsylvanian System.   Thickness   Depth

	Thickness	Depth
Soil	65	65
Shale	95	160
Sandstone	40	200
Shale	10	210
Sandstone blue	20	230
Shale, (gas 300)	95	325
Shale	100	425
Sandstone, gray	20	445
Shale	10	455
Sandstone, blue	70	525
Sandstone, (gas 562)	70	595
Shell	5	600
Sandstone, (water 612)	155	755
Coal, (gas 876-906)	10	765
Sandstone	60	825
Shale and shell	43	868
Sandstone	38	906

Mississippian System.

	Thickness	Depth
Shale, broken (gas 926-966, 1,500,00 cu. ft.)	1	907
Sandstone (Maxon)	62	969
Total depth		969

Log No. 240

S. May, No. 1, lessor.   Pennagrade Oil & Gas Co., lessee.   Location: Mouth of Wilson Creek.   Completed: January, 1920.

Strata.

Pennsylvanian System.	Thickness	Depth
Soil	8	8
Limestone	90	98
Shale	60	158
Sandstone	18	176
Shale	90	266
Sandstone	60	326
Shale	50	376
Limestone	15	391
Sandstone	50	441
Shale	60	501
Shale, hard, shelly, (gas 550)	50	551
Sandstone	69	620
Shale	8	628
Sandstone, (salt water 735)	140	768
Sandstone, limy	30	798
Sandstone, (salt water 820)	62	860
Shale and shell, (gas and oil show 969)	104	964
Sandstone	48	1,012
Shale, (gas pay 1027-1072)	15	1,027
Mississippian System.		
Sandstone (Maxon)	45	1,072
Shell	118	1,190
Limestone and shale	15	1,205
Shale	12	1,217
Limestone (Little Lime)	23	1,240
Limestone (Big Lime), (water)	190	1,430
Total depth		1,430

Log No. 241

H. May, No. 1, lessor.   Pennegrade Oil & Gas Co., lessee.   Location: Right Beaver Creek.   Completed: January, 1921.

Strata.

Pennsylvanian System.	Thickness	Depth
Soil	5	5
Sandstone	20	25
Shale	14	39

A BLUE GRASS DUSTER.

This shallow, Louis C. Weber No. 1, well drilled in the Summer of 1920 on Benson Creek in Franklin County is illustrative of the most recent of the large number of unsuccessful attempts to secure "Trenton" oil in Central Kentucky. The log of the well appears on page 169.

Pennsylvanian System.

	Thickness	Depth
Shale and shell	112	151
Sandstone	39	190
Shale	30	220
Sandstone	165	385
Sandstone	120	505
Shale, (salt water 603)	287	792
Shale, black	4	796
Shale and shell	14	810
Shale	8	818
Sandstone	2	820
Shale, brown	16	836
Sandstone	45	881

Mississippian System.

Shale, sandy, red, Mauch Chunk	64	945
Shale, Mauch Chunk	30	975
Shale, sandy, red, Mauch Chunk	25	1,000
Shale, sandy, (oil show 1005)	12	1,012
Shale	18	1,030
Sandstone	45	1,075
Limestone	30	1,105
Total depth		1,105

Log No. 242

S. May, No. 1, lessor. Ky. Coke Co., lessee. Location: S. May Branch, 2000' from Wilson Creek.

Strata.

Pennsylvanian System.

	Thickness	Depth
Soil	22	22
Sandstone	28	50
Shale	200	250
Sandstone	30	280
Shale, hard, shelly	20	300
Shale	56	356
Sandstone	14	370
Limestone	20	390
Shale	20	410
Sandstone	15	425
Shale	65	490
Sandstone	40	530
Shale	10	540
Sandstone	190	730

Pennsylvanian System.	Thickness	Depth
Shale	8	738
Sandstone	122	860
Limestone	25	885
Shale	10	895

Mississippian System.		
Sandstone (Maxon)	94	989
Shale	16	1,005
Limestone, blue	20	1,025
Sandstone, limy	35	1,060
Shale	10	1,070
Shale, sandy, red	15	1,085
Shale	18	1,103
Total depth		1,103

Log No. 243

J. H. Allen, lessor. Pennagrade Oil & Gas Co., lessee Location: Maytown. Completed: July 8, 1920. Production: Open flow from Maxon, 985, 250,000 cu. ft. gas.

Strata.

Pennsylvanian System.	Thickness	Depth
Soil	40	40
Sandstone	100	140
Coal	10	150
Sandstone	70	220
Shale	30	250
Sandstone	10	260
Shale	100	360
Sands, limy, (gas 505-510)	115	475
First salt	95	570
Shale	5	575
Second salt, (water 650)	170	745
Shale, (water 760)	15	760
Shale, hard, gray, (gas 785)	55	815

Mississippian System.		
Shale, sandy, red, (Maxon) (gas 825)	10	825
Sandstone (Maxon)	25	850
Shale sandy, red	15	865
Limestone (Little Lime)	5	870
Shale, sandy, red	30	900
Shale	20	920

155

Mississippian System. ·   Thickness Depth
   Limestone   10   930
   Shale, sandy, red   30   960
   Shale, blue, (Maxon) (gas 985)   25   985
   Sandstone, (water 1,035)   50   1,035
   Sandstone   16   1,051
      Total depth   1,051

Log No. 244

S. May, No. 1, lessor. Pennagrade Oil & Gas Co., lessee. Location: 1,000 ft. up first right hand branch of Wilson Creek. Completed: September 29, 1920. Production: 2,500,000 cu. ft. gas from Maxon.

   Strata.

Pennsylvanian System.   Thickness Depth
   Sandstone and shale   545   545
   Sandstone, (gas 570)   245   790
   Shale, black, (salt water)   50   840
   Shale, green, (gas 870)   10   850
   Shale, sandy   11   861
   Sandstone   40   901
   Shale, sandy   9   910
   Shale, blue   2   912

Mississippian System.
   Shale, sandy, red   4   916
   Sandstone (Maxon)   41   957
      Total depth   957

Log No. 245

K. Moore, No. 1, lessor. Pennagrade Oil & Gas Co., lessee. Location: Right Beaver Creek, 1,300 feet above R. R. tunnel.

   Strata.

Pennsylvanian System.   Thickness Depth
   Soil   45   45
   Shale   73   118
   Sandstone   55   173
   Shale   140   313
   Sandstone, blue   60   373
   Sandstone, (gas 495, 515)   357   730

Pennsylvanian System.   .   Thickness Depth

	Thickness	Depth
Shale and shell	70	800
Sandstone, (salt water 525)	35	835
Shale, blue, (gas 828)	18	853
Sandstone	71	924
Shale, blue, (salt water 932)	28	952
Limestone shell	7	959
Shale, blue	20	979

Mississippian System.

Shale, red, sandy	1	980
Sandstone (Maxon), (gas 980-997)	17	997
Total depth		997

Log No. 246

W. R. Crisp, No. 1, lessor. Keystone Oil & Gas Co., lessee. Location: 1 mile up Turkey Creek. Completed: July 25, 1918. Production: Gas, 535 cu. ft. open flow, 60 qts. shot.

Strata.

Pennsylvanian System.   Thickness Depth

	Thickness	Depth
Soil	18	18
Sandstone	28	46
Shale	66	112
Sandstone	30	142
Shale	10	152
Sandstone	18	170
Shale	12	182
Sandstone	12	194
Shale	35	229
Shale	5	234
Sandstone	35	269
Shale	91	360
Sandstone	10	370
Shale, (gas 475-495, 100,000 cu. ft.)	11	381
Sandstone, (salt water 675)	370	751
Coal	2	753
Sandstone	15	768
Shale, black	9	777
Shale, sandy	4	781
Shale, light	30	811
Shale, black	16	827

Mississippian System.	Thickness	Depth
Shale, red, sandy, (pay gas 837-872, came in 260 M.)	10	837
Sandstone (Maxon)	35	872
Shale, black	22	894
Total depth		894

Log No. 247

J. P. Akers, No. 1, lessor. Pennagrade Oil & Gas Co., lessee. Location: Maytown. Completed: January, 1921.

Strata.

Pennsylvanian System.	Thickness	Depth
Shale	50	50
Limestone	30	80
Coal	3	83
Fire clay	67	150
Shale	50	200
Limestone, (gas show 225)	50	250
Shale	50	300
Limestone, (gas show 330)	50	350
Shale and shell	75	425
Sandstone, limy	75	500
Sandstone	165	665
Shale	25	690
Sandstone	107	797
Shale	3	800
Sandstone	25	825
Shale	15	840
Limestone	5	845
Shale	5	850
Sandstone, limy	10	860
Shale	20	880
Limestone	10	890
Sandstone	20	910
Shale	8	918
Limestone	32	950
Mississippian System.		
Shale, sandy, red and Sandstone (Maxon)	65	1,015
Shale and shell	55	1,070
Limestone	18	1,088
Limestone and shale	24	1,112
Sandstone, limy	8	1,120
Limestone (Little Lime), (oil show 1,130)	20	1,140

Mississippian System.	Thickness	Depth
Sandstone, limy	56	1,196
Limestone (Big Lime), (oil show 1,256)	204	1,400
Shale, sandy, red, (gas show 1,380)	35	1,435
Shale, black	40	1,475
Sandstone, limy	50	1,525
Shale, black	33	1,558
Sandstone (Wier)	66	1,624
Shale, dark	6	1,630
Limestone	20	1,650
Shale	85	1,735
Limestone	70	1,805
Shale, black	725	2,530
Limestone (Corniferous)	80	2,610
Total depth		2,610

NOTE—Well stopped in sulphur gas in Corniferous.

Log No. 248

A. Ratliffe, No. 1, lessor. Ky. Coke Co., lessee. Location: Wilson Creek. Completed: April, 1921.

Strata.

Pennsylvanian System.	Thickness	Depth
Soil	42	42
Sandstone	8	50
Coal	2	52
Shale	18	70
Coal	5	75
Shale	15	90
Sandstone	40	130
Limestone	20	150
Shale	50	200
Sandstone	40	240
Limestone	40	280
Shale	50	330
Sandstone	50	380
Shale	20	400
Limestone, black	30	430
Shale	20	450
Limestone	20	470
Sandstone	20	490
Shale and shell, (gas 620)	100	590
Sandstone, (gas 700)	240	830
Shale, (salt water 810)	10	840

Pennsylvanian System.	Thickness	Depth
Sandstone	72	912
Shale and sandstone	23	935
Limestone	10	945
Mississippian System.		
Shale, red, sandy	15	960
Sandstone (Maxon), (gas & oil 1,060)	60	1,020
Shale ....:	12	1,032
Sandstone, (Maxon)	43	1,075
Shale ....,.........................	30	1,105
Sandstone, (salt water)	75	1,180
Total depth ....,.............		1,180

Log No. 249

N. Martin, No. 1, lessor. Kentucky Coal Co., lessee. Location: Wilson Creek. Production: Gas in Maxon sand.

Strata.

Pennsylvanian System.	Thickness	Depth
Soil	33	33
Sandstone	15	48
Shale and sandstone	177	225
Sandstone	55	280
Shale and sandstone	165	445
Sandstone, blue, (gas 590-610)	100	545
Sandstone, (salt water 675)	265	810
Shale	10	820
Limestone shell	15	835
Shale	20	855
Mississippian System.		
Shale, sandy, red	3	858
Sandstone (Maxon)	117	975
Total depth		975

Log No. 250

C. B. Webb, No. 1, lessor. Keystone Oil & Gas Co., lessee. Location: 3,600 feet south Maytown, west of main road, Right Fork of Beaver Creek. Completed: October 23, 1919. Open flow: 3,214,000. Rock pressure: 250 lbs. Casing head: 685,610. Production: 3,214,-000 cubic feet gas.

Strata.

Pennsylvanian System.	Thickness	Depth
Soil	20	20
Sandstone	40	60

Pennsylvanian System.	Thickness	Depth
Shale ....,....	60	120
Sandstone	55	175
Shale	40	215
Sandstone, (salt water 220)	40	255
Shale, (gas 345,321,000 cu. ft.)	115	370
Sandstone, gray (shale gas 540,130,000)	225	595
Sandstone, white, (salt water flooded 600)...	8	603
Sandstone, black	6	609
Sandstone, gray	71	680
Sandstone, white	80	760
Sandstone, dark	8	768
Shale, (gas 6-5/773)	30	798
Shale, white	11	809
Shale, dark	6	815
Sandstone, gray	4	819

Mississippian System.		
Shale, yellow (trace of red rock)	6	825
Sandstone (Maxon)	54	879
Total depth		879

Log No. 251

Jonah Webb, No. 1, lessor. Keystone Oil & Gas Co., lessee. Location: ½ mile above Wilson Creek on Beaver Creek. Completed: May 8, 1918. Open flow: 1,267,000 cubic feet gas. Casing head: 826.19. Production: Gas, 100,000 cubic feet.

Strata.

Pennsylvanian System.	Thickness	Depth
Soil	18	18
Sandstone	82	100
Shale and sand	35	135
Shale ....:	20	155
Sandstone dark gray	11	166
Shale, white	17	183
Shale and sand, (case in well, little gas)	20	203
Shale	57	260
Sandstone	35	295
Shale	55	350
Sandstone	15	365
Shale	74	439
Sand, (salt) (gas 439-605) (water 733-820)..	461	900
Shale, (gas 6-5/8 905)	52	952

Mississippian System.

	Thickness	Depth
Shale, red, sandy	2	954
Shale	4	958
Sandstone (Maxon), (gas pay 964-974)	52	1,010
Shale	2	1,012
Total depth		1,012

Log No. 252

T. J. Webb, No. 1, lessor. Keystone Oil & Gas Co., lessee. Location: 1 mile above right fork of Beaver Creek on Henry Branch. Completed: 1918. Open flow: 550,000. Casing head: 707.83.

Strata.

Pennsylvanian System.

	Thickness	Depth
Soil	24	24
Shale, light	17	41
Shale, black	17	58
Sandstone	30	88
Shale, black	25	113
Shale, gray	27	140
Shale, black	15	155
Sandstone	12	167
Shale, sandy (show of gas)	23	190
Sandstone, light	10	200
Shale, light	50	250
Sandstone	35	285
Shale, dark	95	380
Sandstone	20	400
Shale	25	425
Limestone	15	440
Sandstone	12	452
Limestone, gray	28	480
Sandstone, salt, (gas 485-495,75,000)	155	635
Shale, sandy, dark, (gas 540-550, 75,000)	8	643
Sandstone, gray	32	675
Sandstone, dark	5	680
Sandstone, white, (salt water flooded 700)	90	770
Sandstone, dark gray	18	788
Shale, black, (case 6½ 794)	6	794
Shale and sand	46	840
Shale and red rock	7	847
Salt sand	13	860

Mississippian System.	Thickness	Depth
Shale and sand	5	865
Sandstone (Maxon)	49	914
Shale	5	919
Total depth		919

Log No. 253

W. R. Crisp, No. 1, lessor. Keystone Oil & Gas Co., lessee. Location: 1 mile up Turkey Creek. Completed: July 25, 1918. Production: Gas, open flow, small. Casing head: A. T. 677.

Strata.

Pennsylvanian System.	Thickness	Depth
Soil	18	18
Sandstone	28	46
Shale	66	112
Sandstone	30	142
Shale	10	152
Sandstone	18	170
Shale	12	182
Sandstone	12	194
Shale	35	229
Sandstone	10	239
Shale	15	254
Sandstone	15	269
Shale	75	344
Sandstone	16	360
Shale	10	370
Salt sand, (gas 475-495, 100,000 cu. ft.)	381	751
Coal	2	753
Sandstone, (salt water flooded hole 675)	15	768
Shale, black	9	777
Shale and sand	4	781
Shale, light	30	811
Shale, black	16	827

Mississippian System.		
Shale, red, sandy	7	834
Shale, yellow	3	837
Sandstone (Maxon)	35	872
Shale, black	22	894
Total depth		894

Log No. 254

J. P. Allen, No. 1, lessor. Keystone Oil & Gas Co., lessee. Location: ¾ mile south of Maytown, off main road 500 feet. Completed: July 24, 1919. Production: Gas, open flow, 3,618,000 cubic feet. Casing head: A. T. 682.5.

Strata.

Pennsylvanian System.	Thickness	Depth
Soil	45	45
Sandstone gray	20	65
Shale	55	120
Sandstone	60	180
Shale	55	235
Sandstone, (gas 270, 25,000)	35	270
Shale, (gas 340, 50,000)	130	400
Sandstone, gray, (gas 25,000)	125	525
Sandstone, gray, (gas 25,000)	20	545
Sandstone, dark	17	562
Sandstone, gray	6	568
Sandstone, white	30	598
Sandstone, gray and salt sand (flooded 680)	116	714
Coal	3	717
Sandstone, dark, (salt sand)	48	765
Shale	40	805
Sandstone	3	808
Mississippian System.		
Shale, red, sandy	6	814
Sandstone (Maxon)	49	863
Total depth		863

Log No. 255

Kentucky Coke Co., (J. M. Osborn, No. 1), lessor. Louisville Gas & Electric Co., lessee. Location: Wilson Creek. Date Drilled: Nov. 1, 1921. Contractor: E. B. Duncan. Orig. Open Flow: 1,150,000 cubic feet gas. Orig. Rock Press.: 275 lbs.

Strata.

Pennsylvanian System.	Thickness	Depth
Soil	20	20
Sandstone	40	60
Coal	1	61
Sandstone	149	210
Shale	35	245

Pennsylvanian System	Thickness	Depth
Shale, calcareous	65	310
Shale	50	·360
Shale, calcareous	20	380
Sandstone	40	420
Shale	20	440
Shale, calcareous	10	450
Sandstone	7	457
Shale, calcareous	6	463
Sandstone	47	510
Shale, hard	225	735
Shale	15	750
Sandstone	5	755
Shale	95	850
Shale, hard	5	855
Shale	7	862
Sandstone	2	864
Shale	13	877
Sandstone	13	890
Shale	55	945
Sandstone	5	950

Mississippian System.		
Limestone	25	975
Sandstone	3	978
Shale	7	985
Sandstone.	5	990
Total depth		990

Log No. 256

Kentucky Coke Co. (S. P. Ratcliffe, No. 1), lessor. Louisville
Gas & Electric Co., lessee. Location: Head of Wilson Creek, May-
town. Date Drilled: Sept. 30, 1921. Contractor: E. B. Duncan. Orig.
Open Flow: 170,000 cubic feet gas. Orig. Rock Press.: 530 lbs.

Strata.

Pennsylvanian System.	Thickness	Depth
Gravel	35	35
Coal	2	37
Clay	13	50
Shale, calcareous	40	90
Shale	45	135
Sandstone	15	150
Shale	20	170

Pennsylvanian System

	Thickness	Depth
Shale, calcareous	2	172
Shale	102	274
Shale and shell	15	289
Shale, hard	11	300
Shale	50	350
Sandstone	90	440
Shale	25	465
Sandstone	88	553
Shale	341	894
Sandstone	8	902
Shale	18	920
Sandstone	5	925
Shale	115	1,040
Sandstone	10	1,050

Mississippian System.

Limestone, black	10	1,060
Sandstone	20	1,080
Shale	10	1,090
Sandstone	10	1,100
Shale	80	1,180
Shale	15	1,195
Limestone	5	1,200
Shale	12	1,212
Limestone	3	1,215
Shale (pencil cave)	13	1,228
Limestone	1	1,229
Limestone (Big Lime), dark	176	1,405
Red rock (Big Injun)	65	1,470
Limestone	25	1,495
Shale	35	1,530
Limestone	50	1,580
Shale, coffee	90	1,670
Limestone	50	1,720
Shale	50	1,770

Devonian System.

Shale, black	30	1,800
Shale	25	1,825
Shale	45	1,870
Shale	30	1,900
Shale, brown	68	1,968
Shale	10	1,978
Total depth		1,978

Log No. 256-A.

Tom Reffet, No. 1, lessor. Beaver Creek Oil & Gas Co., lessee. Location: Pitts Fork of Left Fork of Middle Creek, Floyd County, Ky. Production: Gas, 3,000,000 cubic feet. Casing head elevation: 860 A. T. Authority: Frank Harmon, Bill Adams. Incomplete Record. Strata.

Pennsylvanian System.	Thickness	Depth
Clay and sandstone	30	30
Coal	5	35
Shale	165	200
Sandstone (Little Dunkard), (1st cow run)	60	260
Shale	115	375
Sandstone (Big Dunkard)	65	440
Shale	20	460
Sandstone, (gas sand)	15	475
Shale	105	580
Sandstone, salt, (1st) (gas)	135	715
Total depth		715

Log No. 256-B.

Lou Ann Wright farm, No. 1, lessor. Beaver Creek Oil & Gas Co., lessee. Location: Pitts Fork of Left Fork of Middle Creek, Floyd County, Ky. Production: Gas. Casing head elevation: 795 A. T. Strata.

Pennsylvanian System.	Thickness	Depth
Sandstone, gray	28	28
Shale, blue	22	50
Sandstone and shale	7	57
Sandstone, gray	3	60
Shale, black	17	77
Coal	5	82
Shale, blue	33	115
Sandstone, gray	62	177
Shale	7	184
Sandstone, white	81	265
Shale, black	60	325
Sandstone, white	55	380
Shale, black	35	415
Sandstone	15	430
Shale, black	155	585
Sandstone	240	825
Shale	6	831
Sandstone	12	843
Shale, black	4	847

Pennsylvanian System	Thickness	Depth
Sandstone, white	48	985
Limestone, black, sandy	11	906
Coal	1	907
Mississippian System.		
Limestone and shale	19	926
Shale, black	41	967
Sandstone (Maxon)	119	1,086
Total depth		1,086

Light show of gas, 605; big flow, 670-715; lot of salt water, 740; show of oil, 1039-1046.

Log No. 256-C.

Colla Allen, No. 2, lessor. Eastern Carbon Co., lessee. Location: On waters of Goslin Branch of Goose Creek, Floyd County, Ky. Contractors: Dial & Meabon.

Strata.

Pennsylvanian System.	Thickness	Depth
Gravel	25	25
Shale	25	50
Limestone, black	10	60
Sandstone	35	95
Shale	35	130
Limestone	20	150
Shale	90	240
Sandstone	110	350
Shale	67	417
Sandstone	23	440
Shale and shells	130	570
Sandstone	10	580
Shale	24	604
Sandstone, salt (1st)	146	750
Shale and shells	40	790
Sandstone, salt (2d)	120	910
Shale	10	920
Mississippian System.		
Limestone, sandy	40	960
Shale	13	973
Limestone, black	22	995
Shale	2	997
Sandstone (Maxon) (1st)	26	1,023
Shale	12	1,035

Mississippian System.

	Thickness	Depth
Sandstone (Maxon) (2nd)	60	1,095
Shale	5	1,100
Sandstone	15	1,115
Shale	23	1,138
Sandstone	18	1,156
Shale	2	1,158
Sandstone	42	1,200
Shale	15	1,215
Sandstone	25	1,240
Shale and shells	45	1,285
Sandstone	11	1,296
Limestone (Little Lime)	27	1,323
Shale (pencil cave)	2	1,325
Limestone (Big Lime)	95	1,420
Shale	4	1,424
Sandstone (Keener)	6	1,430
Total depth		1,430

Casing left in hole 10″ 24-4.
Casing left in hole 8¼″ 208-2.

Log No. 256-D.

Colla Allen, No. 3, lessor. Eastern Carbon Co., lessee. Location: On Goose Creek, Floyd County, Ky. Completed: January 18, 1922. Contractors: Dial & Meabon. Production: Gas, 3½ million cubic feet oil, about 5 bbls. per day.

Strata.

Pennsylvanian System.

	Thickness	Depth
Soil	55	55
Sandstone	15	70
Shale	15	85
Sandstone	40	125
Shale, (fresh water 180)	157	282
Coal	4	286
Sandstone	19	305
Shale	30	335
Sandstone	150	485
Shale	105	590
Sandstone	30	620
Coal	4	624
Shale	96	720
Sandstone	15	735
Shale	88	823
Sandstone, salt (salt water 892, big water 1050)	294	1,117
Shale	5	1,122

Mississippian System.

	Thickness	Depth
Limestone	43	1,165
Sandstone	72	1,237
Shale	5	1,242
Limestone	9	1,251
Sandstone and limestone, shelly	14	1,265
Sandstone	33	1,298
Shale	5	1,303
Sandstone	5	1,308
Shale	8	1,316
Sandstone	94	1,410
Shale, red, and limestone, shelly	20	1,430
Shale	37	1,467
Sandstone	10	1,477
Shale and shells	8	1,485
Limestone, shelly	15	1,500
Sandstone	48	1,548
Total depth		1,548

Show of oil, 1350.
Show of gas, 1393.
Gas started to pay, 1500, and payed to 1546.
Amount of gas, 3½ million cubic feet.
Oil pay at 1546, to 1548.
About 5 bbls. per day.
Casing record:
10 in. casing, 21 ft.
8¼ in. casing, 253 ft.
6¼ in. casing, 1294 ft.
3 in. tubing, 1548 ft.
6¼ in. tubing packer set at 1341.

FRANKLIN COUNTY.

Production: **Small gas.** Producing sands: **Unnamed of Trenton age (Ordovician).**

Log No. 257

Louis C. Weber, No. 1, lessor. Dr. J. S. Goodrich and B. G. Pratte, lessees. Location: Near Devil's Hollow Pike on Benson Creek, above falls. Commenced and completed: Summer, 1920. Production: Oil and gas shows only. Authority: W. T. Congleton, driller, 346 Aylesford Place, Lexington, Ky.

Strata.

Ordovician System.

	Thickness	Depth
Soil	30	30
Shale	20	50
Limestone and shale	350	400
Limestone ''sand''	25	425
Limestone, ''salt water sand''	2	427
Total depth		427

CHAPTER IV.

FULTON COUNTY.

Production: Neither oil or gas to date. Producing Sands: None recognized to date.

Log No. 258

Roney, Mitchell & Bruer, Hickman, Ky., owners and operators. Location: 150 yards S. E. Bondrant Station on C. M. & G. R. R. which is 8 miles S. W. Hickman, 1 mile N. Reelfoot Lake. Drilled with rotary machine. Driller: De Orman. Stratigraphic interpretations by W. R. Jillson. Authority: J. W. Roney. Production: Oil and gas show only to Dec. 7, 1921. Well incomplete and drilling.

Strata.	Thickness	Depth
Quaternary System.		
Soil	15	15
Sand	105	120
Clay	15	135
Tertiary System.		
Sand, Pliocene or Miocene	20	155
Gumbo, Pliocene or Miocene	95	250
Gumbo and gravel, (10" casing), La Grange..	50	300
Sand and gravel, La Grange	50	350
Sand, brown, La Grange	100	450
Sand, hard, La Grange	480	930
Sand rock, La Grange	70	1,000
Brown water sand, La Grange	100	1,100
Sand, hard, La Grange	240	1,340
Shale, black, and gumbo, (8" casing), Porter's Creek	105	1,445
Brown rock, (gas show), Porter's Creek	6	1,451
Black gumbo, Porter's Creek	369	1,820
Cretaceous System.		
Shale, hard, and hard sand and gravel, showing some oil and gas all the way, Ripley-McNairy and Selma)	94	1,914
Shale hard, and sand, (Ripley-McNairy and Selma)	71	1,985
Limestone, hard, gray, with layers of chalk, (Ripley-McNairy and Selma)	285	2,270
Limestone, hard, gray (6" casing)	130	2,400
Limestone, hard, brown, green and red	300	2,700

Cretaceous System.   Thickness Depth

	Thickness	Depth
Limestone, hard black, pyrites and silica	200	2,900
Chalk rock, white	50	2,950
Limestone, hard, gray, sand and brown shale, (oil show)	230	3,180
Incomplete depth, Dec. 7, 1921		3,180

NOTE—The computed thickness of the entire Mississippian Series regarded as present beneath the embayment series has been placed at from 1800 to 2300 feet. Accepting the base of the Cretaceous as 2120 the base of the Mississippian and the top of the Devonian here is probably about 4,300 feet below the surface. For purposes of comparison in this little "wild catted" section of extreme western Kentucky, the record of three recent wells, all drilled near to Reelfoot Lake, in Obion and Lake Counties, Tenn., are given as follows:

OBION COUNTY, TENNESSEE.

Log No. 259

Roger Well, No. 1, lessor. Reelfoot Ranger Oil Co., lessee. Location: 3 miles east of Walnut Log, in Obion County, Tennessee. Collaborated authorities: J. S. Hudnall, collector of cuttings, supplied by Tennessee Geological Survey; Wilbur A. Nelson, stratigraphic divisions; and C. H. Richardson, mineralogical and lithological determinations. This log compiled from actual cuttings of rotary drill.

Strata.

Quaternary System.

	Thickness	Depth
Clay, loess, calcareous, yellowish gray	70	70
Clay, loess ferruginous	20	90
Clay, calcareous, yellow	10	100
Tertiary System.		
Gravel, coarse, river water rounded, Pliocene or Miocene	40	140
Sand and gravel, river rounded, fine, Pliocene or Miocene	5	145
Gravel, coarse, ferruginous, Pliocene or Miocene	10	155
Unknown (no sample collected, Pliocene or Miocene)	145	300
Sand, silicious, fine, gray, La Grange	25	325
Sand or shale, fine, light gray, La Grange....	5	330
Sand, dolomitic and calcareous, and shale La Grange	25	355
Unknown, La Grange	5	360

Tertiary System.   Thickness Depth

Shale, dolomitic, manganiferous and carbonaceous, La Grange	30	390
Sand, fine, with carbonaceous matter, La Grange	20	410
Gravel, fine, light gray, La Grange	25	435
Sand and gravel (break), angular, carbonaceous, La Grange	30	465
Sand, fine, and gravel, light gray, La Grange	30	495
Sand, fine, and gravel, light gray, La Grange..	45	540
Gravel ferruginous, coarse, La Grange	30	570
Sand, mostly white, fine, La Grange	5	575
Sand and gravel, La Grange	10	585
Shale, light colored, fine quartz sand, La Grange	15	600
Sand, with little gravel, fine, and shale, La Grange	20	620
Sand and gravel, slightly dolomitic, mollusca, La Grange	25	645
Sand, coarse, subangular (break), La Grange..	10	655
Sand, shale and gravel, small fossil, Porter's Creek	25	680
Sand, shale and flat limonite gravel, Porter's Creek	70	750
Sand and gravel, coarse and fine, Porter's Creek	20	770
Sandstone, fine and coarse, Porter's Creek ...	100	870
Sand and gravel, some clay, Porter's Creek ...	30	900
Sand, white, and ferruginous gravel, Porter's Creek	10	910
Sand, mostly white, Porter's Creek	10	920
Sand and gravel, flat and angular, pea size, Porter's Creek	30	950
Gravel, ferruginous, (break), some sand, Porter's Creek	115	1,065
Unrecorded, Porter's Creek	140	1,205
Sand and gravel, Porter's Creek	20	1,225
Gravel and clay, Porter's Creek	10	1,235
Sand and gravel, coarse, Porter's Creek	40	1,275
Sand and gravel and clay, Porter's Creek ...	45	1,320
Sand, gravel, sand clear quartz, Porter's Creek	120	1,440
Sand, gravel, mostly clear quartz sand, Porter's Creek	45	1,485
Sand, quartz and gravel of sandstone, Porter's Creek	20	1,505
Sand and gravel, mostly white quartz, Porter's Creek	95	1,600
Sand, very fine, Porter's Creek	25	1,625

Tertiary System.

	Thickness	Depth
Sand and gravel, shaly, Porter's Creek	60	1,685
Sand, gravel and bluish shale, Porter.'s Creek ,	40	1,725
Shale, bluish gray, alumina and silica, Porter's Creek	15	1,740
Total depth		1,740

Log No. 260

O. T. Wollaston, No. 1, lessor. Reelfoot Ranger Oil Co., lessee.
Location: Walnut Log, Obion County, Tennessee. Authority: Tenn.
Geological Survey. Stratigraphic divisions by Wilbur A. Nelson, State
Geologist, Tenn.

Strata.

Quaternary System.

	Thickness	Depth
Surface soil, ...	3	3
Clay, silt and sand, (River fill)	17	20
Quicksand, (River fill)	70	90
Gravel, river water worn, (River fill)	95	185

Tertiary System.

	Thickness	Depth
Clay, silt and sand, Pliocene or Miocene	10	195
Sand, (water), Pliocene or Miocene	20	215
Gravel, clay and artesian flow, Pliocene or Miocene	45	260
Clay, Pliocene or Miocene	15	275
Sand and clay, Pliocene or Miocene	20	295
Sand, clay and rock, Pliocene or Miocene	5	300
Sand and clay, La Grange	10	310
Sand and gravel, La Grange	20	330
Quicksand, La Grange	5	335
Gravel, La Grange	5	340
Gravel and sand, La Grange	5	345
Sand and clay, La Grange	25	370
Sand and gravel, La Grange	40	410
Gravel, sand, flint, chalk rock, La Grange	20	430
Clay, blue, fine, sticky, La Grange	19	449
Sand and flint, La Grange	31	480
Gravel, La Grange	91	571
Clay, sticky, and sand, La Grange ....:.....	29	600
Gravel and sand, La Grange	40	640
Clay, sticky, and sand, La Grange	15	655
Sand, La Grange	65	720
Sandstone, hard, some gas, La Grange	5	725
Gumbo and sand, La Grange	45	770
Sand, La Grange	10	780

Tertiary System.

	Thickness	Depth
Sand and gumbo, La Grange	20	800
Sand, La Grange	40	840
Sand and gumbo, La Grange	15	855
Sand, (asphalt), La Grange	5	860
Clay, sticky, and sand, La Grange	65	925
Sand, La Grange	30	955
Clay, fine, and sand, La Grange	20	975
Sand, La Grange	35	1,010
Clay, sticky, and sand, La Grange	65	1,075
Total depth		1,075

LAKE COUNTY, TENNESSEE.

Log No. 261

Reelfoot Dome Oil Co., lessor. Location: northwest side of Reelfoot Lake, at Proctor City. Authority: De Armand, driller. Stratigraphic division by Wilbur A. Nelson, State Geologist, Tennessee Geological Survey. Selma fossils found in bottom of well.

Strata.

Quaternary System.

	Thickness	Depth
Soil	10	10
Sand and gravel	135	145
Unknown, (no sample), Pliocene or Miocene ..	55	190
Clay, blue gray, sticky, Pliocene or Miocene ..	20	210
Sand and clay, like buttermilk, with wood, some reddish, Pliocene or Miocene	15	225
Quicksand	75	300
Sand, blue, little clay, La Grange	80	380
Sand, gray, La Grange	103	483
Sand, La Grange	45	528
Gumbo, La Grange	37	565
Sand, hard, La Grange	20	585
Sand, brown, coarse, La Grange	200	785
Sand, hard, and gravel, La Grange	115	900
Sand rock La Grange	50	950
Shale, black, La Grange	70	1,020
Sand, hard, coarse, La Grange	60	1,080
Gumbo, gray, La Grange	60	1,140
Sand, brown, coarse, La Grange	125	1,265
Gumbo, sandy, Porters Creek	210	1,475
Shale, black, Porters Creek	25	1,500
Gumbo, sandy, (show of oil) Porters Creek	80	1,580

Tertiary System.	Thickness	Depth
Shale, black, Porters Creek	20	1,600
Shale, hard, yellow, fine shells, Porters Creek ..	20	1,620
Gumbo, sandy, Porters Creek	30	1,650
Cretaceous System.		
Shale, black, with blue lime shells and white flint, Selma—McNairy and Ripley	70	1,720
Shale, blue, with hard shells of flint and pyrite, Selma—McNairy and Ripley	230	1,950
Shells and hard sandstone, Selma—McNairy and Ripley	24	1,974
Limestone, Selma—McNairy and Ripley	101	2,075
Total depth		2,075

GREEN COUNTY.

Production: **Oil and Gas.** Producing **Sands: Corniferous (Devonian);
Niagaran (Silurian).**

Log No. 262

Cashdollar, No. 1, lessor. Location: Gowan, near Russell Creek,
7 miles southwest of Greensburg.

Strata.		
Mississippian System.	Thickness	Depth
Soil	8	8
Limestone, blue, hard, (water 50)	204	212
Shale, gray, ...	32	244
Devonian System.		
Shale, black	43	287
Limestone (cap rock)	2	289
Limestone, (oil "sand")	6	295
Total depth		295

122'—6½ casing.

Drilled into water. Some came with the oil.

Log No. 263

J. E. Thompson, No. 1, lessor. George H. Carson, lessee. Mahan
Bros., drillers. Location: 2 miles east of Coakeley. Completed: September, 1920. Production: ½ bbl. green oil.

Strata.		
Mississippian System.	Thickness	Depth
Soil	18	18

Mississippian System.

	Thickness	Depth
Limestone, brown	29	47
Caves and crevices	24	71
Gravel and water	7	78
Limestone, blue, hard	23	101
Limestone, gray	8	109
Limestone, broken	41	150
Limestone, gray	20	170
Limestone, broken	63	233
Limestone, gray, hard, flinty	12	245
Limestone, gray	29	274
Limestone, blue, hard	20	294
Limestone, broken	14	308
Limestone, gray	32	340
Limestone, broken	8	348
Limestone, gray, hard	25	373
Shale, blue	3	376
Limestone, gray	57	433
Limestone, gray, hard	17	450
Limestone, broken	57	507
Shale, green	1	508

Devonian System.

	Thickness	Depth
Shale, black	59	567
Limestone (cap rock)	14	581
Pay sand	½	581½
Limestone, light gray, sandy	17½	599
Total depth		599

80 ft. 6¼ casing.

Log No. 264

Vance, No. 1, lessor. Molloy & Gardner, lessees. Location: 3 miles southwest of Greensburg.

Strata.

Mississippian System.

	Thickness	Depth
Soil	4	4
Limestone	198	202

Devonian System.

	Thickness	Depth
Shale, brown	57	259
Shale, green	4	263
Shale, black	47	310
Limestone (cap rock)	2	312
Limestone sand	11½	323½
Total depth		323½

BARGING EASTERN KENTUCKY OIL.

The view is in the Kentucky River at Frankfort and shows one method of transporting petroleum to the Ohio River refineries.

Log No. 265 ·

J. N. Nagle, No. 1, lessor. M. B. Cooley Oil & Gas Co., lessees.
Location: 6 miles south of Greensburg, Ky., near Newt Thurlow.
Strata.

Mississippian System.	Thickness	Depth
Soil	3	3
Limestone	62	65
Shale (Waverly), shaly limestone and sand ..	10	75
Limestone, brown, (gas)	15	90
Limestone, shelly	10	100
Limestone, broken, (gas)	15	115
Limestone, shelly	10	125
Limestone, shelly, (gas)	5	130
Limestone, brown	95	225
Limestone, shelly	5	230
Limestone, broken	10	240
Limestone, gray, hard	188.	428
Limestone, brown	20	448
Shale, green	4	452
Devonian System.		
Shale, black	40	492
Limestone ("cap" and "sand"), white	8	500
Shale, gray, soft, fire clay and yellow clay	35	535
Total depth		535

Casing, approx. 234.

Log No. 266

John Risen, No. 1, lessor. Location: Summerville. Commenced:
July 10, 1919. Completed: September 6, 1919. Drilling contractors:
Houser and Mootheart. Authority: The Atlantic Oil Producing Co.
Strata.

Mississippian System..	Thickness	Depth
Soil, yellow, soft	30	30
Limestone, gray, hard	98	128
Shale, blue, soft	22	150
Limestone, brown, hard	226	376
Shale, black, hard	1	377
Limestone, gray, hard, coarse	21	398
Limestone, gray, hard, fine	52	450
Devonian System.		
Shale, black, soft	55	505
Limestone, black and white, hard cap	3	508
Sand, gray, medium	1½	509½
Limestone, light gray, soft	26½	536
Total depth		536

Log No. 267

William Turner, No. 1, lessor. Location: ¼ mile north of Highland School House. Production: Encountered several small pockets of gas, and a small showing of gas on top of pay sand.

Strata.

Mississippian System.	Thickness	Depth
Soil	10	10
Limestone, blue	209	219
Limestone, broken	80	299
Devonian System.		
Shale, brown	48	347
Limestone (cap rock)	4	351
Shale, limy, (sand)	8	359
Shale, limy, (salt sand)	6	365
Total depth		365

Casing head el. above sea level, 690 ft.
Base of black shale, el. 343 ft. above sea level.

Log No. 268

J. H. Kessler, No. 1, lessor. S. W. Meals, et al., Pittsburg, Pa., lessees. Completed: August 21, 1920. Production: 5—10 bbls. oil.

Strata.

Mississippian System.	Thickness	Depth
Soil	8	8
Limestone, hard	198	206
Limestone, broken	41	247
Devonian System.		
Shale, black	44	291
Limestone (cap rock)	2	293
Limestone (pay sand)	5	298
Total depth		298

Remarks: Oil showed in cap rock. Small amount of salt water showed at 298 feet, and drilling was stopped.

Log No. 269

A. H. Akin, No. 1, lessor. Location: 5 miles southwest of Greensburg. Completed: September 15, 1919. Shot July 6, 1921. Production: small oil. Drilled: Mallort and Godden. Authority: G. B. Taylor.

Strata.

Mississippian System.	Thickness	Depth
Soil	30	30

Mississippian System.

	Thickness	Depth
Limestone	88	118
Limestone, brown	101	219
Shale, green	3	222

Devonian System.

Shale, black (Chattanooga)	145	267
Limestone (cap rock)	3	270
Limestone "sand," (good)	5	275
Limestone "sand," white	3	278
Total depth		278

Log No. 270

Blakeman, No. 1, lessor. G. B. Taylor, et al., lessees. Location: 3 miles northeast of Greensburg. Completed: July, 1920.
Strata.

Mississippian System.

	Thickness	Depth
Soil	8	8
Limestone, hard	183	191

Devonian System.

Shale, black (Chattanooga)	45	236
Limestone (cap rock)	24	260
Limestone "sand," (1½ million cu. ft. gas)	20	280
Limestone, broken, (salt water 290)	40	320
Shale, green	20	340
Shale, pink	14	354
Shale, very brown	28	382
Limestone, gray	420	802
Total depth		802

Log No. 271

Blakeman, No. 2, lessor. G. B. Taylor, et al., lessees. Location: 3 miles northeast of Greensburg. Completed: March, 1921. Production: ½ million cu. ft. gas. Authority: G. B. Taylor.
Strata.

Mississippian System.

	Thickness	Depth
Soil	30	30
Limestone, hard, blue	68	98
Limestone, hard, flinty	58	156
Limestone, brown	12	168

Devonian System.

Shale, black (Chattanooga)	48	216
Limestone (cap rock)	10	226
Sandstone and limestone	27	253
Total depth		253

Log No. 272

J. B. Cook, No. 1, lessor. Location: 10 miles southwest of Greensburg. Drilled by P. O. Johnson. Completed: September 10, 1919. Production: Tested on ½ million cu. ft. gas. Well is capped. Authority: G. B. Taylor.

Strata.

Mississippian System.	Thickness	Depth
Soil	4	4
Limestone, blue	16	20
Limestone, gray	249	269
Shale, gray	20	289

Devonian System.		
Shale, black (Chattanooga)	44	333
Limestone (cap rock)	2	335
Limestone "sand," (oil show 349)	25	360
Limestone, broken	7	367
Shale, pink	15	382
Shale, green	1	383
Total depth		383

Log No. 273

Gowen, No. 1, lessor. J. W. Cashdollar, et al., lessees. Location: 7 miles southwest Greensburg on the Little Russell Creek. Completed: August 14, 1919. Production: 120 ft. of oil after the first 12 hours. Well is not being pumped. Authority: G. B. Taylor.

Strata.

Mississippian System.	Thickness	Depth
Soil	10	10
Limestone, hard	206	216
Shale, gray	32	248

Devonian System.		
Shale, black (Chattanooga)	43	291
Limestone (cap rock)	2	293
Limestone ("Irvine sand")	3	296
Total depth		296

Log No. 274

W. L. Hicks, No. 1, lessor. G. B. Taylor, et al, lessees. Location: 1¾ miles north of Greensburg. Contractor: G. B. Taylor. Completed: June 1, 1921. Authority: G. B. Taylor.

Strata.

Mississippian System.	Thickness	Depth
Soil	2	2
Limestone, gray, hard	89	91
Limestone, gray, flinty	50	141
Limestone, gray, broken	13	154
Devonian System.		
Shale, black (Chattanooga)	46	200
Limestone (cap rock)	11	211
Limestone "sand," brown, tight	9	220
Limestone "sand," brown, broken	16	236
Ordovician System.		
Shale, gray, and mud, blue	7	243
Shale, gray	5	248
Limestone, "salt sand," (salt water)	2	250
Total depth		250

Gas 54, 76, 125, 142, 211 and 226 feet. Salt water found in bottom of hole rose approximately 35 feet in hole.

Log No. 275

R. A. White, No. 3, lessor. Green River Gas Co., lessee. Location: on Meadow Creek about 6,000 feet directly north of R. A. White, No. 1, and about 2,500 feet southeast of Whitewood Station. Commenced: January 1, 1921. Completed: February 14, 1921. Production, by Pitot Tube, 2,740,608 cu. ft. gas. Contractors: More and Moss.

Strata.

Mississippian System.	Thickness	Depth
Soil, yellow	3	3
Gravel, creek bed	2	5
Shale, gray	2	7
Limestone, gray	14	21
Limestone, flinty, gray	25	46
Limestone, blue	19	65
Sand, brown	3	68
Limestone, hard, flinty	8	76
Limestone, white	12	88
Shale, blue	16	104
Limestone, hard, shelly	14	118
Limestone, white	6	124
Devonian System.		
Shale, black (Chattanooga)	49	173
Limestone, white	2	175

Devonian System.	Thickness	Depth
Sand, limy, light gray	1	176
Sand, fine, light brown ....,......	1	177
Sand, dark gray	1	178
Limestone, gray, sandy	1	179
Sand, gray	1	18(
Sand, limy, gray	2	182
Limestone, sandy, dark gray	3	185
Limestone, sandy, dark gray	2	187
Limestone, sandy, light gray	2	189
Silurian System.		
Sand, gray, limy (gas show, steel line).......	1½	190½
Sand gray, limy (gas show increasing)	2½	193
Limestone, light gray, sandy	3	196
Sand, limy, light gray (gas increases to half million)	3	199
Sand, limy, light gray (gas increases to 600,000)	-	200
Sand, coarse, gray, limy (gas increases to 2,500,000)	10	210
Sand, coarse, gray, limy (gas increases to 3,000,000)	5	215
Total depth		215

Rock pressure, 37 lbs.
Casing: 8¼″ 10′
         6¼″ 70′

Log No. 276

W. M. Price, No. 1, lessor. Cutler and Wallis, Inc., lessees. Location: near Crab. Drilled by Mahan Bros. Commenced: March 17, 1921. Authority: G. B. Taylor.

Strata.

Mississippian System.	Thickness	Depth
Soil and loose rock	8	8
Limestone, hard, gray, non-cryst, no fossils, (water 27)	44	52
Crystals, rusty	2	54
Limestone, hard, gray	13	67
Limestone, light gray, hard, rusty	8	75
Limestone, gray	2	77
Limestone, gray, (water 88)	11	88
Limestone, blue, gray, broken	11	99
Limestone, hard, blue	9	108
Limestone, dark blue, broken	7	115
Limestone, blue and gray, (some gas)	9	124
Limestone, blue gray, white specks	8	132

Mississippian System.	Thickness	Depth
Limestone, blue gray, broken	76	208
Limestone, dark blue, hard in spots	12	220
Limestone, blue gray, massive	15	235
Limestone, blue gray ....._..............	10	245
Limestone, gray	17	262
Limestone, blue, soft	9	271
Limestone, gray, hard	29	300
Limestone, blue gray, (gas 314)	14	314
Limestone, hard, gray, blue	33	347
Limestone, hard, gray, (little gas)	10	357
Limestone, gray, blue	59	416
Devonian System.		
Shale, black (Chattanooga)	46	462
Limestone (cap rock)	1	463
Limestone ''sand,'' hard, white, (show of oil		
464)	13	476
Limestone, white, sandy	8	484
Shale, gray, limy	2	486
Incomplete depth		486

NOTE—This is an incomplete record of this well, which was drilled deeper.

Log No. 277

Porter Turner, No. 1, lessor. Location: 4 miles north of Greensburg on Big Pitman Creek. Completed: February, 1919. Production: Gas well; the gas is used for domestic purposes. Authority: G. B. Taylor.

Strata.

Mississippian System.	Thickness	Depth
Soil	7	7
Limestone, shelly	2	9
Gravel	2 .	11
Limestone, hard, blue	139	150
Shale, hard, black	43	193
Shale, black	10	203
Shale, green	9	212
Devonian System.		
Shale, brown (Chattanooga)	48	260
Limestone ''sand,'' brown	3	263
Limestone (cap rock)	4	267
Limestone ''sand''	11	278
Total depth		278

Log No. 278

M. P. Vaughn, No. 1, lessor. Location: 6 miles southwest of Greensburg. Drilled by S. W. Neal, et al. Production: Flush 12 bbls. oil, but not being pumped. Authority: G. B. Taylor.

Strata.		
Mississippian System.	Thickness	Depth
Soil	10	10
Limestone, hard	229	239
Devonian System.		
Shale, black, (Chattanooga)	43	282
Limestone (cap rock)	1	283
Limestone ''sand''	9	292
Total depth		292

Log No. 279

A. V. Walker, No. 1, lessor. Location: 2½ miles southwest of Greensburg. Drilled by Mallory and Godden. Completed: July, 1919. Production: small oil. Not under pump. Authority: G. B. Taylor.

Strata.		
Mississippian System.	Thickness	Depth
Limestone, hard	298	298
Devonian System.		
Shale, black (Chattanooga)	48	346
Limestone (cap rock)	2	348
Limestone ''sand''	10	358
Total depth		358

Log No. 280

F. G. Yankey, No. 1, lessor. Completed: January 31, 1921. Authority: G. B. Taylor.

Strata.		
Mississippian System.	Thickness	Depth
Soil	2	2
Limestone, hard	198	200
Limestone, broken	20	220
Devonian System.		
Shale, black (Chattanooga)	48	268
Limestone (cap rock)	20	288
Limestone ''sand''	37	325
Ordovician System.		
Limestone (salt sand)	27	352
Limestone, blue, broken	26	378
Shale, pink	11	389
Shale, green	13	402
Limestone, brown	10	412

Ordovician System.	Thickness	Depth
Limestone, blue, broken, with hard streaks ..	41	453
Limestone, broken	147	600
Total depth		600

Small amount of gas at 290 feet
Set 115 feet with 8¼ inch casing.
Set 356 feet with 6¼ inch casing.

GREENUP COUNTY.

Production: **Oil and gas shown only to date.** Producing **Sands: None** recognized.

Log No. 281

Geo. F. Bradley, No. 1, lessor. United Fuel Gas Co., Transylvania Oil & Gas Co., lessees. Location: Big White Oak Creek, Greenup County, Ky. Completed: June 6, 1918.

Strata.

Mississippian System.	Thickness	Depth
Soil, gravel, etc. (water at 12)	12	12
Limestone (Big Lime)	75	87
Clay, blue	53	140
Shale and shells	165	305
Sandstone	45	350
Shale	65	415
Limestone	133	548
Devonian System.		
Shale, black	33	581
Coal	19	600
Shale, brown, (cased 794'—8¼")	385	985
Shale, white	80	1,065
Limestone, (show of gas)	7	1,072
Limestone (Ragland ''sand''), (water 1,115)	48	1,120
Silurian System.		
Limestone (Niagara)	300	1,420
Ordovician System.		
Shale, white	10	1,430
Shale, sandy, red (cased 1,520'—6⅝")	120	1,550
Limestone, (oil show 1,629)	100	1,650
Shale	17	1,667
Total depth		1,667

Casing record:

10" 32 lbs., 100' pulled.
8¼" 24 lbs., 794' left in well.
6⅝" 17 lbs., 1,520' pulled.

Log No. 282

Sanford Bradley, No. 1, lessor. United Fuel Gas Co., Transylvania Oil & Gas Co., lessees. Location: Big White Oak Creek. Completed: December, 1918.

Strata.

	Thickness	Depth
Mississippian System.		
Surface, gravel, etc. (fresh water 20)	20	20
Limestone	35	55
Shale	45	100
Clay, blue	200	300
Shale and limestone	125	425
Sand	10	435
Limestone	90	525
Devonian System.		
Shale, black	75	600
Shale, white	75	675
Limestone and black shale	50	725
Shale, brown	90	815
Limestone shell	10	825
Shale, brown	100	925
Shale, light	70	995
Limestone, light, hard	320	1,315
Shale, light	10	1,325
Ordovician System.		
Limestone, red, shaly	125	1,450
Shale, white	35	1,485
Limestone, red, shaly	15	1,500
Shale, blue	10	1,510
Limestone	25	1,535
Shale, blue	40	1,575
Shale and shells	35	1,610
Limestone, red, shaly	20	1,630
Shale	125	1,755
Limestone	10	1,765
Shale and limestone shells	536	2,301
Total depth		2,301

Water at 432.

Show of oil and gas, 1,000.

Water, 3 bailers per hr., 1,015.

Water, hole full, 1,080.

Cave, 1,375 to 1,425.

Casing record:

      13" conductor 13½".

      10" casing 106' pulled.

      8¼" casing 500' pulled.

      6⅝" casing 1,330' pulled.

NOTE—The Corniferous limestone occurs in the upper part of the 320 feet of limestone above 1,315 feet in depth. The base of the Devonian and the top of the Silurian is also within this 320 feet of limestone.

HANCOCK COUNTY.

Production: Oil and gas. Producing Sands: ''Pellville'' and ''Tar Springs'' (Chester-Mississippian).

Log No. 283

Breckinridge Cannel Coal Co., England, owner and operator. Location: Victoria Post Office. Drilled in spring of 1921. Driller, Albert MacGarvey. Stratigraphic interpretation by Prof. Arthur M. Miller, Lexington, Ky. Casing head, 550 feet, A. T. Standard Rig. Casinghead strata: Top of Chester.

Strata.

Mississippian System.

	Thickness	Depth
Surface materials	14	14
Shale, light colored, (Buffalo Wallow)	5	19
Limestone, white, (Buffalo Wallow)	6	25
Shale, gray, (Buffalo Wallow)	8	33
Limestone, gray, (Buffalo Wallow)	4	37
Shale, gray, (Buffalo Wallow)	13	50
Limestone, white to gray, (Buffalo Wallow)	32	82
Shale, mainly, light to dark, (Buffalo Wallow)	38	120
Sandstone and dark shale (Tar Springs)	23	143
Limestone dark, (Glen Dean)	1	144
Shale, calcareous, (Glen Dean)	14	158
Limestone, dark, (Glen Dean)	4	162
Shale, dark gray, (Glen Dean)	1	163
Limestone, dark gray, (Glen Dean)	5	168
Shale, dark gray, (Glen Dean)	29	197
Limestone, dark, crystalline, (Glen Dean)	3	200
Shale, dark, (Glen Dean)	2	202
Limestone, dark, crystalline, (Glen Dean)	15	217
Sandstone and shale (Hardinsburg)	11	228
Limestone, dark to light, (Golconda)	37	265
Shale, (Golconda)	19	284
Limestone, white to gray, (Golconda)	52	336

Mississippian System.	Thickness	Depth
Shale, dark to light, (Golconda)	22	358
Limestone, slaty, (Golconda)	12	370
Sandstone with shale (Cypress)	62	432
Limestone, (Casper)	24	456
Sandstone, white, (Casper)	13	469
Limestone, white to dark, (Casper)	231	700
Limestone, oolitic, white, (show of oil), (St. Genevieve)	170	870
Limestone, varying in color, and of varying degrees of purity (St. Louis, Warsaw and Upper Waverly)	820	1,690
Shale, greenish (New Providence)	30	1,720
Devonian System.		
Shale, black (Ohio-Chattanooga)	198	1,918
Limestone, white	52	1,970
Silurian System.		
Limestone, yellow to white	170	2,140
Ordovician System.		
Limestone, of varying colors and textures, at bottom, compact like Highbridge lime- stone	1,005	3,145
Total depth		3,145

Log No. 284

R. C. Jett Farm. Location: 2 miles S. E. of Pellville. Completed: Sept. 1921. Authority: C. Tobin Johnson.

Strata.

Pennsylvanian System.	Thickness	Depth
Soil	8	8
Sandrock	117	125
Sandstone, broken, and shale	185	310
Fire clay	20	330
Shale, muddy	5	335
Mississippian System.		
Limestone, brown	15	350
Shale	5	355
Limestone, gray, and shale	37	392
Shale	4	396
Limestone and shale	8	404

Mississippian System.	Thickness	Depth
Shale	44	448
Limestone, gray	8	456
Shale	8	464
Sand (gas pay) (Tar Springs)	3	467
Sand (oil pay) (Tar Springs)	16	483
Shale	2	485
Total depth		485

30' of 10" casing.

210' of 8" casing.

Shot with 60 qts. Shows for 25 buls.

Sand brown and medium soft.

Drilled by Oak Oil Co.

HARDIN COUNTY.

Production: Neither oil or gas. Producing Sands: None recognized.

Log No. 285

Stuart, No. 1, lessor. Frank X. Piatt, lessee. Location: near Colesburg. Commenced: December 29, 1920. Completed: January 20, 1921. Production: Salt water.

Strata.

Mississippian System.	Thickness	Depth
Soil, clay	4	4
Shale, blue, sticky	62	66
Devonian System.		
Shale, black (Chattanooga)	79	145
Limestone, gray	17	162
Limestone, brown	23	185
Limestone, sandy (salt water)	35	220
Shale, "fire clay"	10	230
Silurian System.		
Limestone, sandy (salt water)	48	278
Limestone, shaly	82	360
Total depth		360

HARRISON COUNTY.

Production: **Neither oil or gas.** Producing Sands: **None recognized.**

Log No. 286

Maybrier, No. 1, lessor. Starts in top of Cynthiana. Gas, 250 to 254; salt water 254. Authority: L. Beckner.

Strata.

Ordovician System.	Thickness	Depth
Limestone	254	254
Limestone, blue gray, hard, (lithograph)	436	690
Limestone, light dove gray, very hard	70	760
Limestone, shaly, dark, almost black, soft (grained almost)	65	825
Limestone, dark pepper dove, very hard fine..	15	840
Limestone, light blue green, very soft	33	873
Limestone, light dove, soft	6	879
Limestone, blue, muddy, very soft	41	920
Limestone, dark pepper and salt, with green shale, hard	45	965
Salt sand, light, dove yellow, St. Peter, very fine crystalline	10	975
Limestone, very light dove yellow crystal	25	1,000
Unrecorded sediments	225	1,225
Limestone, fine, sandy, dark yellow	65	1,290
Limestone, fine, white, sandy	3	1,293
Limestone, fine, light, sandy, wet	22	1,315
Limestone, fine, light, sandy, wet	10	1,325
Limestone, fine, dark, sandy	20	1,345
Limestone, fine, light, sandy, (mineral water)	7	1,352
Limestone, sandy, very coarse, and white magnesite	6	1,358
Limestone, sandy, coarse, with less magnesite, but small pyrite crystals	6	1,364
Total depth		1,364

HART COUNTY.

Production: **Small oil and gas.** Producing Sands: **Unnamed.**

Log No. 287

Elizabeth Gaddie, No. 1, lessor. New Domain Oil & Gas Co., lessee. Location: ¾ mile south of Boiling Springs Church, at bend of Green River. Completed: February 12, 1919. Production: filled up with salt water within 30 ft. of top ½ hr. Well abandoned. Casing: 234—6¼. Authority: New Domain Oil & Gas Co.

Strata.

Mississippian System.	Thickness	Depth
Soil, mud	20	20
Limestone, gray	140	160
Limestone, blue	375	535

Devonian System.		
Shale, black (Chattanooga)	55	590
Limestone, white	60	650
Shale, blue	10	660
Sand, gray	60	720
Total depth		720

Log No. 288

J. C. Nunn, No. 1, lessor. New Domain Oil & Gas Co., lessee. Location: 1½ miles northwest Boiling Springs Church. Completed: May 10, 1919. Authority: New Domain Oil & Gas Co.

Strata.

Mississippian System.	Thickness	Depth
Gravel, red	7	7
Limestone, white	233	240
Shale, black	5	245
Limestone, black	6	251
Shale	8	259
Limestone	551	810

Devonian System.		
Shale, black (Chattanooga)	66	876
Limestone, gray	16	892
Shale, black	2	894
Limestone, white	42	936
Shale, black	5	941
Limestone	31	972
Limestone (salt water)	14	986
Total depth		986

Filled 600 ft. south.
Salt water 2 ft. in sand.
Casing 267—8¼.
      876—6¼.

NOTE—This hole filled with salt water when the tools were pulled out for the last 14 feet. Well abandoned.

Log No. 289

H. L. Richardson, No. 1, lessor. New Domain Oil & Gas Co., lessee. Location: ¼ mile north Boiling Springs Church, about 5 miles northeast of Munfordville. Completed: December 5, 1918. Production: Well dry and abandoned. Authority: New Domain Oil & Gas Co.

Strata.

Mississippian System.	Thickness	Depth
Clay	15	15
Limestone, gray	150	165
Limestone, black	461	626
Devonian System.		
Shale, black (Chattanooga)	60	686
Limestone, gray	60	746
Limestone ''sand,'' gray	39	785
Silurian System.		
Limestone ''sand'' (salt water)	25	810
Limestone, black, (6¼" casing, 855)	190	1,000
Shale and limestone	424	1,424
Limestone, gray	76	1,500
Limestone, black	80	1,580
Total depth		1,580

HENDERSON COUNTY.

Production: Oil and gas. Producing sands: Sebree and Pottsville (Pennsylvanian). The Tar Springs and Cypress sands (Mississippian) are also untried possibilities.

Log No. 289A.

O'Nan Heirs, No. 1, lessors. Union County Syndicate, Union County, Ky., lessee. Location: 500 yards northeast of Highland Creek, and about 500 yards south of the Illinois Central Railroad right of way. This well is 1 mile southeast of Proctor, No. 1, well (Union County). Commenced: March 4, 1922. Completed: April 1, 1922. Authority: Ivyton Oil & Gas Co. Production: Salt water; well plugged and abandoned.

Strata.
Pennsylvanian System.

		Thickness	Depth
Drift		* 125	125
Shale and slate	Lisman Formation, Conemaugh Series	47	172
Fire clay		2	174
Lime, flinty		4	178
Fire clay		2	180
Coal (No. 11)		4	184
Slate		61	245
Slate		55	300
Shale, hard		12	312
Slate		38	350
Coal		22	372
Fire clay		3	375
Sand, dark		20	395
Slate		25	420
Coal		1	421
Slate, dark	Carbondale Formation (composed of DeKoven and Mulford), Allegheny Series.	64	485
Coal and slate		4	489
Slate		4	493
Sandy shell		4	497
Slate, sandy		53	550
Slate, dark		15	565
Coal		1	566
Shale, black		3	569
Shale, light		17	586
Slate, dark		27	613
Sand, gritty, dark		8	621
Slate, dark		5	626
Slate, hard		.	630

Pennsylvanian System.   Thickness Depth

		Thickness	Depth
Fire clay and light shale,	Carbondale Forma-tion (composed of DeKoven and Mulford), Allegheny Series.	34	664
Sand, white		31	695
Sand, salt water. Sebree		15	710

Total depth   710

NOTE—Full representation of Caseyville and Tradewater formations of Pottsville Series undrilled. Estimated thickness about 600 feet in this locality. The Pottsville Series was not drilled.

HOPKINS COUNTY.

Production: Small oil and gas. Producing Sands: Unnamed (Pennsylvanian), unnamed (Mississippian).

Log No. 290

Pools, No. 3, lessor. Moss Hill Oil & Gas Co., lessee. Location: 2 miles south of White Plains, and ½ mile from well No. 2 on this farm. Completed: in 1918. Production: at first was about 5 bbls. per day; oil is in this well now, but is not being pumped out, August, 1920. Authority: L. E. Littlepage.

Strata.

Pennsylvanian System.   Thickness Depth

	Thickness	Depth
Clay	3	3
Clay and gravel	7	10
Clay, sandy	17	27
Shale, hard, limy	1	28
Fire clay	7	35
Shale, soft	25	60
Shale	22	82

Pennsylvanian System.	Thickness	Depth
Shale, hard	2	84
Fire clay	12	96
Shale	7	103
Shale, hard	6	109
Shale, sandy	36	145
Shale, soft	41	186
Shale, hard, limy	2	188
Shale	12	200
Shale, hard, limy	3	203
Shale, soft	5	208
Sand rock, gray	42	250
Shale, soft	4	254
Shale, hard	3	257
Fire clay	8	265
Shale	75	340
Limestone and shale	10	350
Sand, (oil)	10	360
Sandstone, white	7	367
Shale	5	372
Limestone and shale	63	435
Sandstone, (water)	5½	440½
Total depth		440½

Log No. 291

Pools, No. 2, lessor. Moss Hill Oil & Gas Co., lessee. Location: 2 miles south of White Plains. Completed: in 1918. Production: Flush 20 bbls. pumped; now the well stands 300 feet in oil, August, 1920. Authority: L. E. Littlepage.

Strata.

Pennsylvanian System.	Thickness	Depth
Clay and soil	19	19
Coal	1	20
Fire clay	13	33
Sand, rock	7	40
Shale	4	44
Shale, hard, limy	3	47
Shale	43	90
Fire clay	6	96
Shale	29	125
Shale, hard, limy	7	132
Shale	37	169
Shale, hard, limy	1	170
Shale	65	235

Pennsylvanian System	Thickness	Depth
Sandstone	9	244
Shale, soft	1	245
Sandstone	5	250
Fire clay	3	253
Shale, hard, limy	1	254
Shale	24	278
Shale, hard, limy	12	290
Shale	27	317
Coal	1	318
Shale	20	338
Shale (cap rock), hard	1	339
Sand, (oil)	3	342
Sand rock, white	1	343
Total depth		343

Log No. 292

Bailey, No. 6, lessor, The Moss Hill Oil & Gas Co., lessee. Location: ⅛ mile north of White Plains. Completed: in 1919. Authority: L. E. Littlepage.

Strata.

Pennsylvanian System.	Thickness	Depth
Soil	15	15
Shale, hard	10	25
Sand	10	35
Shale, gray	25	60
Sand and shale	60	120
Shale, shelly	4	124
Shale, brown	51	175
Sand	15	190
Shale	50	240
Sand	20	260
Shale	90	350
Shale, shelly	5	355
Sand, (oil)	5	360
Shale	20	380
Limestone	**15**	**395**
Shale, brown (pencil cave)	155	550
Sand, (water)	155	705
Shale	10	715
Sand, broken	15	730
Shale, brown	20	750
Sand	5	755

Pennsylvanian System.	Thickness	Depth
Limestone, brown, and shells	20	775
Shale	10	785
Limestone (cap rock)	3	788
Sand, white, (oil)	8	796
Total depth		796

JACKSON COUNTY.

Production: **Oil and Gas.** Producing **Sands: Unnamed (Mississipian);**
Corniferous (Devonian).

Log No. 293

Sereno Johnson, No. 1, lessor. Wheeling-Kentucky Development
Co., lessee. Location: Moore's Creek. Authority: E. A. Meade, contractor, through L. Beckner.
Strata.

Pennsylvanian System.	Thickness	Depth
Soil	8	8
Shale	8	16
Shale, hard gray, (water 33, water and gas 80)	64	80
Shale, hard, gray	50	130
Sandstone	20	150
Shale	50	200
Sand	20	220
Shale	10	230
Sand, (water 300)	95	325
Shale, black	15	340
Shale, gray, hard	30	370
Mississippian System.		
Shale, red, sandy	40	410
Shale, gray, hard	90	500
Limestone (Big Lime)	200	700
Shale	20	720
Shale, red, sandy	10	730
Sandstone, (little gas)	5	735
Shale, hard	35	770
Shell, very hard	5	775
Shale, hard	295	1,070
Devonian System.		
Shale, black	207	1,277
Limestone, brown, hard, gritty	5	1,282
Limestone, brown, sandy	15	1,297

	Thickness	Depth
Devonian System.		
Shale, white (turning green)	78	1,375
Shale, very red	33	1,408
Shale, gray, hard	10	1,418
Total depth		1,418

8 inch casing in well, 40 feet.
6⅝ inch casing in well, 431 feet and 5 inches.
6⅝ inch casing in water well 21 feet.
Depth of water well 37 feet.

NOTE—The base of the Devonian and the top of the Silurian occur in the 78 feet of shale above 1,375 feet.

JEFFERSON COUNTY.

Production: Neither oil or gas.   Producing Sands: None recognized.

Log No. 294

William Yann, No. 1, lessor.   Buechel Oil & Mineral Co., operators. Location: Buechel, Ky.   Commenced: Oct., 1919.   Completed June, 1920.   Contractor: J. H. Wolfe.   Elevation: 495.   Casing head strata: Base of Silver Creek horizon of Sellersburg limestone (base of the Devonian).

Strata.

Devonian, Silurian & Ordovician Systems.	Thickness	Depth
Conductor	6	6
Limestone, blue, hard, (water 18)	85	91
Fireclay, light	5	96
Limestone, gray, hard	44	140
Limestone and shale, shelly, (water 160, 320)	230	370
Limestone, dark, (8¼ in. casing)	130	500
Limestone and shale, shelly	315	815
Limestone, gray	85	900
Limestone, dark	70	970
Limestone, light gray, hard	15	985
Limestone, dark gray	30	1,015
Limestone, (cavernous to about 1,325)	70	1,085
Limestone, gray, hard	115	1,200
Limestone, light, hard	75	1,275
Limestone, dark, hard	155	1,430
Limestone, gray, hard, (water 1,570)	140	1,570
Limestone, (water "sand")	52½	1,622½
Total depth		1,622½

Log No. 295

Sam R. Armstrong, No. **1**, lessor. Caldwell, **et al.**, lessees. Location: Fairdale, Jefferson Co. Casinghead strata: basal Mississippian. Authority: Joseph Howard.

Strata.

Mississippian System.	Thickness	Depth
Clay and soapstone	56	56
Devonian System.		
Shale, black, (sulphur water 145)	89	145
Limestone, white	23	163
Limestone, white, hard	106	274
Silurian System.		
Shale, green	37	311
Sand, white	19	330
Shale, red, or limestone	3	333
Limestone, gray	42	375
Limestone, gray, hard	12	387
Shale, green	4	391
Ordovician System.		
Sand, white, (fresh water)	22	413
Limestone, gray	18	431
Limestone, white, (fresh water)	9	440
Shale, gray, (set 476 ft. 8" casing)	35	475
Limestone, white	5	480
Shale, gray, and limestone	130	610
Limestone, gray	290	900
Shale, gray	72	972
Limestone, gray, fine	138	1,110
Shale, blue, and limestone, white, hard	35	1,145
Limestone rock, brown	20	1,165
Sand, brown, hard	2	1,167
Limestone rock, brown, hard, fine	28	1,195
Limestone or shale, gray, hard	7	1,202
Sand, brown, hard	3	1,205
Limestone rock, brown	5	1,210
Shale and limestone, gray, fine	15	1,225
Limestone rock, brown	30	1,255
Limestone, brown and white	5	1,260
Limestone rock, brown, hard	11	1,271
Limestone, brown, soft	30	1,301
Limestone, brown, soft	5	1,306
Limestone, gray and brown	140	1,446
Limestone, gray	10	1,456
Limestone, gray, fine	20½	1,476½

Ordovician System.

	Thickness	Depth
Limestone, gray	23½	1,500
Limestone, brown	25	1,525
Limestone, brown and white	5	1,530
Limestone, brown	10	1,540
Limestone, dark brown	20	1,560
Limestone, light brown and white	5	1,565
Limestone, very dark brown	5	1,570
Limestone, gray and bluish	13	1,583
Limestone, brown, (sample lost)	25	1,608
Limestone, gray and bluish	77	1,685
Shale, blue, (salt water)	5	1,690
Sand, salt water	20	1,710
Sand, brown	15	1,725
Sand, gray, and rock	5	1,730
Total depth		1,730

JESSAMINE COUNTY.

Production: Neither oil or gas.   Producing Sands: None recognized.

Log No. 296

William Hoover, No. 1, lessor.   J. T. Acker, Broadway, Va., and L. C. Wilson, Buffalo, N. Y., lessees, and drillers.   Location: ¼ mile south of Nicholasville.   Elevation: about 940.   Commenced: October 28, 1918.   Completed: November 11, 1918.   Production: Dry.

Strata.

Ordovician System.

	Thickness	Depth
Soil	13	13
Limestone	3	16
Limestone, gray, fine (water 45, 55, 90)	578	594
Limestone, hard	16	610
Limestone, soft, (sulphur water 702)	190	800
Limestone (sand), (black sulphur 820)	40	840
Limestone	210	1,050
Limestone (sand)	30	1,080
Limestone	10	1,090
Limestone (sand)	60	1,150
Limestone	10	1,160
Limestone (sand)	10	1,170
Limestone, black	30	1,200
Limestone (sand)	20	1,220
Limestone	35	1,255
Limestone (sand)	10	1,265
Limestone	110	1,375

Ordovician System.

	Thickness	Depth
Limestone (sand)	10	1,385
Limestone	40	1,425
Limestone (sand), (water)	6	1,431
Limestone	23	1,454
Limestone, dark	26	1,480
Limestone (sand)	20	1,500
Limestone	15	1,515
Limestone (sand)	23	1,538
Limestone, white	20	1,558
Limestone, dark	15	1,573
Limestone, real white	12	1,585
Sand, white	15	1,600
Limestone, brown	38	1,638
Limestone	42	1,680
Limestone, hard, gritty	20	1,700
Limestone, real	15	1,715
Pebbles, white	5	1,720
Blue water	5	1,725
Limestone, white	20	1,745
Limestone, black	20	1,765
Sand, white	20	1,785
Limestone, blue	60	1,845
Limestone, blue	3	1,848
Limestone, brown	12	1,860
Limestone, white	13	1,873
Sand, hard	8	1,881
Limestone, black	14	1,895
Sand, white	5	1,900
Limestone, dark	15	1,915
Limestone, white	13	1,928
Sand, white	12	1,940
Limestone, brown	15	1,955
Sand, white	45	2,000
Limestone, dark	25	2,025
Limestone, white	25	2,050
Water sand	25	2,075
Limestone, hard	25	2,100
Sand, white	50	2,150
Limestone, dark	50	2,200
Sand, white	40	2,240
Limestone, brown	20	2,260
Limestone, white	40	2,300
Limestone, dark	35	2,335
Limestone (15 feet), white (sand), (salt water)	40	2,375
Limestone, brown	25	2,400

Ordovician System.	Thickness	Depth
Limestone, white (sand)	25	2,425
Limestone, white (sand)	25	2,450
Limestone, brown, (black skim on water)	25	2,475
Limestone, white, very hard ....;	25	2,500
Sand (25 feet), white, foam (more salt water, strong)	50	2,550
Limestone ....,...............	20	2,570
Sand (5 feet)	20	2,590
Limestone, hard	10	2,600
Limestone, very hard	10	2,610
Limestone, hard, (could not make bits stand)	9	2,619
Limestone, black	16	2,635
Limestone (sand)	15	2,650
Limestone, white	10	2,660
Limestone (sand) ....·................	20	2,680
Limestone, white	10	2,690
Limestone, brown	10	2,700
Limestone, light	8	2,708
Limestone asphalt tar	9	2,717
Limestone, brown, sandy	7	2,724
Limestone, light	11	2,735
Limestone asphalt tar ....·....	2	2,737
Limestone, brown	6	2,743
Limestone, brown, sandy ....:.......·...	7	2,750
Limestone dark asphalt tar	5	2,755
Limestone, white	6	2,761
Limestone, brown, sandy	5	2,766
Sand, white, (looks like water sand)	5	2,771
Sand, white, (looks like water sand)	9	2,780
Limestone, brown	5	2,785
Limestone, white	15	2,800
Shale	10	2,810
Limestone, brown	70	2,880
Limestone, gray	55	2,935
Shale (pencil cave), (caving)	8	2,943
Limestone, brown	17	2,960
Shale	5	2,965
Limestone	15	2,980
Shale	5	2,985
Limestone, sandy	10	2,995
Limestone	7	3,002
Shale	5	3,007
Limestone	6	3,013
Shale ....·....	10	3,023
Limestone, brown	6	3,029
Limestone, pink, shaly, (caving)	8	3,037

Ordovician System.	Thickness	Depth
Limestone, soft, from 3,031	32	3,069
Shale, pink, (set casing)	6	3,075
Limestone, gray, hard	6	3,081
Shale, chocolate color, (caving) ......•.....	4	3,085
Limestone, shaly, black	36	3,121
Shale, soft, black	64	3,185
Total depth		3,185

NOTE—The limestone rocks were filled with water from top to bottom and the well was cased about twenty-eight times in an effort to get shut of this water. First 8 in. casing at 475 feet. The drill went through limestone rock at 2,935 feet into shale, at which level the water drained off completely. The lower part of this record is undoubtedly in the upper Cambrian, but the line of demarkation cannot be made because of insufficient data.

JOHNSON COUNTY.

Production: Oil and Gas. Producing Sands: Big Lime, Big Injun, Wier and Berea (Mississippian).

Log No. 297

Dan Hitchcock, No. 1, lessor. Ken-Mo Oil & Gas Co., lessee. Location: on headwaters of Barnett's Creek.

Strata.

Pennsylvanian System.	Thickness	Depth
Soil	14	14
Sand	41	55
Shale	89	130
Sand	144	274
Shale	157	331
Sand	183	514
Shale	9	523
Sand	17	540
Sand and shale	25	565
Mississippian System.		
Limestone, black	5	570
Shale, muddy	20	590
Limestone (Big Lime)	27	617
Shale, break	2	619
Limestone	58	677
Shale	2	679
Sandstone (Big Injun)	6	685
Shale	165	850
Shale, dark	86	936
Shale, white	12	948

A KENTUCKY RIVER ANTICLINE.

The view shows a low group of Ordovician Limestones just below the Twin Chimneys, Mercer County, Kentucky River gorge. The axis of this slight fold is nearly north and south.

Mississippian System.

	Thickness	Depth
Sand, gas, 200,000 ft.	92	1,040
Shale (Sunbury)	12	1,052
Sand, soft (Berea)	91	1,143
Total depth		1,143

NOTE—Not on pump, but shows for small producer.

Log No. 298

Coon Conley, No. 1, lessor. John G. White, lessee. Location: Head of Pigeon Creek, 1 mile southeast of Win P. O.

Strata.

Pennsylvanian System.

	Thickness	Depth
Soil (conductor)	16	16
Shale	134	150
Sand, gray	35	185
Sand, white	180	365
Shale	115	480
Sand, white	50	530
Shale	5	535
Sand, black	9	544
Limestone (Little Lime)	6	550
Shale	15	565
Limestone (Big Lime)	70	635
Shale (Waverly)	75	810
Shale	65	875
Sand	20	895
Sand, loose	28	923
Total depth		923

NOTE—Well No. 1 and No. 2 gauged day shot and produced 1½ million feet of gas.

Log No. 299

Ross Well, No. 1. South West Pet. Co. & Cliff Pet. Co., lessees. Location: Flat Gap P. O. Production: Slight show oil.

Strata.

Pennsylvanian System.

	Thickness	Depth
Quicksand	21	21
Sand, settling	64	85
Sand, hard	165	250

Pennsylvanian System.	Thickness	Depth
Shale	5	255
Sand	15	270
Shale	30	300
Sand	20	320
Shale	5	325

Mississippian System.		
Limestone ·(Little Lime)	10	335
Sand, settling, and water	40	375
Sandstone (Maxon)	10	385
Sand, pink, limestone and shale	15	400
Limestone (Big Lime)	110	510
Sandstone (Big Injun)	25	535
Shale, sandy	95	630
Shale	35	665
Sand	10	675
Shale	15	690
Sand	30	720
Shale and shell	80	800
Shale, black	60	860
Top of grit	44	904
Limestone and shale	61	965
Total depth		965

Log No. 300

George Conley, No. 1, lessor.   Bedrock Oil Co., lessee.   Location: On Pigeon Fork.   Elevation: 936

Strata.

Pennsylvanian System.	Thickness	Depth
Soil	12	12
Shale	6	18
Shell	6	24
Shale	76	100
Sandstone, hard	10	110
Shale	10	120
Sand	225	345
Shale, sandy	85	430
Sand	30	460

Mississippian System.		
Shells	10	470
Limestone (Big Lime)	130	600
Limestone, sandy	200	800

Mississippian System.

	Thickness	Depth
Shale, sandy	25	825
Sand (gas at 825)	50	875
Shale (gas at 880)	5	880
Sand	90	970
Shale, black	10	980
Limestone, sandy	56	1,036
Total depth		1,036

Log No. 301

Tom Cantrill, No. 1, lessor. Mid South Gas Co., lessee. Location: Hargis Creek. Elevation: 840. Production: 2,000,000 feet gas.

Strata.

Pennsylvanian System.

	Thickness	Depth
Soil	18	18
Sand	6	24
Shale	56	80
Sand	175	255
Shale	125	380
Sand	60	440
Shale	5	445
Limestone (Little Lime)	5	450
Shale	25	475
Limestone (Big Lime)	90	565
Shale	50	615
Limestone, sandy	167	782
Sand, gas	53	835
Shale	10	845
Shells, broken	4	849
Total depth		849

Blew out mercury 1.

Log No. 302

C. II. Williams, No. 1, lessor. Red Bush Syndicate, lessee. Location: Near Red Bush P. O. Elevation: 811.

Strata.

Pennsylvanian System.

	Thickness	Depth
Soil and mud	30	30
Shale, black	40	70
Sand	150	220
Shale	33	255
Sand, settling	48	301

Mississippian System.	Thickness	Depth
Shale	7	308
Limestone, black	5	313
Shale	6	319
Sand	7	326
Sand	21	347
Limestone, black	23	370
Limestone, white	100	470
Sand	12	482
Shale	221	703
Sandstone (Wier)	33	736
Shale	3	739
Limestone, hard	3	742
Shale	6	748
Shale and shell	6	754
Limestone, hard	4	758
Shale	39	797
Shale, brown (Sunbury)	20	817
Sandstone (Berea)	90	907
Shale	2	909
Total depth		909

Log No. 303

A. J. Tackett, No. 4, lessor. Location: Near Win P. O. Elevation: 1125, approx. Production: Oil, 1050-1060. Water, 1065-1075.

Strata.

Pennsylvanian System.	Thickness	Depth
Soil	14	14
Sand	224	238
Coal	1	239
Sand	3	242
Shale	12	254
Sand	215	469
Shale	4	473
Coal	3	476
Sand	58	534
Coal	2	536
Sand	19	555
Shale	45	600
Sandstone, gray, white, black	14	614
Shale	32	646
Sandstone, gray	10	656

Pennsylvanian System.

	Thickness	Depth
Sand, settling	40	696
Shale, muddy	10	706
Sand	12	718

Mississippian System.

Shale	7	725
Limestone (Big Lime)	84	809
Limestone, light	176	985
Limestone, dark	50	1,035
Sandstone (Wier)	40	1,075
Shale, dark	38	1,113
Sand	34	1,147
Shale, blue	16	1,163
Shale, white	24	1,187
Shale, brown (Sunbury)	17	1,204
Sandstone (Berea)	41	1,245
Shale	2	1,247
Total depth		1,247

Log No. 304

A. J. Tackett, No. 1, lessor. Location: On Hargis Ck., near Win P. O. Elevation: 881.

Strata.

Pennsylvanian System.

	Thickness	Depth
Soil	40	40
Sandstone, brown	10	50
Sand, gray	50	100
Shale, blue	60	160
Sand, gray	70	230
Shale, blue	112	342
Sand, white	32	374
Shale, blue	26	400

Mississippian System.

Limestone and pencil cave	30	430
Limestone (Big Lime), gray and white	160	590
Sand, broken	70	660
Shale, gray and black	108	768
Sand, gray, strong flow gas	35	803
Sandstone, broken	32	835
Total depth		835

Log No. 305

A. J. Tackett, No. 3, lessor. Location: Near Win P. O. Elevation: 935? Started: January 17, 1920. Finished: February 11, 1920.

Strata.

Pennsylvanian System.	Thickness	Depth
Soil	25	25
Sand, soft	83	108
Sand	198	306
Shale	134	440
Sand, salt	55	495
Sand and shale	15	510
Shale, mud	10	520
Mississippian System.		
Shale	11	531
Limestone, (Big Lime)	29	560
Limestone, break, (Big Lime)	5	565
Limestone (Big Lime)	41	606
Sandstone (Big Injun)	164	770
Sand, dark, white limestone	25	795
Sandstone (Wier), show oil 795	123	918
Total depth		918

Log No. 306

Bud Conley, No. 3, lessor. Location: Pigeon Creek. Elevation: 945. Commenced: December 19, 1919. Finished: February 6, 1920.

Strata.

Pennsylvanian System.	Thickness	Depth
Soil	18	18
Shale	22	40
Sand	40	80
Shale	68	148
Sand	92	240
Shale	7	247
Sand	73	320
Shale	95	415
Sand	35	450
Shale	7	457

Mississippian System.

	Thickness	Depth
Limestone, (Little Lime)	7	464
Shale	18	482
Limestone (Big Lime)`	108	590
Shale	80	670
Shale, sandy	60	730
Shale	77	807
Sand, gas	28	835
Shale	16	851
Sand	19	870
Shale	17	887
Shale	15	902
Total depth		902

Log No. 307

Bud Conley, No. 2, lessor. Location: Pigeon Creek. Elevation: 1020?

Strata.

Pennsylvanian System.

	Thickness	Depth
Soil	22	22
Shale	136	158
Sand	40	198
Shale	12	210
Sand	135	345
Shale	98	443
Sand	14	457
Shale	15	472
Sand	47	519
Shale	15	534

Mississippian System.

	Thickness	Depth
Limestone	5	539
Shale	14	553
Limestone	85	638
Shale, sandy	137	775
Limestone, sandy	71	846
Shale	14	860
Sand, gas	30	890
Shale	7	897
Sand, gas	5	902
Shale, sandy	56	958
Total depth		958

Log No. 308

John Cochran, No. 1, lessor.   Location: Below mouth Oil Branch at Little Paint Creek.   Elevation: 730.

Strata.

Pennsylvanian System.	Thickness	Depth
Sandstone	40	40
Shale	90	130
Sand, settling	45	175
Shale	5	180
Mississippian System.		
Limestone (Little Lime)	5	185
Shale (pencil cave)	5	190
Limestone (Big Lime)	153	343
Shale, green, sandy	60	403
Shale, dark, sandy	134	537
Sandstone (Wier)	21	558
Shale, sandy	85	643
Shale (Sunbury)	15	658
Sandstone (Berea)	18	676
Shale and sandstone	39	715
Total depth		715

Berea, 60 quarts.

Wier, 40 quarts.

Log No. 309

Bud Conley, No. 1, lessor.   Location: Pigeon Creek.   Elevation: 1020 approx.

Strata.

Pennsylvanian System.	Thickness	Depth
Soil	8	8
Sand	16	24
Sand	6	30
Shale	102	132
Sand	217	349
Shale	121	470
Sand	45	515

Mississippian System.

	Thickness	Depth
Limestone (Little Lime)	9	524
Shale (Pencil cave)	24	548
Limestone (Big Lime)	92	640
Shale, sandy	95	735
Limestone, sandy	120	855
Shale	5	860
Limestone, sandy	25	885
Shale	17	902
Sand, gas	10	912
Shale	15	927
Total depth		927

Log No. 310

Auxier Oil Company, No. 1, lessor. Location: Glade Farm, near Glade's Branch. Started: January 10, 1921. Completed: March 29, 1921. Production: 5 bbls oil. Authority: C. E. Bales.

Strata.

Pennsylvanian System.

	Thickness	Depth
Sandstone	130	130
Shale, sandy	190	320
Sand, settling	107	427
Limestone, black	24	451
Sand (Maxon), oil	11	462
Limestone (Big Lime)	46	508
Limestone, sandy, (Big Lime)	18	526
Sandstone, oil (Big Injun)	16	542
Sandstone (Big Injun)	20	562
Shale, blue, sandy	12	574
Shale, black, sandy	15	589
Shale, gray, sandy	86	675
Sand (Wier), oil	40	715
Shale, gray, sandy	83	798
Shale	100	898
Shale (Sunbury)	9½	907½
Sandstone, cap rock	3	910½
Sandstone (Berea), oil	50	960½
Sandstone, shaly	15	975½
Shale, black	16	991½
Total depth		991½

Log No. 311

John Wright, No. 1, lessor. Pulaski-Johnson Oil & Gas Co., lessee. Location: Near Barn Rock P. O. Started: August 23, 1920. Completed: September 22, 1920. Production: 500,000 cubic feet. Authority: C. E. Bales.

Strata.

Pennsylvanian System.	Thickness	Depth
Soil	50	50
Sandstone, dark	130	180
Sandstone, white	65	245
Shale, white	10	255
Limestone	20	275
Sandstone, white	40	315
Shale, blue	40	355
Mississippian System.		
Limestone, blue	35	390
Limestone (Big Lime), gas 430-431, 470-475	125	515
Shale, white	35	550
Limestone, blue, and shale	100	650
Sandstone, soft	70	720
Shale, blue	22	742
Sand (Wier), gas 747-750	33	775
Shale, blue	5	780
Limestone, blue	30	810
Shale, white	5	815
Shale, black	7	822
Limestone, blue	53	875
Shale, black	50	925
Shale, black	58	983
Total depth		983

Log No. 312

A. J. Spradlin, No. 1, lessor. Location: Hargis Creek. Elevation: 905. Drilled: April 18, 1919.

Strata.

Pennsylvanian System.	Thickness	Depth
Soil	15	15
Shale, blue	50	65
Sand, salt, water at 230	175	240
Shale, black	110	350

Mississippian System.	Thickness	Depth
Limestone, black	17	367
Sand (Maxon), oil	36	403
Shale, black	6	409
Limestone (Little Lime), white	18	427
Shale, black (Pencil Cave)	17	444
Limestone (Big Lime), gray	142	586
Shale	20	606
Sandstone, (Big Injun)	110	716
Shale, blue	18	734
Sandstone (Squaw), gas at 746	52	786
Shale, blue	38	824
Sandstone (Wier), gas	8	832
Limestone, dark, gritty	18	850
Shale, black, hard	7	857
Shale, brown (Sunbury)	23	880
Total depth		880

Log No. 313

A. J. Spradlin, No. 2, lessor. Location: Hargis Creek. Elevation: 1095.

Strata.

Pennsylvanian System.	Thickness	Depth
Soil	12	12
Sandstone, hard	48	60
Shale	75	135
Sand	25	160
Shale	90	250
Sand	190	440
Shale	105	545
Sand	57	602
Shale	3	605
Shale, shelly	8	613
Shale	8	621
Shale, shelly	3	624
Shale	26	650

Mississippian System.	Thickness	Depth
Limestone (Big Lime)	80	730
Shale, sandy	95	825
Limestone, sandy	95	920
Shale, sandy	51	971
Sand, gas	45	1,016

Mississippian System.	Thickness	Depth
Shale, sandy	34	1,050
Sand, gas	33	1,083
Shale	40	1,123
Shale (Sunbury)	18	1,141
Sandstone (Berea)	19	1,160˙
Limestone, sandy	18	1,178
Total depth		1,178

Log No. 314

A. J. Spradlin, No. 3, lessor.   Location: Hargis Creek.

Strata.

Pennsylvanian System.	Thickness	Depth
Soil	25	25
Sand	30	55
Shale	51	106
Sand	14	120
Shale	25	145
Sand	255	400
Shale	6	406
Shale, muddy	12	418
Shale	82	500
Sand	50	550

Mississippian System.	Thickness	Depth
Limestone, dark, sandy	37	587
Shale, muddy (pencil cave)	19	606
Limestone (Big Lime)	60	666
Sand and shale	3	669
Limestone	7	676
Sand, blue	184	860
Limestone shell	40	900
Limestone and shale	36	936
Sand (some gas)	40	976
Shale, hard	25	1,001
Sand	17	1,018
Shale, white	28	1,046
Limestone, brown, hard	9	1,055
Shale, blue	15	1,070
Shale, brown (Sunbury)	17	1,087
Sandstone (Berea)	44	1,131
Shale and shell	14	1,145
Total depth		1,145

Log No. 315

A. J. Spradlin, No. 4, lessor. Location: Hargis Creek. Elevation: 980.

Strata.

Pennsylvanian System.

	Thickness	Depth
Soil	13	13
Sandstone	25	38
Shale, sandy	32	70
Shale	15	85
Shale, hard	28	113
Sandstone, (show oil 248-768; little water 256-268)	174	287
Shale	131	418
Sand, settling	47	465
Shale	3	468

Mississippian System.

Limestone (Little Lime), mud and shale	18	486
Shale, muddy	16	502
Limestone (Big Lime)	72	574
Sandstone (Big Injun)	181	755
Sand, shaly	70	825
Sand (Wier), gas at 16-30 feet	38	863
Shale	48	911
Total depth		911

Log No. 316

A. J. Spradlin, No. 5, lessor. Location: Hargis Creek. Elevation: 1095.

Strata.

Pennsylvanian System.

	Thickness	Depth
Soil	20	20
Sand	20	40
Shale	90	130
Sand	20	150
Shale	100	250
Sand	197	447
Shale	123	570
Sand	57	627

Mississippian System.	Thickness	Depth
Limestone (Big Lime)	96	723
Shale, sandy	95	818
Limestone, sandy	96	914
Shale, sandy	55	969
Sand (Wier), gas	46	1,015
Shale	15	1,030
Total depth		1,030

Log No. 317

A. J. Spradlin, No. 6, lessor. Location: Hargis Creek. Elevation: 1020.

Strata.

Pennsylvanian System.	Thickness	Depth
Soil	16	16
Shale	39	55
Sandstone	47	102
Shale, sandy, mud	48	150
Sandstone	25	175
Shale	10	185
Sandstone	170	355
Shale	134	489
Sandstone	59	548

Mississippian System.	Thickness	Depth
Limestone (Little Lime)	32	580
Limestone (Big Lime)	27	607
Shale	10	617
Limestone (Big Lime)	33	650
Sand, shaly	8	658
Sand, shaly	172	830
Sand, dark	30	860
Total depth		860

Log No. 318

A. J. Spradlin, No. 7, lessor. Location: Hargis Creek. Elevation: 1,000 feet. Started: September 30, 1919. Completed: October 21, 1919.

Strata.

Pennsylvanian System.	Thickness	Depth
Soil	18	18
Shell, hard	5	23

IRREGULARLY CEMENTED POTTSVILLE SANDS.

The Pottsville Sands of Western Kentucky are thick ideal oil "sands" surrounded by thick bituminous shales. This outcrop occurs south of Sebree, on the Sim Sutton farm, Webster County, on the south limb of the Rough Creek Anticline.

Pennsylvanian System.	Thickness	Depth
Clay, soft, caving	5	28
Shale, shelly	4	32
Shale	38	70
Sand	23	93
Salt	1	94
Sand	3	97
Coal	3	100
Shale	3	103
Sand	194	297
Sandstone, salt water	3	300
Shale, muddy	14	314
Shale	89	403
Sand (salt water)	47	450
Shale	12	462
Sandstone	5	467
Shale	7	474

Mississippian System.		
Limestone	4	478
Shale	3	481
Sand	2	483
Shale, muddy (Pencil Cave)	12	495
Limestone (Big Lime)	65	560
Shale, Injun blue	175	735
Shale	91	826
Sandstone (Wier)	33	859
Shale	27	886
Sand	15	901
Shale, black, and shell	23	924
Total depth		924

Water at 285 and 425.
Show black oil at 220.
Little gas close to top and bottom.

Log No. 319

A. J. Spradlin, No. 8, lessor. Location: Hargis Creek. Elevation: 1100.

Strata.

Pennsylvanian System.	Thickness	Depth
Soil	15	15
Sand	23	38
Shale	65	103

Pennsylvanian System.

	Thickness	Depth
Sand	21	124
Shale	35	159
Sand	56	215
Shale (oil 200-220)	14	229
Sand	36	265
Coal	1½·	266½
Sand	51½	318
Shale	2	320
Sand (oil 345)	106	426
Shale	54	480
Sand	9	489
Shale	83	572
Sand, settling	43	615
Coal	1	616
Sand	6	622
Shale	5	627
Sand	5	632

Mississippian System.

	Thickness	Depth
Limestone	5	637
Limestone	17	654
Limestone	74	728
Shale	162	890
Total depth		890

Log No. 320

.H. M. Rice, No. 2, lessor. Emden Oil Company, lessee. Location: Near Barnett's Creek, at mouth of Grassy Fork of Barnett's Creek. Started: October 26, 1920. Completed: February 19, 1921. Production: 6 bbls. oil. Authority: C. E. Bales.

Strata.

Pennsylvanian System.

	Thickness	Depth
Soil	115	115
Sandstone	255	370
Sand, settling	76	446

Mississippian System.

	Thickness	Depth
Limestone (Big Lime)	115	561
Shale, dark	356	917
Sandstone (Wier), oil	12	929
Shale, dark	2	931
Sandstone (Wier), oil	19½	950½
Shale, hard	8	958½
Total depth		958½

Log No. 321

David Conley, No. 1, lessor. Mid-South Oil Co. (D. T. Evans, Pres., Huntington W. Va.) lessee. Location: on Litteral Fork, Mg. Co. Elevation: 960. Commenced: May 17, 1920. Completed: June 12, 1920.

Strata.

Pennsylvanian System.	Thickness	Depth
Soil	20	20
Clay and shale	26	46
Sandstone, very hard	9	55
Coal	4	59
Shale, hard	126	182
Sand	42	224
Shale and clay	2	226
Sand	164	390
Shale and clay	61	451
Sand	35	486
Shale and clay	4	490
Mississippian System.		
Limestone	10	500
Shale	22	522
Limestone, very hard	5	527
Shale (Pencil Cave)	19	546
Limestone (Big Lime)	54	600
Shale	210	810
Shale, dark	56	866
Sandstone (Wier)	43	909
Sand, oil 22' pay at top	27	936
Shale, dark	12	948
Sand, gas at top	13	961
Total depth		961

Log No. 322

Lindsay Conley, No. 1, lessor. Eastern Imperial Co., lessee. Location: ¼ mile southeast of I. G. Rice farm well by church and school-house, 2½ miles northwest of Paintsville. Completed: May, 1919. Production: 1 bbl. oil natural, and good flow of gas.

Strata.

Pennsylvanian System.	Thickness	Depth
Soil	50	50
Shale, bluish	12	62
Sandstone and shale	390	452

Mississippian System.

	Thickness	Depth
Limestone (Big Lime)	98	540
Shale, grayish	275	815
Shale, blue	75	890
Shale (Sunbury)	25	915
Sandstone (Berea), shale streaks	80	995
Shale	5	1,000
Total depth		1,000

NOTE—The 75 feet above recorded as shale, blue, is the correct position of the Wier sand, which evidently was not recognized by the drillers.

Log No. 323

Andy Jayne, No. 1, lessor. Gibson Petroleum Co., lessee. Location: near Forks of Big Paint Creek, 1 mile south of Elna P. O.

Strata.

Pennsylvanian System.

	Thickness	Depth
Soil	10	10
Sandstone (Mountain), oil show	60	70
Shale	100	170
Sand, settling	60	230
Shale	10	240

Mississippian System.

	Thickness	Depth
Limestone (Little Lime)	10	250
Shale (pencil cave)	10	260
Limestone (Big Lime), gas	100	360
Sandstone, shale, oil show	100	460
Shale, black	40	500
Shale	30	530
Sandstone (Big Injun), oil show	80	610
Shale, black	22	632
Total depth		632

Log No. 324

I. G. Rice, No. 1, lessor. Va.—Ky. Oil Co., lessee. Location: ¼ mile above Paint Creek on Ruel Branch 2½ miles northwest of Paintsville. Elevation: 625 approx. Production: Estimated at 4 bbls. oil. Drilled in June 11, 1920.

Strata.

Pennsylvanian System.

	Thickness	Depth
Soil	25	25
Shale, bluish	8	33
Sandstone	375	408

Mississippian System.

	Thickness	Depth
Limestone (Big Lime)	90	498
Shale, bluish	175	673
Shale, red	10	683
Shale, gray	106	789
Sandstone	8	797
Shale, bluish	75	872
Shale (Sunbury)	18	890
Sandstone (Berea)	30	920
Total depth		920

Shot 60 qts, 15 bbls.
Salt water bailed off.

Log No. 325

Jesse Stafford, No. 1, lessor. Nitro Oil & Gas Co., Huntington, W. Va., lessee. Location: on North Fork of Paint Creek, 3 miles west of Paintsville. Completed: June 25, 1918. Production: 1 barrel oil.

Strata.

Pennsylvanian System.

	Thickness	Depth
Soil	20	20
Shale	18	38
Sandstone	380	418

Mississippian System.

	Thickness	Depth
Limestone (Big Lime)	102	520
Shale, bluish	175	695
Shale, grayish	110	805
Sandstone	4	809
Shale, bluish	40	849
Shale (Sunbury)	11	860
Sandstone (Berea)	60	920
Total depth		920

Log No. 326

Jesse Lyons, No. 1, lessor. Keaton Oil Co., lessee. Location: ¼ mile up Keaton Creek from Blaine Creek, on right hand side. Production: 38 barrels oil.

Strata.

Pennsylvanian System.

	Thickness	Depth
Sandstone, (35 ft. below surface a heavy flow of water)	202	202
Sand	167	369

Mississippian System.

	Thickness	Depth
Limestone (Big Lime), salt water	148	517
Sandstone (Big Injun)	7	524
Shale	210	734
Sandstone	6	740
Shale, blue	36	776
Sandstone (Wier), 38 bbls. oil	42	818
Total depth		818

Log No. 327

Jesse Lyons, No. 2, lessor. Practically the same as Log No. 1. No. 1 produces 38 bbls., and No. 2, is estimated to produce about the same. No. 2 located slightly below No. 1 on Keaton Creek.

Log No. 328

Joe Hamilton, No. 1, lessor. Wheeler-Watkins Co., lessee. Location: on Mine Fork just above the mouth of Little Paint Creek. Authority: J. J. Baker.

Strata.

Pennsylvanian System.

	Thickness	Depth
Soil	30	30
Shale, blue	35	65
Sand (show oil)	30	95
Shale and sand	51	146
Sand, dark	6	152

Mississippian System.

	Thickness	Depth
Limestone (Big Lime)	53	205
Sand, gray (Keener), show oil	165	370
Shale	35	405
Sand, dark gray (Big Injun), gas	50	455
Shale, blue	12	467
Sand (Squaw), some gas	12	479
Shale	36½	515½
Sand, gray (Wier), some oil	24	539½
Sand, dark	5	544½
Shale, dark, sandy	20	564½
Sandstone	15	579½
Shale	21½	601
Shale (Sunbury)	25	626
Sandstone (Berea)	61	687
Total depth		687

Log No. 329

Joe Hamilton, No. 2, lessor.   Wheeler-Watkins Oil Co., lessee.
Location: on Mine Fork just above the mouth of Little Paint Creek.
          Strata.

Pennsylvanian and Mississippian Systems.	Thickness	Depth
Sandstones, shales, and limestones	504	504
Sandstone (Wier)	40	544
Shale and sand, broken	91	635
Sandstone (Berea)	54	689
Total depth		689

NOTE—This record is very incomplete, but is reported to have
been practically the same as J. H. No. 1 above the Wier sand.

Log No. 330

H. M. Rice, No. 1, lessor. Emden Oil Company, lessee. Location: on Road Fork of Barnett's Creek, about 2 miles N. E. of Oil
Springs, and 8 miles west of Paintsville. Started: Aug. 1, 1920. Completed: Sept. 28, 1920. Initial Production: 15 bbls, oil. Authority:
C. E. Bales.
          Strata.

Pennsylvanian System.	Thickness	Depth
Soil	8	8
Sandstone (oil show 75 in.)	487	495
Mississippian System.		
Limestone (Big Lime), strong gas	100	595
Sandstone, shaly, green to gray	349	944
Sandstone ("Berea Grit"), oil	25	969
Shale and sandstone	46	1,015
Total depth .:......................		1,015

Log No. 331

Will Turner, No. 1, lessor. Mid-South Oil Co., lessee. Location:
Little Mine. Elevation: 860 approx.
          Strata.

Pennsylvanian System.	Thickness	Depth
Soil	12	12
Shale	14	26
Sand, gray	20	46
Limestone, brown	12	58
Shale, blue :...........................	75	133
Sand, gray	84	217
Shale, gray	50	267
Sandstone (Maxon), white	65	332

Mississippian System.	Thickness	Depth
Limestone, gray, dark	63	395
Shale, green (pencil cave)	5	400
Limestone (Big Lime), white	82	482
Shale, green	26	508
Sand, gray	15	523
Shale, sandy	4	527
Shale (Waverly)	97	624
Shale, light gray	100	724
Sand (Wier), gray, hard	56	780
Shale, brown (Sunbury)	10	790
Sand, gray (Berea)	66	856
Devonian System.		
Shale, white and black (Chattanooga)	269	1,125
Limestone sandy, light brown (Corniferous)..	85	1,210
Limestone, light red	180	1,390
Limestone, gray, hard	20	1,410
Shale gray blue and green	40	1,450
Limestone, blue, hard	30	1,480
Limestone, gray and light brown	85	1,565
Limestone, light blue, hard	65	1,630
Limestone dark gray	80	1,710
Limestone, black grannett	20	1,730
Limestone, blue	40	1,770
Limestone, blue, hard	20	1,790
Shale, big red	205	1,995
Total depth		1,995

NOTE—This is a very poorly kept record, especially in its lower part.

PARTIAL RECORDS.

Log No. 332

Felix Fyffe, No. 1, lessor. Location: Big Lick. Production: Gas, 500.000 ft. Commenced: August 1916. Completed: April. 1917. Depth to sand, 638. Total depth, 672. Feet sand, 34.

Log No. 333

A. M. Lyon, No. 1, lessor. Union Oil & Gas Co., lessee. Location: Big Lick. Production: Gas, 500,000 ft. Commenced: May, 1917. Completed: Aug., 1917. Depth to sand, 605. Total depth, 645. Feet sand, 40. Not shot.

Log No. 334

A. M. Lyon, No. 2, lessor. Location: Big Lick. Production: Gas, 250,000 ft. Commenced: Sept., 1917. Completed: Oct., 1917. Depth to sand, 705. Total depth, 755. Feet sand, 50.

Log No. 335

Steve Fyffe, No. 2, lessor. Location: Big Lick. Production: Gas, 500,000 ft. Commenced: Oct., 1918. Completed: Nov., 1918. Depth to sand, 760. Total depth, 800. Feet sand, 40.

Log No. 336

Henry Fyffe, No. 3, lessor. Location: Big Lick. Production: Gas, 300,000 ft. Commenced: July, 1919. Completed: Sept. 3, 1919. Depth to sand, 730. Total depth, 768. Feet sand, 38.

Log No. 337

A. M. Lyon, No. 3, lessor. Location: Big Lick. Production: Gas, 300,000 ft. Commenced: Oct., 1919. Completed: Dec. 19, 1919. Depth to sand, 650. Total depth, 690. Feet sand, 40. Not shot.

Log No. 338

Jim Evans, No. 1, lessor. Location: Upper Laurel Creek. Production: Gas, 250,000 feet. Commenced: Aug. 6, 1919. Completed: Sept. 19, 1919. Depth to sand, 635. Total depth, 666. Feet sand, 31. Not shot.

Log No. 339

Jim Evans, No. 2, lessor. Location: Upper Laurel Creek. Production: Gas, 250,000 ft. Commenced: Sept. 28, 1918. Completed: Oct. 22, 1919. Depth to sand, 678. Total depth, 718. Feet sand, 40. Not shot.

Log No. 340

J. S. Young, No. 1, lessor. Location: Upper Laurel Creek. Commenced: Dec., 1918. Completed: Jan. 28, 1920. Depth to sand, 664. Total depth, 700. Feet sand, 36.

CHAPTER V.

KNOTT COUNTY.

Production: Oil and Gas. Producing Sands: Pottsville (Pennsylvanian); Maxton and Big Lime (Mississippian).

Log No. 341

Greenville Sloan, No. 1, lessor. Ohio Fuel Oil Co., lessee. Location: 2 miles from mouth of Caney Creek of Right Beaver Creek. Completed: September 15, 1914. Authority: The Eastern Petroleum Co.

Strata. Pennsylvanian System.	Thickness	Depth
Sandstone	18	18
Shale, sandy	12	30
Shale, hard	10	40
Coal	3	43
Shale, hard, and shells	57	100
Sandstone	60	160
Shale, hard, and shells	60	220
Coal	2	222
Shale, hard	18	240
Shale, hard, gray	15	255
Shale, hard, and shells	57	312
Sandstone, (gas show 312)	23	335
Shale, hard, and shells	60	395
Sandstone	105	500
Shale, hard, and shells	150	650
Sandstone	110	760
Shale, hard	80	840
Shale, sandy	45	885
Shale, hard	55	940
Sandstone (salt water 995)	120	1,060
Shale, hard	15	1,075
Sandstone, white (gas 1080, 1,000,000 feet)..	30	1,105
Shale, hard	25	1,130
Shale, hard, limy	12	1,142
Shale, hard, and shells	16	1,158
Sandstone	12	1,170
Shale, hard	5	1,175
Sandstone (salt sand), (small gas 1195 to 1205, show of oil 1238, 1 bailer salt water per hour at 1255)	95	1,270
Shale, hard, and shells	70	1,340

Mississippian System.	Thickness	Dep
Sandstone, (a little gas 1340)	10	1,350
Shale, hard, and shells	76	1,426
Total depth		1,426

Hole plugged at 435 and 1335 feet.

NOTE—This record principally in the Pottsville. The Maxton sal should be near the 10 feet of sandstone above 1350 feet in depth.

Log No. 342

Joseph Hall, No. 1, lessor. Location: Mouth of Dry Creek of Rig Beaver Creek. Casing head: 801 feet A. T. Completed: October 1 1904. Authority: The Eastern Gulf Oil Co.

Strata.

Pennsylvanian System.	Thickness	Dep
Soil	18	18
Sandstone, gray (fresh water)	10	28
Shale hard	92	120
Sandstone (fresh water)	8	128
Shale, hard	112	240
Sandstone	35	275
Shale, hard, shelly	105	380
Sandstone, white	24	404
Shale, hard, shelly, (little gas 435-440)	41	445
Shale, hard, black	85	530
Sandstone, gray	220	750
Shale, hard, black	60	810
Sandstone, gray	75	885
Shale, hard, black	65	950
Sandstone, gray (salt water 981)	40	990
Shale, hard, pebble shell	12	1,002
Sandstone, white (salt water 1123)	153	1,155
Coal	5	1,160
Sandstone (salt water flooded hole 1190)	83	1,243
Shale, hard	12	1,255
Sandstone	8	1,263
Mississippian System.		
Shale, hard	27	1,290
Sandstone (Maxon), (oil at 1390 to 1396) ..	108	1,398
Shale, hard, limy	39	1,437
Limestone	10	1,447
Shale, hard, limy	31	1,478

Mississippian System.

	Thickness	Depth
Sand, limy	22	1,500
Limestone (Big Lime)	178	1,678
Sandstone, reddish (Big Injun)	14	1,692
Sandstone, white, fine (Big Injun)	29	1,721
Shale, red	34	1,755
Shale, hard, black	93	1,848
Total depth		1,848

KNOX COUNTY.

Production: **Oil and Gas**. Producing Sands: Pottsville (**Pennsylvanian**); Maxton, **Big Lime** and **Big Injun** (**Mississippian**).

Log No. 343

Jim Walker, No. 1, lessor. E. J. Wyrick, No. 1, lessee. Location: On Omandas Branch of the road fork of Stinking Creek. Commenced: January, 1920. Completed: March 31, 1920. Authority: The Associated Producers Oil Co.

Strata.

Pennsylvanian System.

	Thickness	Depth
Soil	47	47
Sand, white	10	57
Shale, blue, hard	103	160
Shale, hard, sandy	40	200
Shale, blue, hard	100	300
Shale, dark, hard	20	320
Shale, blue, hard	20	340
Shale, hard, and limestone shells	70	410
Shale, hard	20	430
Shale, black, hard	35	465
Shale and limestone shells, hard	35	500
Shale, blue, hard	10	510
Shale, black, hard	25	535
Shale, limy, sandy	10	545
Sandstone (salt sand), white, (oil 550 and 830)	515	1,060
Shale, limy, black, hard	10	1,070
Shale, blue, hard	10	1,080
Sand, white (Beaver)	175	1,255
Shale, limy, hard	10	1,265
Shale, black, hard	5	1,270
Shale, gray, limy	10	1,280
Shale, black, hard	5	1,285
Sand, white (salt)	75	1,360
Shale, gray, limy	10	1,370

Mississippian System.	Thickness	Depth
Shale, red, sandy	50	1,420
Shale, red, shelly	20	1,440
Shale, red, limy	5	1,445
Sand, blue (Maxon)	55	1,500
Limestone, red	5	1,505
Shale, red, sandy	15	1,520
Sand, white (Maxon)	55	1,575
Shale, black, hard	40	1,615
Limestone (Little Lime), dark	105	1,720
Shale (pencil cave)	3	1,723
Limestone (Big Lime), white	127	1,850
Sand, white, (Big Injun)	70	1,920
Sandstone, red, hard, (Big Injun)	10	1,930
Shale, red, sandy, (Big Injun)	15	1,945
Sand, red, (Big Injun)	50	1,995
Limestone, blue, (Big Injun)	5	2,000
Shale, hard, and limestone shells	100	2,100

Devonian System.		
Shale, black (Chattanooga)	344	2,444
Limestone (Irvine sand), (1st 10. feet gritty, then mostly limestone, gas 2449)	53	2,497

Silurian System.		
Limestone	20	2,517
Shale, hard, and limestone shells	60½	2,577½
Total depth		2,577½

Log No. 344

North Jellico Coal Co., lessor. Louisville Cement Co., lessee. Location: Near Wilton, Knox Co., Ky.

Strata.

Pennsylvanian System.	Thickness	Depth
Shale, sandy	17	17
Coal	.8	17.8
Shale, sandy	4.4	22
Sandstone	42.8	64.8
Coal	.11	65.7
Shale, dark	.11	66.6
Sandstone	2	68.6
Sandstone, dark	73.6	142
Sandstone z	2	144

Pennsylvanian System.   Thickness Depth

	Thickness	Depth
Shale, soft	6	150
Shale, dark	44.4	194.4
Coal	.10	195.2
Shale, dark	3.10	199
Sand shale	3	202
Shale, dark	6	208
Sandstone, shale parting	57	265
Shale, dark	20	285
Sand shale	25	310
Shale, dark	44.4	354.4
Coal	.2	354.6
Sand shale	41.6	396
Coal	1.6	397.6
Sandstone and shale	12.6	410
Shale, dark	1	411
Sandstone	80.3	491.3
Coal	.6	491.9
Sand shale	3.3	495
Sandstone	92	587
Shale, dark	21	608
Sandstone	87	695
Shale, black	2.10	697.10
Coal	2.4	700.2
Coal and shale mixed	1.8	701.10
Sandstone, shale partings	25.2	727
Sandstone	47	774
Sandstone conglomerated	6	780
Sandstone	103	883
Shale, sandy	10	893
Shale, dark	7	900
Sandstone	4	904
Shale, sandy	18	922
Shale, dark	2	924
Limestone, sandy	5	929
Shale, dark	1	930
Sandstone	11.4	941.4
Coal	.3	941.7
Shale, gray	1.11	943.6
Shale, sandy	12.6	956
Limestone, sandy	3	959
Shale	7	966
Limestone, sandy	3	969
Shale, gray	4	973
Sandstone, shaly	11	984
Shale, dark	4	988

Pennsylvanian System.	Thickness	Depth
Shale, sandy	2	990
Shale, blue	7	997
Shale, black	3	1,000
Total depth		1,000

NOTE—This well finished in the Pottsville, but is undoubtedly close to the top of the Mississippian Series.

LAUREL COUNTY.

Production: **Oil and Gas.**   Producing Sand: Corniferous (Devonian).

Log No. 345

Hiram Watkins, No. '1' lessor.   Atlanta Oil & Gas Co., lessee. Location: ¼ mile from Atlanta P. O. Production: Dry; well abandoned.

Strata.

Pennsylvanian System.	Thickness	Depth
Soil	1	1
Clay	4	5
Shale	6	11
Sand (show of coal)	10	21
Shale	15	36
Sand, shale and coal	6	42
Sand	100	142
Shale	6	148
Coal	2	150
Shale, brown	140	290
Sand	55	345
Shale, white	3	348
Coal show	2	350

Mississippian System.	Thickness	Depth
Limestone and shale	15	365
Limestone	20	385
Limestone, blue	5	390
Limestone, white, and shale	5	395
Shale	7	402
Shale and limestone	7	409
Shale, pink, and limestone	4	413
Limestone and shale	13	426
Shale, pink	35	461
Shale, white	10	471
Shale, blue	20	491
Limestone	6	497

Mississippian System.	Thickness	Depth
Limestone	7	504
Shale, white	6	510
Limestone, blue and gray	45	555
Limestone, brown	7	562
Limestone, (oil at 705)	143	705
Limestone, soft	61	766
Shale	12	778
"Sand," green (New Providence)	15	793
Devonian System.		
Shale, brown (Chattanooga)	47	840
Limestone "sand," (oil show)	60	900
Shale, gray and blue, with white noles	200	1,100
Limestone black	135	1,235
Shale, white	69	1,304
Limestone, red, and sand	13	1,317
Shale	17	1,343
Total depth		1,343

NOTE—The Devonian-Silurian contact occurs in the 60 feet of limestone above 900 feet, the Silurian-Ordovician contact in the 200 feet just above 100 feet. The well finished in the Ordovician.

LAWRENCE COUNTY.

Production: **Oil and Gas.** Producing Sands: **Pottsville (Pennsylvanian),** Wier **and** Berea **(Mississippian).**

Log No. 346

L. S. Alley, No. 1, lessor. Ohio Fuel Oil & Gas Co., lessee. Location: Lower Louisa Township. Commenced: May 1, 1919. Completed: June 14, 1919. Production: 2½ bbls. per day after shot.

Strata.

Pennsylvanian System.	Thickness	Depth
Soil	12	12
Shale	83	95
Sand	145	240
Shale	45	285
Sand	24	310
Coal	2	312
Shale	298	610
Coal	5	615
Shale	75	690
Salt sand, (water 720)	130	820
Shale, white	30	850
Second sand	160	1,010
Shale and mud	20	1,030

Mississippian System.

	Thickness	Depth
Limestone (Big Lime)	145	1,175
Sandstone (Big Injun)	90	1,265
Shale (break)	5	1,270
Limestone	50	1,320
Shale and shell	388	1,708
Shale, brown (Sunbury)	20	1,728
Sandstone (Berea)	26½	1,754½
Total depth		1,754½

Log No. 347

L. S. Alley, No. 3, lessor. Ohio Fuel Oil & Gas Co., lessee. Commenced: September 5, 1919. Completed: October 6, 1919. Production: 2½ bbls. daily. Well shot October 10, 1919, 30 qts.

Strata.

Pennsylvanian System.

	Thickness	Depth
Sub-soil and mud	40	40
Sand	40	80
Shale	120	200
Sand, buff	80	280
Shale	60	340
Limestone	22	362
Shale	38	400
Sand	75	475
Shale	30	505
Limestone	35	540
Shale and shells	110	650
Sand (salt), (water 675)	175	825
Shale and shells	115	940

Mississippian System.

	Thickness	Depth
Sandstone (Maxon)	23	963
Shale (pencil cave)	2	965
Limestone (Big Lime)	150	1,115
Sandstone (Big Injun)	84	1,199
Shale and shells	451	1,650
Shale, brown (Sunbury)	28	1,678
Sandstone (Berea)	26	1,704
Total depth		1,704

Log No. 348

L. S. Alley, No. 3, lessor   Ohio Fuel Oil & Gas Co., lessee. Commenced: November 3, 1919. Completed: December 3, 1919. Shot
December 3, 90 qts. Production: 2 bbls. per day.

Strata.

Pennsylvanian System.	Thickness	Depth
Sub-soil	13	13
Shale	47	60
Sand	50	110
Shale	125	235
Sand	25	260
Shale	350	610
Sand	60	670
Shale	15	685
Sand	75	760
Shale	20	780
Sand	30	810
Shale	15	825
Sand	95	920
Shale	25	945
Sand	15	960
Shale	10	970
Mississippian System.		
Limestone (Little Lime)	20	990
Shale (pencil cave)	2	992
Limestone (Big Lime)	148	1,140
Sand	20	1,160
Shale	2	1,162
Sand	58	1,220
Shale	462	1,682
Shale, brown (Sunbury)	25	1,707
Sand (Berea), (pay 1,707-1,726)	22	1,729
Total depth		1,729

Log No. 349

L. S. Alley No. 6, lessor. Ohio Fuel Oil & Gas Co., lessee. Commenced: January 20, 1920. Completed: February 21, 1920. Shot
Feb. 23, 1920, 80 qts. Production: 4 bbls. per day.

Strata.

Pennsylvanian System.	Thickness	Depth
Sub-soil	15	15
Sand	15	30
Shale	20	50

Pennsylvanian System.

	Thickness	Depth
Sand	30	80
Shale	10	90
Sand	40	130
Shale	15	145
Sand	15	160
Shale	400	560
Sand	175	735
Shale	40	775
Sand	135	910
Shale	20	930

Mississippian System.

Limestone (Big Lime)	160	1 090
Sandstone (Big Injun)	122	1,212
Shale	2	1,214
Shale and shell	439	1,653
Shale, brown (Sunbury)	23	1,676
Sand (Berea), (pay 1,677-1,697)	25	1,701
Total depth		1,701

Log No. 350

W. F. Austin, No. 1, lessor. Ohio Fuel Oil & Gas Co., lessee. Location: Lower Louisa Township. Commenced: April 7, 1919. Completed: May 6, 1919. Shot May 10, 1919, 60 qts.

Strata.

Pennsylvanian System.

	Thickness	Depth
Sub-soil	4	4
Sand	24	28
Coal	2	30
Shale	5	35
Sand	15	50
Coal	3	53
Sand	25	78
Shale	7	85
Sand	8	93
Shale	52	145
Sand	40	185
Shale	50	235
Sand	35	270
Shale	60	330
Sand	20	350
Shale	15	365

Pennsylvanian System.

	Thickness	Depth
Sand	13	378
Shale	37	415
Sand	17	432
Shale	8	440
Sand	25	465
Shale and shells	55	520
Sand	55	575
Shale	3	578
Sand (salt)	122	700
Shale and shells	40	740
Sand	25	765
Mud	3	768
Sand	62	830
Shale	3	833

Mississippian System.

	Thickness	Depth
Sand (Maxon)	12	845
Shale and mud	23	868
Shale (pencil cave)	4	872
Limestone (Big Lime)	148	1,020
Clay, white	3	1,023
Sandstone (Big Injun)	112	1,135
Shale	3	1,138
Limestone	77	1,215
Shale and shells	370	1,585
Shale, brown (Sunbury)	24	1,609
Sand (Berea)	24	1,633
Total depth		1,633

Log No. 351

W. F. Austin, No. 2, lessor. Ohio Fuel Oil & Gas Co., lessee. Commenced: August 6, 1919. Completed: September 5, 1919. Shot September 5, 1919, 30 qts. Production: 5 bbls. per day.

Strata.

Pennsylvanian System.

	Thickness	Depth
Sub-soil	7	7
Shale, hard	7	14
Mud	44	58
Sand	7	65
Coal	2	67
Shale	18	85
Sand	115	200

Pennsylvanian System.

	Thickness	Depth
Shale and shell	115	315
Sand	25	340
Mud	35	375
Sand, shelly	115	490
Shale	110	600
Shells	55	655
Sand (salt)	25	680
Shale (break)	3	683
Sand, (big water)	217	900
Shells	45	945
Shale, black	5	950

Mississippian System.

	Thickness	Depth
Limestone (Little Lime)	5	955
Shale (pencil cave)	5	960
Limestone (Big Lime)	160	1,120
Sandstone (Big Injun)	90	1,210
Shale and shells	444	1,654
Shale, brown (Sunbury)	20	1,674
Sand (Berea), (pay 1st 10 feet)	22	1,696
Total depth		1,696

Log No. 352

W. F. Austin, No. 3, lessor. Ohio Fuel & Gas Co., lessee. Commenced: October 10, 1919. Completed: November 12, 1919. Shot November 13, 1919, 40 qts. Production: 5 bbls. per day.

Strata.

Pennsylvanian System.

	Thickness	Depth
Sub-soil	12	12
Shale hard	6	18
Shale	17	35
Mud	25	60
Sand	50	110
Shale	40	150
Sand	90	240
Shale	20	260
Sand	60	320
Shale	40	360
Sand	220	580
Mud	20	600
Sand	40	640
Shale, (big water)	35	675

Pennsylvanian System.

	Thickness	Depth
Sand (salt)	110	785
Mud, black	65	850
Sand	105	955

Mississippian System.

	Thickness	Depth
Shale (pencil cave)	4	959
Limestone (Big Lime)	160	1,119
Sandstone (Big Injun)	80	1,199
Shale and shell	471	1,670
Shale, brown (Sunbury)	20	1,690
Sand (Berea), (pay first 10 feet)	21	1,711
Total depth		1,711

Log No. 353

W. F. Austin, No. 4, lessor. Ohio Fuel Oil & Gas Co., lessee. Commenced: December 27, 1919. Completed: January 29, 1920. Shot January 30, 1920, 40 qts. Production: 4 bbls. per day.

Strata.

Pennsylvanian System.

	Thickness	Depth
Sub-soil	3	3
Sandstone	12	15
Shale soft	140	155
Sandstone, buff	30	185
Shale, soft	5	190
Sandstone	20	210
Shale	40	250
Sandstone	15	265
Shale, soft	10	275
Sandstone, white	15	290
Shale	10	300
Sandstone	40	340
Sandstone, shelly	5	345
Sandstone	30	375
Shale	25	400
Sandstone	10	410
Coal	2	412
Shale	3	415
Sandstone	10	425
Shale, soft	25	450
Coal	7	457
Shale, soft	13	470
Sandstone	10	480

Pennsylvanian System.	Thickness	Depth
Shale	65	545
Shale, black, caving	5	550
Shale	35	585
Sandstone	74	659
Shale and shells	61	720
Sandstone (salt)	120	840
Shale, muddy	45	885
Sand (salt)	65	950
Shale	10	960
Sandstone	10	970

Mississippian System.	Thickness	Depth
Limestone	5	975
Sand (Maxon)	20	995
Limestone (Little Lime), black	18	1,013
Shale (pencil cave)	5	1,018
Limestone (Big Lime)	155	1,073
Sandstone (Big Injun)	105	1,178
Shale and shells	459	1,637
Shale, brown (Sunbury)	24	1,661
Sand (Berea), (pay from 1,757 to 1,767)	20½	1,681½
Total depth		1,681½

Log No. 354

R. Blankenship, No. 1, lessor. Ohio Fuel Oil & Gas Co., lessee. Location: 2½ miles northwest of Busseyville. Commenced: September 9, 1913. Completed: October 28, 1913. Shot November 7, 1913, 30 quarts. Production: Pumping water.

Strata.

Pennsylvanian System.	Thickness	Depth
Gravel	52	52
Sand	108	160
Shale	110	270
Coal	8	278
Shale	122	400
Sand	35	435
Shale	80	515
Sand (salt)	15	530
Shale	170	700
Sand	35	735
Shale	65	800
Limestone, black	15	815

Mississippian System.

	Thickness	Depth
Sand (Maxon)	35	850
Shale	15	865
Limestone (Big Lime)	17	1.040
Shale	5	1,045
Sandstone (Big Injun)	30	1,075
Shale	5.	1,080
Sand	55	1,135
Shale and shell	422	1,557
Shale, brown (Sunbury)	20	1,577
Sand (Berea)	53	1,630
Shale	4	1,634
Total depth		1,634

Log No. 355

Raish Blankenship, No. 3, lessor.   Ohio Fuel Oil & Gas Co., lessee. (Partial Record).

Strata.

Pennsylvanian System.

	Thickness	Depth
Soil	20	20
Shale	40	60
Sandstone	50	110
Shale	25	135
Sandstone	40	175
Shale	110	285
Sandstone	35	320
Shale and shells	70	390
Sandstone	25	415
Shale	95	510
Sandstone	20	530
Shale	35	565
Sandstone	170	735
Shale	10	745
Sandstone	145	890
Shale	5	895
Sandstone	15	910
Shale	5	915

Mississippian System.

	Thickness	Depth
Sandstone (Maxon)	20	935
Shale	10	945
Limestone (Big Lime)	135	1,080
Sandstone	85	1,165
Shale	15	1,180
Incomplete at		1,180

Log No. 356

Well not completed when recorded.

Arthur Blankenship, No. 2, lessor.   Ohio Fuel Oil & Gas Co., lessee.
Commenced: June 6, 1917.   Completed: July 3, 1917.   Production: 3
bbls. per day.   Shot July 5, 1917, 100 quarts.   After shot, 4 bbls per
day.

Strata.

Pennsylvanian System.	Thickness	Depth
Shale, soft	40	40
Shale	50	90
Sandstone	95	185
Shale	115	300
Sandstone	50	350
Shale	70	420
Sandstone	20	440
Shale	40	480
Sandstone	20	500
Shale	50	550
Sandstone	25	575
Shale	55	630
Sandstone	60	690
Shale	30	720
Sandstone (water 800)	210	930
Shale	20	950
Sandstone	85	1,035
Shale	35	1,070
Sandstone	10	1,080
Shale	10	1,090

Mississippian System.		
Limestone (Big Lime)	145	1,235
Sandstone (Big Injun)	115	1,350
Shale and shell	440	1,790
Shale, brown (Sunbury)	31	1,821
Sandstone (Berea), (oil)	37½	1,858½
Total depth		1,858½

Log No. 357

A. Blankenship, No. 3, lessor. Ohio Fuel Oil & Gas Co., lessee. Commenced: November 15, 1919. Completed: March 10, 1920. Shot March 11, 1920, 60 quarts. Production: 3 bbls per day.

Strata.

Pennsylvanian System.	Thickness	Depth
Soil	10	10
Mud	25	35
Shale	20	55
Sand	10	65
Shale	40	105
Sand	95	200
Shale	115	315
Sand	50	365
Shale	70	435
Sand	20	455
Shale	40	495
Sand	20	515
Shale	50	565
Sand	25	590
Shale	55	645
Sand	60	705
Shale	30	735
Sand, (water 820)	215	950
Shale	15	965
Sand	85	1,050
Shale	35	1,085

Mississippian System.		
Sand (Maxon)	10	1,095
Shale	10	1,105
Limestone (Big Lime)	145	1,250
Sandstone (Big Injun)	115	1,365
Shale and shells	461	1,826
Shale, brown (Sunbury)	24	1,850
Sandstone (Berea), (pay 1850-1865)	22	1,872
Total depth		1,872

Log No. 358

Arthur Blankenship, No. 6, lessor. Ohio Fuel Oil & Gas Co., lessee. Commenced: March 29, 1920. Completed: April 28, 1920. Well shot April 28, 1920, 60 qts. Production: 6 bbls. per day.

Strata.

Pennsylvanian System.	Thickness	Depth
Clay	60	60
Shale and shell	180	240
Coal	2	242
Shale, (water 250)	8	250
Mountain sand	60	310
Shale and shell	470	780
Sand (salt), (water)	175	955
Shale	57	1,012
Mississippian System.		
Sand (Maxon)	58	1,070
Shale and shells	45	1,115
Limestone (Big Lime)	140	1,255
Sandstone (Big Injun)	135	1,390
Shale and shell	430	1,820
Shale, brown (Sunbury)	21	1,841
Sandstone (Berea), (pay 1,846-1,859)	18	1,859
Total depth		1,859

Log No. 359

T. H. Burchett, No. 1, lessor. Ohio Fuel Oil & Gas Co., lessee. Commenced: January 7, 1914. Completed: February 7, 1914. Shot February 12, 1914, 150 qts. Well first produced 5 bbls. in 24 hrs. Production: 2 bbls. oil per day.

Strata.

Pennsylvanian System.	Thickness	Depth
Gravel	15	15
Sand	10	25
Shale	75	100
Sand	13	113
Coal	3	116
Shale	129	245
Sand	30	275
Shale	70	345
Coal	5	350

Pennsylvanian System.

	Thickness	Depth
Shale	130	480
Sand	40	520
Shale	160	680
Shell sand	50	730
Shale	15	745
Sand, (hole full of water)	130	875
Shale	45	920

Mississippian System.

	Thickness	Depth
Sand (Maxon)	40	960
Shale	30	990
Limestone (Little Lime), black	12	1,002
Limestone (Big Lime)	208	1,210
Shale	5	1,215
Sandstone (Big Injun)	93	1,308
Shale and shell	400	1,708
Shale, brown (Sunbury)	20	1,728
Sandstone (Berea)	52	1,780
Shale	2	1,782
Total depth		1,782

Log No. 360

Thos. H. Burchett, No. 2, lessor. Ohio Fuel Oil & Gas Co., lessee. Commenced: January 25, 1916. Completed: February 22, 1916. Shot: February 22, 1916, 120 qts. Production: 2 bbls. oil per day.

Strata.

Pennsylvanian System.

	Thickness	Depth
Soil	16	16
Sandstone	109	125
Shale	70	195
Sandstone	100	295
Shale	70	365
Coal	3	368
Shale	77	445
Limestone	40	485
Shale	35	520
Sandstone	70	590
Shale	30	675
Shale and shells	75	750
Sand (salt)	180	930
Shale and shell	75	1,005

Mississippian System.	Thickness	Depth
Sand (Maxon)	30	1,035
Shale	18	1,053
Shale (pencil cave)	3	1,056
Limestone (Big Lime)	152	1,208
Sandstone (Big Injun)	89	1,297
Shale and shells	419	1,716
Shale, brown (Sunbury)	24	1,740
Sand (Berea), (pay 1,742-1,769)	50	1,790
Total depth		1,790

Log No. 361

Thos. H. Burchett, No. 3, lessor. Ohio Fuel Oil & Gas Co., lessee. Commenced: March 11, 1916. Completed: April 6, 1916. Shot April 14, 1916, 120 qts. Production: 4 bbls. oil per day.

Strata.

Pennsylvanian System.	Thickness	Depth
Soil	16	16
Sandstone	133	149
Shale	70	219
Sandstone	100	319
Shale	70	389
Coal	3	392
Shale	77	469
Limestone	40	509
Shale	35	544
Sandstone	70	614
Shale	55	669
Shale	30	699
Shale and shell	75	774
Sand (salt)	180	954
Shale and shells	75	1,029
Mississippian System.		
Sand (Maxon)	30	1,059
Shale	18	1,077
Shale (pencil cave)	3	1,080
Limestone (Big Lime)	152	1,232
Sandstone (Big Injun)	89	1,321
Shale and shells	419	1,740
Shale, brown (Sunbury)	24	1,764
Sand (Berea), (pay 1,764-1,791)	45	1,809
Total depth		1,809

Log No. 362

Thos. H. Burchett, No. 5, lessor. Ohio Fuel Oil & Gas Co., lessee. Commenced: December 8, 1919. Completed: January 28, 1920. Shot: January 29, 1920, 40 qts. Production: 5 bbls. oil per day.

Strata.

Pennsylvanian System.

	Thickness	Depth
Soil	9	9
Shale and shells	231	240
Sand, buff	100	340
Shale and shells	385	725
Sand	75	800
Shale	30	830
Sand (salt)	165	995
Shale	25	1,020

Mississippian System.

Sand (Maxon)	30	1,050
Shale (pencil cave)	30	1,080
Limestone (Big Lime)	180	1,260
Sandstone (Big Injun)	98	1,358
Shale	5	1,363
Sandstone	40	1,403
Shale	404	1,807
Sand (Berea), (pay 1,808-1,818)	20	1,827
Total depth		1,827

Log No. 363

J. C. Short, No. 1, lessor. Ohio Fuel Oil & Gas Co., lessee. Commenced: July 30, 1917. Completed: Sept. 1, 1917. Shot: Sept. 4, 1917, 100 qts. Production: ¾ bbl. per day.

Strata.

Pennsylvanian System.

	Thickness	Depth
Soil	40	40
Shale and shells	160	200
Sandstone, buff	50	250
Shale	70	320
Shale, hard	58	378
Shale and shells	82	460
Sandstone	60	520
Shale and shells	255	775
Sandstone (salt), (water flood 800)	135	910
Shale and shells	40	950

Mississippian System.	Thickness	Depth
Sandstone (Maxon)	70	1,020
Shale	45	1,065
Sandstone (Maxon)	14	1,079
Shale (pencil cave)	1	1,080
Limestone (Big Lime)	160	1,240
Sandstone (Big Injun)	85	1,325
Shale and shells	463	1,788
Shale, brown (Sunbury)	20½	1,808½
Sandstone (Berea), (oil 1,833)	34½	1,843
Total depth		1,843

Log No. 364

J. C. Short, No. 2, lessor. Ohio Fuel Oil & Gas Co., lessee. Commenced: Sept. 21, 1917. Completed: Oct. 24, 1917. Shot: Oct. 26, 1917, 60 qts. Production: 3 bbls. oil per day.

Strata.

Pennsylvanian System.	Thickness	Depth
Soil	8	8
Clay	12	20
Sandstone	20	40
Shale, soft	35	75
Shale, hard	20	95
Shale, red, sandy	45	140
Shale	75	215
Sandstone	65	280
Shale and shells, (little water 310)	120	400
Sandstone	75	475
Shale, black	10	485
Shale, dark	45	530
Shale, black	60	590
Sandstone (Cow Run)	50	640
Shale, black	40	680
Limestone, sandy	45	725
Shale, (big water 780)	5	730
Sandstone (salt)	150	880
Shale, black	25	905
Sand, (gas)	20	925
Shale, white	105	1,030
Mississippian System.		
Limestone (Big Lime)	160	1,190
Sandstone (Big Injun)	107	1,297
Shale and shells	459	1,756
Shale, brown (Sunbury)	20	1,776
Sandstone (Berea)	29	1,805
Total depth		1,805

Log No. 365

J. C. Short, No. 3, lessor. Ohio Fuel Oil & Gas Co., lessor. Commenced: Feb. 17, 1919. Completed: March 20, 1919. Shot: March 24, 1919, 60 qts. Production: 3 bbls. per day.

Strata.

Pennsylvanian System.	Thickness	Depth
Soil	14	14
Sandstone	6	20
Shale	70	90
Coal	2	92
Shale	58	150
Sandstone	65	215
Shale	55	270
Shale, hard	30	300
Shale	40	340
Sandstone	60	400
Shale	30	430
Sandstone	60	490
Shale	60	550
Shale, hard	50	600
Shale and shells	130	730
Sandstone (salt)	145	875
Shale and shells	125	1,000

Mississippian System.		
Sandstone (Maxon)	10	1,010
Shale	7	1,017
Shale (Pencil Cave)	3	1,020
Limestone (Big Lime)	170	1,190
Sandstone (Big Injun)	90	1,280
Sandstone and shale, hard	51	1,331
Shale and shells	389	1,720
Shale, brown (Sunbury)	25	1,745
Sandstone (Berea)	`11	1,756
Sand and shale	5	1,761
Sand and shale, (1st oil 1,745)	15	1,776
Total depth		1,776

Log No. 366

Jas. Short, No. **1**, lessor. Ohio Fuel Oil & Gas Co., lessee. Location: near Louisa. Commenced: Apr. **19, 1917.** Completed: May **17, 1917.** Shot: May **18, 1917, 100** qts. Production: 2½ bbls. per day.

Strata.

Pennsylvanian System.

	Thickness	Depth
Soil	30	30
Shale, red, sandy	188	218
Coal	2	220
Shale	30	250
Sandstone	55	305
Coal	3	308
Shale	144	452
Sandstone	28	480
Shale	40	520
Clay, (little gas)	3	523
Shale	17	540
Shale, shelly	20	560
Sandstone	25	585
Shale	90	675
Sandstone	75	750
Shale	30	780
Shale, shelly and sandy	25	805
Shale	2	807
Sandstone (salt), (water)	153	960
Shale, shelly	70	1,030
Sandstone	20	1,050
Shale	40	1,090

Mississippian System.

Sandstone (Maxon)	15	1,105
Shale	2	1,107
Limestone, gritty	27	1,134
Shale (pencil cave)	2	1,136
Limestone (Big Lime)	145	1,281
Sandstone (Big Injun)	119	1,400
Shale and shells	425	1,825
Shale, brown (Sunbury)	20	1,845
Sandstone (Berea)	35	1,880
Total depth		1,880

Log No. 367

Jas. Short, No. 2, lessor. Ohio Fuel Oil & Gas Co., lessee. Commenced: Nov. 21, 1917. Completed: Jan. 9, 1918. Shot: Jan. 15, 1918, 100 qts. Very small show of oil, small well after shot. Production: 3 bbls. well.

Strata.

Pennsylvanian System.	Thickness	Depth
Soil, shale, soft	45	45
Shale	135	180
Coal	3	183
Sandstone, buff	67	250
Shale	75	325
Sandstone	85	410
Shale and shells	90	500
Shale, shelly	157	657
Shale and shells	80	737
Shale, hard	50	787
Shale (break)	3	790
Sandstone (salt)	150	940
Shale	2	942
Sandstone	60	1,002
Mississippian System.		
Shale and shell	68	1,070
Sandstone (Maxon)	18	1,088
Shale (pencil cave)	2	1,090
Limestone (Big Lime)	150	1,240
Sandstone (Big Injun)	100	1,340
Shale	5	1,345
Limestone	40	1,385
Shale and shells	407	1,792
Shale, brown (Sunbury)	20	1,812
Sandstone (Berea)	37	1,849
Total depth		1,849

Log No. 368

Jas. Short, No. 3, lessor. Ohio Fuel Oil & Gas Co., lessee. Commenced: June 2, 1920. Completed: June 26, 1920. Shot: June 28, 1920, 60 qts. Production: 4 bbls. oil per day.

Strata.

Pennsylvanian System.	Thickness	Depth
Soil	8	8
Shale	52	60

Pennsylvanian System.	Thickness	Depth
Sandstone	45	105
Shale, blue	15	120
Sandstone, (water 125)	35	155
Shale, blue, and shell, (water flood 300)	300	455
Sandstone (cow run)	15	470
Shale, blue	90	560
Sandstone (salt)	200	760
Shale, blue	40	800
Sandstone	45	845
Shale, blue	40	885
Mississippian System.		
Sandstone (Maxon)	15	900
Shale (pencil cave)	5	905
Limestone (Big Lime)	140	1,045
Sandstone (Big Injun)	75	1,120
Shale, blue	5	1,125
Shale, shelly	35	1,160
Shale, blue, and shells	8	1,168
Shale, black	422	1,590
Sandstone	8	1,598
Shale, brown (Sunbury)	21	1,619
Sandstone (Berea), (pay 1,620-1,638)	23	1,642
Total depth		1,642

Log No. 369

Mollie Burton, No. 1, lessor. Ohio Fuel Oil & Gas Co., lessee. Location: near Twin Branch. Commenced: June 4, 1919. Completed: July 3, 1919. Shot: July 4, 1919, 60 qts. Production: Gas well, 200,000 ft.

Strata.

Pennsylvanian System.	Thickness	Depth
Soil	20	20
Sandstone	80	100
Shale	20	120
Sandstone	25	145
Shale (water)	150	295
Coal	4	299
Shale	101	400
Shale and shells	280	680
Sandstone (salt)	195	875
Shale	40	915

Mississippian System.

	Thickness	Depth
Limestone (Big Lime)	160	1,075
Sandstone (Big Injun)	82	1,157
Shale and shells	462	1,619
Shale, brown (Sunbury)	16	1,635
Sand (Berea)	36	1,671
Total depth		1,671

Log No. 370

Joe Carter, No. 2, lessor. Ohio Fuel Oil & Gas Co., lessee. Commenced: Feb. 18, 1920. Completed: Apr. 30, 1920. Shot: May 1, 1920, 60 qts. Production: 1½ bbls. per day, 20 ft. in sand.

Strata.

Pennsylvanian System.

	Thickness	Depth
Clay	13	13
Shale and shells	51	64
Shale, hard	11	75
Shale and shell	25	100
Coal	1	101
Shale	18	119
Shale, hard	6	125
Shale and shells	95	220
Sandstone, buff	100	320
Shale and shells	50	370
Shale, hard	10	380
Shale	20	400
Shale and shells, (show gas)	320	720
Sandstone	20	740
Shale	20	760
Shale, hard	10	770
Shale	20	790
Shale, hard	20	810
Sandstone (salt), (hole full gas and water) ..	110	920
Shale	5	925
Sandstone	20	945
Shale	5	950
Shale, hard	20	970
Shale	20	990
Sandstone	30	1,020

Mississippian System.

	Thickness	Depth
Limestone	10	1,030
Sand (Maxon)	30	1,060
Shale, black	15	1,075

Mississippian System.	Thickness	Depth
Shale (pencil cave)	3	1,078
Limestone (Big Lime)	202	1,280
Sandstone (Big Injun)	50	1,330
Shale and sandstone	100	1,430
Shale	170	1,600
Sandstone and limy shells	100	1,700
Shale	75	1,775
Shale, black (Sunbury)	18	1,793
Sandstone (Berea)	20	1,813
Total depth		1,813

Log No. 371

Elizabeth Pigg, No. 1, lessor. Ohio Fuel Oil & Gas Co., lessee. Location: Busseyville District. Commenced: Oct. 4, 1912. Completed: Oct. 28, 1912. Shot: October 28, 1912, 60 qts. Production: 3 bbls. per day.

Strata.

Pennsylvanian System.	Thickness	Depth
Clay	30	30
Shale	50	80
Sandstone	50	130
Shale	120	250
Sandstone	45	295
Shale	45	340
Sandstone	60	400
Shale	160	560
Sandstone (salt)	240	800
Shale	15	815
Sandstone	130	945

Mississippian System.	Thickness	Depth
Limestone (Little Lime), black	10	955
Limestone (Big Lime), white	165	1,120
Shale (break)	15	1,135
Sandstone (Big Injun)	65	1,200
Shale, shelly	360	1,560
Shale	19	1,579
Sandstone (Berea)	16	1,595
Shale (break)	14	1,609
Sandstone (Berea)	23	1,632
Shale	13	1,645
Total depth		1,645

Log No. 372

Elizabeth Pigg, No. 2, lessor. Ohio Fuel Oil & Gas Co., lessee. Commenced: Dec. 26, 1912. Completed: Jan. 18, 1913. Production: 3 bbls. per day.

Strata.

Pennsylvanian System.	Thickness	Depth
Sand	120	120
Shale	60	180
Sand	45	225
Shale	255	480
Sand	60	540
Shale	50	590
Sand (salt)	350	940
Shale	40	980
Mississippian System.		
Limestone (Big Lime)	120	1,100
Shale	20	1,120
Sandstone (Big Injun)	60	1,180
Shale	295	1,475
Limestone, shelly	105	1,580
Shale	23	1,603
Sandstone (Berea)	62	1,665
Total depth		1,665

Log No. 373

Elizabeth Pigg, No. 3, lessor. Ohio Fuel Oil & Gas Co., lessee. Commenced: Aug. 1, 1917. Completed: Aug. 23, 1917. Shot: Aug. 27, 1917, 60 qts. Production: 2½ bbls. oil.

Strata.

Pennsylvanian System.	Thickness	Depth
Shale soft	12	12
Sandstone	8	20
Shale and shells	485	505
Sandstone	30	535
Shale, shelly	65	600
Sand (salt), (water 610)	131	731
Sand (salt)	149	880
Shale	5	885
Sand (salt)	65	950
Shale	1	951

Mississippian System.	Thickness	Depth
Sandstone	3	954
Shale	1	955
Limestone (Little Lime)	5	960
Limestone (Big Lime)	125	1,085
Sandstone (Big Injun)	50˙	1,135
Shale and shells	443	1,578
Shale, brown (Sunbury)	20	1,598
Sandstone (Berea), (oil)	12	1,610
Shale (break)	3	1,613
Sandstone (Berea)	17	1,620
Sandstone and shale	10	1,630
Total depth		1,630

Log No. 374

Elizabeth Pigg, No. 4, lessor. Ohio Fuel Oil & Gas Co., lessee.
Commenced: Nov. 29, 1917. Completed: Jan. 12, 1918. Shot: Jan.
14, 1918, 60 qts. Production: 3 bbls. oil.

Strata.

Pennsylvanian System.	Thickness	Depth
Soil	9	9
Surface and shale	36	45
Sandstone	35	80
Shale	5	85
Sandstone	5	90
Shale	25	115
Sandstone	33	148
Shale, soft	10	158
Sandstone	15	173
Shale and shell	67	240
Coal	4	244
Shale, soft	6	250
Sandstone	60	310
Shale, shelly	80	390
Sandstone	15	405
Shale	125	530
Sandstone	6	536
Shale	54	590
Shale, hard	22	612
Shale and shells	3	615
Sand (salt)	320	935
Shale	6	941

Pennsylvanian System.

	Thickness	Depth
Shale, hard	5	946
Shale	4	950
Sandstone	30	980

Mississippian System.

	Thickness	Depth
Shale	2	982
Limestone (Big Lime)	140	1,122
Sandstone (Big Injun)	60	1,182
Shale	50	1,232
Sandstone, fine, hard	55	1,287
Shale, shelly	185	1,472
Sandstone, fine, hard	15	1,487
Shale, shelly	115	1,602
Shale, brown (Sunbury)	25	1,627
Sandstone (Berea)	7	1,634
Shale	2	1,636
Sand and shale	14½	1,650½
Total depth		1,650½

Log No. 375

Lornad Adams, No. 1, lessor. Ohio City Gas Co., lessee. Location: Sand Branch. Production: Dry hole.

Strata.

Pennsylvanian System.

	Thickness	Depth
Soil	20	20
Shale, hard, white	10	30
Coal	5	35
Limestone, blue	30	65
Shale	5	70
Sandstone	20	90
Shale	20	110
Sandstone	130	240
Shale	10	250
Sandstone	25	275
Shale	40	315
Sandstone (salt)	130	445
Shale	30	475
Sandstone	5	480
Shale	40	520
Sandstone	55	575
Shale	5	580

Mississippian System.	Thickness	Depth
Sandstone (Maxon)	65	645
Shale	60	705
Limestone (Little Lime)	77	782
Limestone (Big Lime), (show gas 936)	154	936
Sandstone (Big Injun), (water 937)	45	981
Shale	172	1,153
Sandstone (Wier)	30	1,183
Shale	145	1,328
Sandstone, gritty	5	1,333
Shale and shell	71	1,404
Shale, brown (Sunbury)	17	1,421
Sandstone (Berea)	47	1,468
Shale (break)	20	1,488
Sandstone (Berea)	19	1,507
Total depth		1,507

Log No. 376

H. H. Gambill, No. 1, lessor. Location: near Blaine. Completed: July 15, 1904. Production: Dry. Well plugged and abandoned. Authority: The New Domain Oil & Gas Co.

Strata.

Pennsylvanian System.	Thickness	Depth
Gravel, loose	69	69
Sandstone, hard, fine	50	119
Sandstone, hard, white	200	319
Shale, black, soft	45	364
Sandstone, hard, white	10	374
Shale, black, soft	25	399
Sandstone, blue, hard	56	455
Sandstone, hard	18	473

Mississippian System.		
Limestone (Big Lime), hard	117	590
Shale, soft	320	910
Shale, hard, black, soft	20	930
Sandstone, hard	60	990
Sand, soft	20	1,010
Shale, hard, white	20	1,030
Shale, black, soft (Chattanooga)	420	1,450
Shale, red, sandy	45	1,495
Shale, black, hard	15	1,510
Shale, white, hard	30	1,540

Mississippian System.

	Thickness	Depth
Shale, black, hard	10	1,550
Shale white, hard	90	1,640
Limestone, soft	16	1,656
Limestone, hard	10	1,666
Limestone, soft	5	1,671

Silurian System.

Limestone, hard	100	1,771
Limestone, soft	20	1,791
Limestone, hard	94	1,885
Total depth		1,885

Log No. 377

J. H. Grambill, No. 1, lessor. Location: on Spring Branch. Commenced: Apr. 20, 1920. Completed: June 12, 1920. Production: Approx. 40 bbls. oil.

Strata.

Pennsylvanian and Mississippian Systems.

	Thickness	Depth
Sandstone, shale and limestone	629	629
Sandstone (stray), (oil show)	17	646
Shale	40	686
Sand (Wier)	37	723
Shale	9	732
Total depth		732

Log No. 378

Jim Bartlett, No. 1, lessor. Holt Shannon Oil Co., lessee. Location: near Irad. Completed: in 1912.

Strata.

Pennsylvanian System.

	Thickness	Depth
Sandstone	55	55
Shale	5	60
Coal	3	63
Shale	7	70
Coal	5	75
Shale	50	125
Sandstone	25	150
Shale	52	202

Pennsylvanian System.	Thickness	Depth
Sandstone	40	242
Shale	43	285
Sandstone	10	295
Shale	55	350
Sandstone	20	370
Shale	20	390
Sandstone (salt)	70	460
Shale (break)	10	470
Sandstone	10	480
Shale	44	524
Sandstone (salt), (salt water flood 524)	111	635
Shale (break)	5	640
Shale, hard, gray	50	690
Shale, soft	25	715
Shale, hard, gray	25	740

Mississippian System.		
Sand and limestone	10	750
Sand (Maxon), (water)	40	790
Limestone (Little Lime)	20	810
Shale (pencil cave)	10	820
Limestone (Big Lime)	198	1,018
Sandstone (Big Injun)	52	1,070
Shale	20	1,090
Shells	370	1,460
Shale, brown (Sunbury)	20	1,480
Sandstone (Berea)	2	1,482
Total depth		1,482

Log No. 379

F. R. Bussey, No. 1, lessor. Venora Oil & Gas Co., of Huntington, W. Va., lessee. Location: near Busseyville.

Strata.

Pennsylvanian System.	Thickness	Depth
Gravel	30	30
Shale black	50	80
Sandstone, white	15	95
Shale, white	30	125
Sandstone, white	20	145
Limestone, black	40	185
Shale, black	15	200
Sandstone, white	30	230

Pennsylvanian System.   Thickness Depth

	Thickness	Depth
Coal, black	15	245
Shale, black	20	265
Sandstone (salt)	4	269
Shale, black, (oil show 455)	186	455
Sandstone, white	30	485
Shale, black	70	555
Sandstone (salt), (water flood 580)	140	695
Shale, black	20	715
Sandstone	80	795
Shale, black	30	825
Sandstone	10	835
Shale, black	30	865

Mississippian System.

	Thickness	Depth
Sand (Maxon)	40	905
Shale, black	30	935
Shale, red, sandy	20	955
Limestone (Little Lime)	15	970
Shale, black	10	980
Limestone (Big Lime)	100	1,080
Shale and shells	215	1,295
Shale and shells, white	255	1,550
Shale, black (Sunbury)	20	1,570
Sand (Berea), white	28	1,598
Total depth		1,598

Log No. 380

F. R. Bussey, No. 2, lessor. Venora Oil & Gas Co., Huntington, W. Va., lessee. Location: Near Busseyville. Commenced: April 30, 1912. Completed: May 25, 1912.

Strata.

Pennsylvanian System.   Thickness Depth

	Thickness	Depth
Clay, yellow	20	20
Sandstone, white	80	100
Shale, brown	40	140
Sandstone, white	80	220
Shale, white	130	350
Shale, gray, hard	8	358
Shale, black	142	500
Sandstone, white	10	510
Shale, black	105	615

Pennsylvanian System.	Thickness	Depth
Sandstone, brown	15	630
Shale, black	10	640
Sandstone, white	375	1,015

Mississippian System.		
Shale, black	2	1,017
Limestone (Big Lime), white	130	1,147
Sandstone (Big Injun), gray	60	1,207
Shale and shell, white	268	1,475
Shale, black (Sunbury)	178	1,653
Sand (Berea), gray	64	1,717
Total depth		1,717

Hole full of water, 645.

Break shale 23-26.

Log No. 381

F. R. Bussey, No. 4, lessor. New Domain Oil & Gas Co., lessee.
Location: Near Busseyville. Commenced: August 15, 1919. Completed: September 29, 1919. Production: 2 bbls. per day, shot 50 quarts.

Strata.

Pennsylvanian System.	Thickness	Depth
Quicksand	52	52
Coal	2	54
Shale and shells	410	464
Sandstone	·388	852

Mississippian System.		
Limestone (Big Lime)	151	1,003
Sandstone (Big Injun)	15	1,018
Shale and shells	442	1,460
Shale, brown (Sunbury)	27	1,487
Sandstone (Berea)	57½	1,544½
Total depth		1,544½

First pay, 1487-1491.

Second pay, 1527-1537.

Log No. 382

F. R. Bussey, No. 5, lessor. New Domain Oil & Gas Co., lessee.
Production: 2 bbls. per day, shot Oct. 13, 1919, 40 quarts.

Strata.

Pennsylvanian System.	Thickness	Depth
Loam	14	14
Sandstone	40	54
Shale	39	92
Coal	2	94
Shale and shell	410	504
Sand (salt)	386	890
Mississippian System.		
Limestone (Big Lime)	118	1,008
Sandstone (Big Injun)	40	1,048
Shells	465	1,513
Shale, brown (Sunbury)	20	1,533
Sandstone (Berea), (oil pay 1581-1597)	64	1,597
Total depth		1,597

Log No. 383

F. R. Bussey, No. 6, lessor. New Domain Oil & Gas Co., lessee.
Production: 3 bbls. oil; shot Nov. 5, 1919, 140 quarts.

Strata.

Pennsylvanian System.	Thickness	Depth
Gravel	30	30
Quicksand	15	45
Sandstone	100	145
Shale, soft, blue	80	225
Shale	180	405
Sand (salt)	325	730
Shale	115	845
Mississippian System.		
Limestone (Big Lime)	160	1,005
Sandstone (Big Injun)	60	1,065
Shale	415	1,480
Shale, brown (Sunbury)	7	1,487
Sandstone (Berea)	63	1,550
Total depth		1,550

Log No. 384

F. R. Bussey, No. 7, lessor.   New Domain Oil & Gas Co., lessee.
Production: 3 bbls. oil per day; shot March 2, 1920, 80 quarts.

Strata.

Pennsylvanian System.	Thickness	Depth
Gravel	20	20
Sandstone	80	100
Shale and shells	475	575
Sand (salt)	359	934

Mississippian System.		
Limestone (Big Lime)	145	1,079
Sandstone (Big Injun)	25	1,104
Shale and shells	479	1,583
Shale, brown (Sunbury)	17	1,600
Sandstone (Berea)	54½	1,654½
Total depth		1,654½

Oil 1660-1668.
Second pay, 1632-1650.

Log No. 385

F. R. Bussey, No. 8, lessor.   New Domain Oil & Gas Co., lessee.
Production: 3 bbls.

Strata.

Pennsylvanian System.	Thickness	Depth
Clay	20	20
Sandstone	185	205
Shale	250	455
Sandstone	170	625
Shale	20	645
Sandstone (salt)	335	980
Shale	60	1,040

Mississippian System.		
Limestone (Big Lime)	155	1,195
Sandstone (Big Injun)	25	1,220
Shale and shells	423	1,643
Shale, brown (Sunbury)	15	1,658
Sandstone (Berea)	60	1,718
Total depth		1,718

Log No. 386

F. R. Bussey, No. 9, lessor. New Domain Oil & Gas Co., lessee.
Shot: October 13, 1919, 40 quarts. Production: 1 bbl. oil.

Strata.

Pennsylvanian System.	Thickness	Depth
Loam	14	14
Sandstone	40	54
Shale	38	92
Coal	2	94
Shale and shells	410	504
Sand (salt)	386	890
Mississippian System.		
Limestone (Big Lime)	118	1,008
Sandstone (Big Injun)	40	1,048
Shells	465	1,513
Shale, brown (Sunbury)	20	1,533
Sandstone (Berea)	64	1,597
Total depth		1,597

Log No. 387

F. R. Bussey, No. 1, lessor. Sullivan-Mayo Oil & Gas Co., lessee.
Commenced: September 21, 1912. Completed: November 14, 1912.
Shot: 60 quarts.

Strata.

Pennsylvanian System.	Thickness	Depth
Shale and sandstone, (water)	470	470
Sandstone (gas)	205	675
Shale, black	40	715
Sandstone (salt)	170	885
Mississippian System.		
Limestone (Big Lime)	149	1,034
Shale	456	1,490
Shale, brown (Sunbury)	38	1,528
Sandstone (Berea)	58	1,586
Shale	14	1,600
Total depth		1,600

NOTE—The above record of the Sullivan-Mayo Oil & Gas Co., and the one following of the Louisa Coal Co., are both F. R. Bussey No. 1 wells of the named lessees. These wells are not to be confused with the F. R. Bussey No. 1 of the Venora Oil & Gas Co., which appears on an earlier page. The three wells are entirely distinct and somewhat separated geographically, though all in Lawrence County.

Log No. 388

F. R. Bussey, No. 1, lessor. Louisa Coal Co., lessee. Commenced: December 20, 1912. Completed: January 23, 1913. 1st shot, 50 quarts; 2nd shot, 200 quarts.

Strata.

Pennsylvanian System.	Thickness	Depth
Soil and clay	36	36
Sandstone white	92	128
Shale, white	2	130
Sandstone, white	20	150
Shale and shell	360	510
Sandstone, white	25	535
Shale, blue	105	640
Sandstone, white	20	660
Shale, white	30	690
Sand (salt), gray	50	740
Shale, white	40	780
Sandstone, white	10	790
Shale, blue	6	796
Sandstone, gray	104	800
Shale, white	5	805
Sandstone, white	154	959
Coal	1	960
Shale, white	10	970

Mississippian System.		
Limestone (Big Lime)	135	1,105
Shale, white	33	1,138
Limestone, black	10	1,148
Shale, white	438	1,586
Shale, coffee (Sunbury)	20	1,606
Sandstone (Berea)	56	1,662
Total depth		1,662

Log No. 389

C. J. Carter, No. 1, lessor. Ohio Fuel Oil & Gas Co., lessee. Location: Near Yatesville. Commenced: October 13, 1919. Completed: November 19, 1919. Shot November 20, 1919, 60 quarts. Production: Gas, 300,000 cubic feet.

Strata.

Pennsylvanian System.	Thickness	Depth
Soil	14	14
Sandstone	86	110

Pennsylvanian System.	Thickness	Depth
Shale	50	150
Sandstone	20	170
Shale	40	210
Sandstone, buff	25	235
Shale	25	260
Sandstone	20	280
Shale	345	625
Sandstone	100	725
Shale	20	745
Sand (salt)	215	960
Shale	12	972

Mississippian System.		
Sand (Maxon)	21	993
Shale	7	1,000
Limestone (Little Lime)	20	1,020
Shale (pencil cave)	3	1,023
Limestone (Big Lime)	137	1,160
Sandstone (Big Injun)	89	1,249
Shale	26	1,275
Shale and limestone shells	25	1,300
Shale	297	1,597
Shale and limestone shells	12	1,609
Shale	125	1,734
Sandstone (Berea)	24½	1,758½
Total depth		1,758½

Log No. 390

C. J. Carter, No. 2, lessor. Ohio Fuel Oil & Gas Co., lessee. Commenced: December 9, 1919. Completed: January 9, 1920. Shot January 12, 1920, 60 quarts. Production: Gas, 130,000 cubic feet.

Strata.

Pennsylvanian System.	Thickness	Depth
Soil	30	30
Sandstone	20	50
Shale	30	80
Coal	3	83
Shale	52	135
Sandstone	65	200
Shale	30	230
Sandstone	15	245

Pennsylvanian System.	Thickness	Depth
Shale	125	370
Sandstone	20	390
Shale	70	460
Sandstone	270	730
Shale	10	740

Mississippian System.		
Sand (Maxon)	40	780
Shale (pencil cave)	3	783
Limestone (Big Lime)	172	955
Sandstone	20	975
Shale	5	980
Sandstone	75	1,055
Shale	170	1,225
Shells	10	1,235
Shale	140	1,375
Shells	8	1,383
Shale	52	1,435
Shale and shells	45	1,480
Sand, brown	22	1,502
Sandstone (Berea), (pay 1502-1514)	16	1,518
Total depth		1,518

Log No. 391

Hester Carter, No. 3, lessor. Ohio Fuel Oil & Gas Co., lessee. Commenced: February 23, 1918. Completed: March 23, 1918. Shot May 1, 1918, 60 quarts. Production: 200,000 gas, ½ bbl. oil.

Strata.

Pennsylvanian System.	Thickness	Depth
Soil	11	11
Sandstone	89	100
Shale	100	200
Sandstone, buff	45	245
Shale	45	290
Sandstone	20	310
Shale	75	385
Sandstone	45	430
Shale and shells	60	490
Limestone and shells	50	540
Shale, black	70	610
Limestone, sandy, (gas 610)	40	650
Shale	50	700

Pennsylvanian System.

	Thickness	Depth
Limestone	50	750
Shale	5	755
Sand, (salt) (water 800)	195	950
Shale (break)	3	953
Sandstone	57	1,010

Mississippian System.

	Thickness	Depth
Shale, shelly	40	1,050
Sand (Maxon)	20	1,070
Shale (pencil cave)	3	1,073
Limestone (Big Lime)	145	1,218
Sandstone (Big Injun)	112	1,330
Shale and shells	446	1,776
Shale, brown (Sunbury)	22	1,798
Sandstone (Berea), (gas and oil)	46	1,844
Total depth		1,844

Log No. 392

Hester Carter, No. 4, lessor. Ohio Fuel Oil & Gas Co., lessee. Commenced: March 5, 1920. Completed: April 3, 1920. Shot April 5, 1920, 40 quarts. Production: 6 bbls. per day.

Strata.

Pennsylvanian System.

	Thickness	Depth
Soil	10	10
Sandstone	20	30
Shale	70	100
Sandstone	50	150
Shale	10	160
Sandstone	55	215
Shale	60	275
Sandstone	15	290
Shale	55	345
Sandstone	25	370
Shale	15	385
Sandstone	55	440
Shale	30	470
Sandstone	10	480
Shale	45	525
Sandstone	15	540
Shale	50	590
Sandstone	365	955

Mississippian System.	Thickness	Depth
Limestone (Big Lime)	140	1,095
Sandstone (Big Injun)	131	1,226
Shale	4	1,230
Limestone	40	1,270
Shale and shells	388	1,658
Shale, brown (Sunbury)	31	1,689
Sandstone (Berea), (oil is 10 feet in sand) ...	24	1,713
Total depth		1,713

Log No. 393

Hester Carter, No. 5, lessor. Ohio Fuel Oil & Gas Co., lessee. Commenced: April 15, 1920. Completed: May 13, 1920. Gas, 300,- 000 cubic feet gas per day. Shot May 14, 1920, 60 quarts. Production: Gas, 400,000 cubic feet.

Strata.

Pennsylvanian System.	Thickness	Depth
Quicksand	65	65
Shale	20	85
Sandstone	35	120
Shale	80	200
Sandstone	25	225
Shale	15	240
Sandstone	50	290
Shale	20	310
Sandstone	20	330
Shale	30	360
Sandstone	15	375
Shale	5	380
Sandstone	55	435
Shale	5	440
Sandstone	328	768
Mississippian System.		
Limestone (Big Lime)	182	950
Sand	70	1,020
Shale and sandstone, hard	52	1,072
Shale and limestone shells	93	1,115
Shale and shells	395	1,510
Sandstone (Berea), (1st pay, 12 feet in sand)..	23	1,533
Total depth		1,533

Log No. 394

Landon Carter, No. 1, lessor. Ohio Fuel Oil & Gas Co., lessee. Location: Twin Branch. Commenced: March 24, 1919. Completed: April 17, 1919. Production: Dry hole.

Strata.

Pennsylvanian System.	Thickness	Depth
Soil	16	16
Sandstone	34	50
Shale	40	90
Gravel	26	116
Shale	4	120
Shale (fire clay)	40	160
Shale	60	220
Shale (fire clay)	17	237
Coal	2	239
Shale	41	280
Shale (fire clay)	20	300
Shale, hard	15	315
Shale	75	390
Shale, hard	10	400
Shale	68	468
Sandstone (Cow Run), gas	32	500
Shale, hard	18	518
Shale	5	523
Shale, shelly	27	550
Shale, sandy	65	615
Shale, (water 635-650)	3	618
Sandstone	32	650
Shale	3	653
Sandstone	75	728

Mississippian System.		
Shale and shells	37	765
Limestone (Little Lime)	15	780
Shale (pencil cave)	2	782
Limestone (Big Lime)	163	945
Sandstone (Big Injun)	100	1,045
Shale	17	1,062
Limestone	18	1,080
Shale	285	1,365
Shells	115	1,480

WEATHER PITTED POTTSVILLE CONGLOMERATE.
The irregular hardness and cross bedding of this important oil "sand" is well shown. This outcrop is just below Natural Bridge in Powell County, Kentucky.

Mississippian System.	Thickness	Depth
Shale, brown (Sunbury)	20	1,500
Sandstone (Berea)	42	1,542
Total depth		1,542

Little gas 1491-1497; little gas and water 1521-1527. Drilled 42 feet in sand.

Log No. 395

Pricey Chapman, No. 1, lessor. Ohio Fuel Oil & Gas Co., lessee. Location: Near Louisa. Commenced: March 9, 1920. Completed: April 15, 1920. Shot April 16, 1920, 90 quarts. Production: 2 bbls. per day.

Strata.

Pennsylvanian System.	Thickness	Depth
Soil	14	14
Shale, white	21	35
Shale, soft	20	55
Coal	2	57
Shale	13	70
Shale	10	80
Shale, soft	30	110
Sandstone, blue, (water 125)	45	155
Shale	45	200
Shale, shelly	15	215
Shale and mud	35	250
Sandstone	40	290
Shale	160	450
Sandstone	25	475
Shale	35	510
Shale, white, hard	15	525
Sandstone	75	600
Shale, shelly	55	655
Sand (salt)	115	770
Shale, soft	20	790
Sandstone	70	860
Shale, sandy	40	900
Shale	25	925

Mississippian System.		
Sand	5	930
Limestone	5	935
Shale (pencil cave)	5	940
Limestone (Big Lime)	160	1,100

Mississippian System.	Thickness	Depth
Sandstone (Big Injun)	105	1,205
Shale and shells	458	1,663
Shale, brown (Sunbury)	22	1,685
Sand (Berea)	27½	1,712½
Total depth		1,712½
Pay 1684-1710.		

Log No. 396

Pricey Chapman, No. 3, lessor. Ohio Fuel Oil & Gas Co., lessee. Commenced: May 18, 1920. Completed June 11, 1920. Shot June 12, 1920. Production: 3 bbls. per day.

Strata.

Pennsylvanian System.	Thickness	Depth
Clay	12	12
Sandstone	18	30
Shale (fire clay), coal	130	160
Sandstone, buff, (water)	30	190
Shale, soft	40	230
Shale and shells	40	270
Limestone, white	15	285
Shale	10	295
Sandstone	60	355
Shale	35	390
Sandstone	25	415
Shale	115	530
Sandstone	40	570
Shale	10	580
Sandstone	70	650
Shale and shells	50	700
Sand (salt)	110	810
Shale (break)	10	820
Sandstone	80	900
Shale, soft	25	925
Sandstone	50	975
Shale	15	990

Mississippian System.	Thickness	Depth
Sand (Maxon)	20	1,010
Shale (pencil cave)	25	1,035
Limestone (Big Lime)	140	1,175
Sandstone (Big Injun)	92	1,267

Mississippian System.

	Thickness	Depth
Shale, shelly	63	1,330
Shale and shells.	370	1,700
Shale, brown (Sunbury)	27	1,727
Sandstone (Berea), (pay 1,739-1,751)	38	1,765
Total depth		1,765

Log No. 397

James L. Clark, No. 2, lessor. Ohio Fuel Oil & Gas Co., lessee. Commenced: July 8, 1918. Completed: Aug. 15, 1918. Shot Aug. 19, 1918, 60 qts. Production: 1 bbl. oil per day.

Strata.

Pennsylvanian System.

	Thickness	Depth
Soil	14	14
Shale, hard	26	40
Sandstone	45	85
Shale	15	100
Sandstone	20	120
Shale	6	126
Sandstone	6	132
Coal	3	135
Shale and mud	75	210
Shale, hard	25	235
Shale, soft	40	275
Shale	10	285
Coal	3	288
Shale	152	440
Sandstone	16	456
Shale	102	558
Sandstone	22	580
Shale, hard	20	600
Shale	6	606
Sand (salt)	159	765
Shale	35	800
Sandstone	45	845
Shale, soft	3	848

Mississippian System.

	Thickness	Depth
Limestone, black	15	863
Shale, red, sandy	4	867
Sand	3	870
Shale, soft, black	3	873
Clay, white	4	877

Mississippian System.   Thickness Depth

	Thickness	Depth
Shale (pencil cave)	5	882
Limestone (Big Lime)	163	1,045
Sandstone (Big Injun)	60	1,105
Limestone shells	40	1,145
Shale and shells	440	1,585
Shale, brown (Sunbury)	25	1,610
Sand (Berea), (1st oil 1,610-1,618)	24½	1,634½
Total depth		1,634½

Log No. 398

William Clark, No. 1, lessor.   Ohio Fuel Oil & Gas Co., lessee.
Commenced: Sept. 8, 1917. Completed: Oct. 31, 1917. Shot Nov. 1,
1917, 80 qts. Production: 4 bbls.

Strata.

Pennsylvanian System.   Thickness Depth

	Thickness	Depth
Soil	20	20
Shale, soft	20	40
Sandstone, bluff	110	150
Shale	50	200
Sandstone	80	280
Shale	85	365
Shale, hard	55	420
Shale	70	490
Shale, hard	68	558
Shale	2	560
Sandstone (salt)	165	725
Shale	15	740
Shale, hard, gray	40	780
Shale and shells	95	875

Mississippian System.

	Thickness	Depth
Sand (Maxon)	22	897
Shale (pencil cave)	3	900
Limestone (Big Lime)	160	1,060
Sandstone (Big Injun)	98	1,158
Shale and shells	433	1,591
Shale, brown (Sunbury)	25	1,616
Sandstone (Berea), (oil pay 1,617-1,627) ...	26½	1,642½
Total depth		1,642½

Log No. 399

William Clark, No. 2, lessor. Ohio Fuel Oil & Gas Co., lessee. Location: near Busseyville. Commenced: Sept. 8, 1917. Completed: Oct. 31, 1917. Shot Nov. 1, 1917, 80 qts. Production: 2 bbls. per day.

Strata.

Pennsylvanian System.	Thickness	Depth
Soil	20	20
Shale, soft	20	40
Sandstone, bluff	110	150
Shale	50	200
Sandstone	80	280
Shale	85	365
Shale, hard	55	420
Shale	70	490
Shale, hard	68	558
Shale	2	560
Sandstone (salt)	165	725
Shale	15	740
Shale, hard	40	780
Shale and shell	95	875
Mississippian System.		
Sand (Maxon)	22	897
Shale (pencil cave)	3	900
Limestone (Big Lime)	160	1,060
Sandstone (Big Injun)	98	1,158
Shale and shells	433	1,591
Shale, brown (Sunbury)	25	1,616
Sandstone (Berea), (pay 1,617-1,627)	26½	1,642½
Total depth		1,642½

Log No. 400

A. Collinsworth, No. 4, lessor. Ohio Fuel Oil & Gas Co., lessee. Commenced: March 12, 1918. Completed: April 10, 1918. Shot April 11, 1918, 60 quarts. Production: 4 or 5 bbls. daily.

Strata.

Pennsylvanian System.	Thickness	Depth
Shale and mud	80	80
Sandstone	4	84
Coal	3	87

Pennsylvanian System.   Thickness Depth

	Thickness	Depth
Shale	78	165
Sandstone	35	200
Shale	10	210
Sandstone	20	230
Shale	15	245
Sandstone and shale	35	280
Coal	4	284
Shale	106	390
Sandstone	10	400
Shale	20	420
Sandstone	15	435
Coal	3	438
Shale	42	480
Sandstone	15	495
Shale	85	580
Coal	4	584
Shale	31	615
Sandstone	11	626
Shale	10	636
Sandstone	14	650
Shale	20	670
Sandstone	70	740
Shale	5	745
Sand (salt)	135	880
Shale	65	945

Mississippian System.

Sandstone	20	965
Shale, soft	45	1,010
Limestone (Little Lime)	25	1,035
Shale, soft	5	1,040
Limestone (Big Lime)	45	1,185
Sandstone (Big Injun)	72	1,257
Limestone, black	58	1,315
Shale and shells	403	1,718
Shale, brown (Sunbury)	22	1,740
Sandstone (Berea), (oil)	16	1,756
Shale (gas 1756)	2	1,758
Shale and sandstone	9	1,767
Total depth		1,767

Log No. 401

A. Collinsworth, No. 5, lessor. Ohio Fuel Oil & Gas Co., lessee. Commenced: June 14, 1918. Completed: July 15, 1918. No record to 610. Production: 4 bbls. per day.

Strata.

Pennsylvanian System.	Thickness	Depth
Unrecorded	610	610
Sand (salt)	225	835
Shale	15	850
Sandstone	10	860
Shale	115	975
Limestone	15	990
Shale	20	1,010
Mississippian System.		
Limestone (Big Lime)	145	1,155
Shale	25	1,180
Sandstone (Big Injun)	50	1,230
Shale	5	1,235
Shale and shells	420	1,655
Shale, brown (Sunbury)	25	1,680
Sandstone (Berea), (pay 1681-1691)	27	1,707
Total depth		1,707

Log No. 402

W. A. Copley, No. 2, lessor. Ohio Fuel Oil & Gas Co., lessee. Location: Near Fallsburg. Commenced: May 8, 1918. Completed: June 8, 1918. Shot June 20, 1918, 60 quarts. Production: Dry hole.

Strata.

Pennsylvanian System.	Thickness	Depth
Soil	5	5
Shale	45	50
Sandstone	40	90
Shale and shell	40	130
Sandstone	15	145
Shale, soft, blue	30	175
Sandstone, white	5	180
Shale and coal	35	215
Limestone, sandy	85	300
Shale	100	400
Sandy shell	50	450

Pennsylvanian System.	Thickness	Depth
Shale	30	480
Sandstone (Cow Run)	35	515
Shale	45	560
Limestone, sandy	20	580
Shale	20	600
Sandstone	80	680
Shale	10	690
Sandstone	160	850
Shale	50	900

Mississippian System.	Thickness	Depth
Sand, shelly	30	930
Shale	40	970
Sand	20	990
Shale, white	55	1,045
Limestone (Big Lime)	135	1,180
Sand, shelly	88	1,268
Sandstone (Big Injun)	80	1,348
Limestone	52	1,400
Shale and shells	323	1,723
Shells, brown (Sunbury)	20	1,743
Sand (Berea)	10	1,753
Shale (break)	2	1,755
Sand, (gas 1768)	13	1,768
Sand	13	1,781
Total depth		1,781

Log No. 403

W. A. Copley, No. 4, lessor. Ohio Fuel Oil & Gas Co., lessee. Production: Gas, 100,000 cubic feet, and 1 bbl. oil.

Strata.

Pennsylvanian System.	Thickness	Depth
Soil (clay)	20	20
Limestone	40	60
Shale	15	75
Sandstone	9	84
Shale	26	110
Shale (fire clay)	5	115
Sandstone	10	125
Shale soft	25	150
Sandstone	13	163
Coal	2	165

Pennsylvanian System.	Thickness	Depth
Shale	30	195
Sandstone (water)	12	207
Shale	33	240
Sandstone	15	255
Shale	65	320
Sandstone, (water)	40	360
Shale	50	410
Sandstone, (water)	35	445
Shale	70	515
Sandstone	25	540
Shale	40	580
Sandstone	25	605
Shale	13	618
Sandstone	44	662
Coal	3	665
Sandstone	81	746
Shale	29	775
Sand (salt)	115	890
Shale and shells	90	980
Mississippian System.		
Sand (Maxon)	12	992
Shale	38	1,030
Limestone, sandy	15	1,045
Limestone (Big Lime)	140	1,185
Sandstone (Big Injun)	15	1,200
Shale (break)	15	1,215
Sand	70	1,285
Shale and mud	25	1,310
Limestone	15	1,325
Shale and shells	414	1,739
Shale, brown (Sunbury)	26½	1,765½
Sand (Berea)	23	1,788½
Total depth		1,788½

Break from 1775-1777.
Pay from 1765½-1775.

Log No. 404

William Crider, No. 1, lessor. Ohio Fuel Oil & Gas Co., lessee. Location: Near Louisa. Commenced: July 2, 1919. Completed: August 4, 1919. Shot August 6, 1919, 60 quarts. Production ½ bbl. oil per day.

Strata.

Pennsylvanian System.	Thickness	Depth
Soil	26	26
Shale	119	145
Sandstone	65	210
Shale and shells	180	390
Limestone, sandy	25	415
Shale and shells	225	640
Limestone, sandy	90	730
Shale	5	735
Sand (salt)	95	830
Shale	5	835
Sandstone	20	855
Shale	45	900
Sandstone	25	925
Shale	20	945

Mississippian System.		
Sandstone	40	985
Shale (pencil cave)	11	996
Limestone (Big Lime)	166	1,162
Sandstone (Big Injun)	93	1,255
Limestone, black	40	1,295
Shale and shells	405	1,700
Shale, brown (Sunbury)	21	1,721
Sandstone (Berea)	36	1,757
Total depth		1,757

Log No. 405

D. W. Diamond, No. 2, lessor. Ohio Fuel Oil & Gas Co., lessee. Commenced: July 24, 1917. Completed: August 25, 1917. Shot August 28, 1917, 100 quarts. Production: 1 bbl. per day.

Strata.

Pennsylvanian System.	Thickness	Depth
Soil	20	20
Shale and shells	70	90
Sandstone	90	180
Shale and shells	100	280
Coal, (water)	3	283
Shale, (water)	337	620
Sand, (salt), (much water 675)	100	720
Shale and shells	100	820
Shale and sand	120	940

Mississippian System.	Thickness	Depth
Sandstone (Maxon)	18	958
Shale (Pencil Cave)	2	960
Limestone (Big Lime)	140	1,100
Sandstone (Big Injun)	110	1,210
Shale and shells	457	·1,667
Shale, brown (Sunbury)	20	1,687
Sandstone (Berea)	·8	1,695
Shale	3	1,698
Sand, (oil 1699-1709)	20	1,718
Total depth		1,718

Log No 406

D. W. Diamond, No. 3, lessor. Ohio Fuel Oil & Gas Co., lessee. Commenced: April 6, 1918. Completed: June 22, 1918. Shot June 22, 1918, 60 quarts. Production: 2½ bbls. per day.

Strata.

Pennsylvanian System.	Thickness	Depth
Soil	14	14
Sandstone	26	40
Shale, red, sandy	20	60
Sandstone	20	80
Shale	40	120
Shale, soft	40	160
Shale and shells	40	200
Sandstone	50	250
Shale	25	275
Sandstone	25	300
Shale	380	680
Shale, shelly	100	780
Sand (salt)	80	860
Shale and sand	90	950
Sand	25	975
Shale, muddy	25	1,000
Mississippian System.		
Sandstone (Maxon)	60	1,060
Shale (Pencil Cave)	20	1,080
Shale, white	20	1,100
Limestone (Big Lime)	158	1,258
Sandstone (Big Injun)	75	1,333
Shale and shells	442	1,775
Shale, brown (Sunbury)	17	1,792
Sandstone (Berea), (oil 1792-1798)	26½	1,818½
Total depth		1,818½

Log No. 407

D. W. Diamond, No. 4, lessor. Ohio Fuel Oil & Gas Co., lessee. Commenced: October 31, 1919. Completed: January 7, 1920. Shot January 9, 1920, 60 quarts. Production: 1 bbl.

Strata.

Pennsylvanian System.	Thickness	Depth
Alluvium	10	10
Shale, hard	30	40
Shale	5	45
Coal	4	49
Shale, hard	31	80
Shale	10	90
Sandstone	45	135
Shale	15	150
Sandstone, (water 180)	50	200
Shale	360	560
Shale, hard	50	610
Shale	20	630
Shale, hard	55	685
Sandstone (salt sand), (water)	140	825
Shale, (water)	25	850
Mississippian System.		
Sand (Maxon)	30	880
Shale	75	955
Limestone (Big Lime)	210	1,165
Sandstone (Big Injun)	65	1,230
Shale	140	1,370
Limestone	30	1,400
Shale and shells	252	1,652
Shale (Sunbury)	27	1,679
Sandstone (Berea), (oil 1679-1684)	21	1,700
Total depth		1,700

Log No 408

J. F. Diamond, No. 2, lessor. Ohio Fuel Oil & Gas Co., lessee. Commenced: March 20, 1920. Completed: June 14, 1920. Shot June 14, 1920, 60 quarts. Production: 8 bbls.

Strata.

Pennsylvanian System.	Thickness	Depth
Loam and quicksand	25	25
Sandstone	155	180

Pennsylvanian System.

	Thickness	Depth
Shale	20	200
Shale, hard	65	265
Coal	5	270
Shale and shells	240	510
Sand (salt)	50	560
Shale	75	635
Sand (salt)	165	800
Shale	25	825
Sand	110	935
Shale	15	950

Mississippian System.

	Thickness	Depth
Limestone (Little Lime)	20	970
Shale (pencil cave)	7	977
Limestone (Big Lime)	153	1,130
Sandstone (Big Injun)	80	1,210
Shale and shells	445	1,655
Shale, brown (Sunbury)	18	1,673
Sand (Berea), (oil 1675-1687)	22½	1,695½
Total depth		1,695½

Log No. 409

J. H. Diamond, No. 1, lessor. Ohio Fuel Oil & Gas Co., lessee. Location: Near Busseyville. Commenced: April 21, 1919. Completed: May 27, 1919. Shot May 28, 1919, 60 quarts. Production: 3 bbls. per day.

Strata.

Pennsylvanian System.

	Thickness	Depth
Soil	11	11
Shale	44	55
Sandstone	5	60
Shale, soft	4	64
Coal	3	67
Shale	48	115
Sandstone	10	125
Shale	10	135
Sandstone, buff	40	175
Shale	10	185
Sandstone	15	200
Shale and shells	400	600
Shale, hard	15	615
Shale	115	730

Pennsylvanian System.	Thickness	Depth
Sandstone	210	940
Shale, soft, black	3	943
Shale, sandy, hard	57	1,000
Shale	10	1,010

Mississippian System.		
Sand (Maxon)	10	1,020
Limestone (Little Lime)	17	1,037
Shale (pencil cave)	3	1,040
Limestone (Big Lime)	160	1,200
Sandstone (Big Injun)	75	1,275
Shale and shells	435	1,710
Shale, brown (Sunbury)	24½	1,734½
Sandstone (Berea)	27	1,761¼
Total depth		1,761¼

Log No. 410

J. H. Diamond, No. 2, lessor.   Ohio Fuel Oil & Gas Co., lessee.
Commenced: January 12, 1920.   Completed: February 23, 1920.   Shot
February 24, 1920, 60 quarts.   Production: 6 bbls. per day.

Strata.

Pennsylvanian System.	Thickness	Depth
Soil	25	25
Shells and blues	135	160
Sand, mountain	60	220
Shells and blues	360	580
Sand, (oil show 625)	70	650
Shale, blue	80	730
Sand (salt)	90	820
Shale, blue	10	830
Sand	15	845
Shale, black, and shell	55	900
Sand	85	985
Shale, black	15	1,000

Mississippian System.		
Limestone (Little Lime)	20	1,020
Shale, blue	5	1,025
Limestone (Big Lime)	145	1,170
Sandstone (Big Injun)	45	1,215
Limestone, shelly	30	1,245

Mississippian System.	Thickness	Depth
Shells	470	1,715
Shale, brown (Sunbury)	22	1,737
Sandstone (Berea), (pay 1738-1753)	26	1,763
Total depth		1,763

Log No. 411

W. I. Diamond, No. 1, lessor. Ohio Fuel Oil & Gas Co., lessee. Commenced: August 27, 1919. Completed: October 2, 1919. Shot October 3, 1919, 40 quarts. Production: 6 bbls.

Strata.

Pennsylvanian System.	Thickness	Depth
Soil	8	8
Sandstone	17	25
Shale	95	120
Sandstone	30	150
Coal	3	153
Clay	27	180
Sandstone, buff	125	305
Shale	20	325
Coal	3	328
Shale	112	440
Sandstone	30	470
Shale	45	515
Sand, shelly	20	535
Shale	15	550
Sandstone	30	580
Shale and shells	70	650
Sandstone	40	690
Shale	70	760
Sand (salt)	200	960
Shale	5	965
Sandstone	95	1,060
Shale, black	10	1,070
Sandstone	10	1,080
Shale, black	5	1,085
Mississippian System.		
Sand (Maxon)	23	1,108
Shale	22	1,130
Limestone (Big Lime)	150	1,280
Sandstone (Big Injun)	70	1,350
Sand, shells	145	1,495

Mississippian System.	Thickness	Depth
Shale	125	1,620
Sand shells	180	1,800
Shale, brown (Sunbury)	23	1,823
Sandstone (Berea), (pay 1823-1829)	21	1,844
Total depth		1,844

Log No. 412

Jas. Grubbs, No. 1, lessor. Ohio Fuel Oil & Gas Co., lessee. Location: Twin Branch District. Commenced: February 3, 1920. Completed: March 25, 1920. Shot March 26, 1920, 60 quarts. . Production: 300,000 cubic feet gas.

Strata.

Pennsylvanian System.	Thickness	Depth
Soil	30	30
Sand and shale	510	540
Sand (salt)	180	720
Sand and shale	80	800
Mississippian System.		
Limestone (Big Lime)	155	955
Sandstone (Big Injun) ......·.............	116	1,071
Shale and shells	418	1,489
Shale, brown (Sunbury)	22	1,511
Sand (Berea) (pay 1512-1522)	17	1,528
Total depth		1,528

Log No. 413

Tom Hays, No. 3, lessor. Ohio Fuel Oil & Gas Co., lessee. Location: Fallsburg District. Commenced: May 19, 1920. Completed: June 14, 1920. Shot June 20, 1920, 80 quarts. Production: 4 bbls. per day.

Strata.

Pennsylvanian System.	Thickness	Depth
Soil	7	7
Limestone	3	10
Shale, soft	80	90
Sandstone	45	135
Coal	4	139

Pennsylvanian System.

	Thickness	Depth
Sandstone	143	282
Coal, (water 285)	3	285
Sandstone	115	400
Shale	40	440
Sandstone	30	470
Shale	30	500
Sandstone, (oil 515)	45	545
Shale	175	720
Sandstone, (water 520)	30	750
Shale, (oil 865)	30	780
Sand (salt)	120	900
Shale	15	915
Sandstone	15	930

Mississippian System.

	Thickness	Depth
Shale, sandy, red	10	940
Sand	80	1,020
Shale	25	1,045
Sand	12	1,057
Shale	8	1,065
Limestone (Big Lime)	140	1,205
Shale	20	1,225
Sandstone (Big Injun)	95	1,320
Shale and shells	445	1,765
Shale, brown (Sunbury)	24	1,789
Sand (Berea), (pay 1790-1800 and 1803-1813)	26	1,815
Total depth		1,815

Log No. 414

Tom Hayton, No. 1, lessor. Ohio Fuel Oil & Gas Co., lessee. Commenced: August 17, 1917. Completed: September 10, 1917. Shot September 11, 1917, 100 quarts. Production: 2 bbls. oil.

Strata.

Pennsylvanian System.

	Thickness	Depth
Shale	100	100
Sandstone	50	150
Shale	230	380
Sandstone	55	435
Shale	150	585
Limestone	27	612
Sandstone	38	650
Shale	6	656

Pennsylvanian System.

	Thickness	Depth
Sandstone	12	668
Shale	52	720
Sand (salt)	105	825
Shale	5	830
Sandstone	54	884
Shale	6	890
Sandstone	10	900
Shale and shells	10	910
Sandstone	12	922
Shale	18	940
Sandstone	35	975

Mississippian System.

	Thickness	Depth
Shale and shells	35	1,010
Limestone (Big Lime)	170	1,180
Sandstone (Big Injun)	61	1,241
Shale, soft	4	1,245
Sandstone	55	1,300
Shale and shells	404	1,704
Shale, brown (Sunbury)	24	1,728
Sandstone (Berea), (oil pay)	36	1,764
Total depth		1,764

Log No. 415

Marion Herd, No. 1, lessor. Ohio Fuel Oil & Gas Co., lessee. Location: Busseyville District. Commenced: May 29, 1917. Completed: June 29, 1917. Shot June 30, 1917, 100 quarts. Production: 6 bbls. oil.

Strata.

Pennsylvanian System.

	Thickness	Depth
Soil	16	16
Sandstone	34	50
Shale	40	90
Coal	3	93
Shale	182	275
Coal, (water)	3	278
Shale	72	350
Sandstone	90	440
Shale	25	465
Coal	3	468
Shale, hard	57	525
Shale and shells	75	600

Pennsylvanian System.

	Thickness	Depth
Sandstone	50	650
Shale	20	670
Shale, gritty, hard	90	760
Shale	2	762
Sand (salt)	118	· 880
Shale and shells	40	920
Sandstone	20	940
Shale	15	955
Sandstone	10	965
Shale, (water 775)	35	1,000

Mississippian System.

	Thickness	Depth
Sand (Maxon)	18	1,018
Shale (pencil cave)	2	1,020
Limestone (Big Lime)	165	1,185
Sandstone (Big Injun)	75	1,260
Shale and shells	454	1,714
Shale, brown (Sunbury)	24	1,738
Sand (Berea) (oil pay)	33	1,771
Total depth		1,771

Log No. 416

Marion Herd, No. 2, lessor. Ohio Fuel Oil & Gas Co., lessee. Commenced: October 29, 1917. Completed: November 26, 1917. Shot November 28, 1917, 80 quarts. Production: 3 bbls oil.

Strata.

Pennsylvanian System.

	Thickness	Depth
Soil and shale	80	80
Coal	2	82
Shale, soft	28	110
Sandstone	100	210
Shale and shells	50	260
Sandstone	25	285
Shale and shells	95	380
Sandstone	20	400
Shale	170	570
Sandstone	22	592
Coal	4	596
Shale	54	650
Sandstone	15	665
Shale	15	680
Sand (salt)	100	780

Pennsylvanian System.	Thickness	Depth
Shale	5	785
Sandstone	5	790
Shale	8	798
Sandstone	37	835
Shale and shells	5	840
Sandstone	18	858
Shale	22	880
Sandstone	40	920
Shale, soft	20	940
Mississippian System.		
Sandstone (Maxon)	25	965
Shale	10	975
Limestone (Big Lime)	220	1,195
Sandstone (Big Injun)	17	1,212
Shale, soft	3	1,215
Limestone	25	1,240
Sand	20	1,260
Shale and shells	408	1,668
Shale, brown (Sunbury)	25½	1,693½
Sandstone (Berea), (oil pay)	26	1,719½
Total depth		1,719½

Log No. 417

A. M. Holbrook, No. 1, lessor.   Completed: August 28, 1904.
Production: Dry.   Authority: The New Domain Oil & Gas Co.

Strata.

Pennsylvanian System.	Thickness	Depth
Gravel and sand	40	40
Shale, hard, black, soft	10	50
Shale, light, soft	10	60
Shells light, hard	5	65
Shale, light, hard	12	77
Sand, light, soft, (water 95)	18	95
Shale, hard, dark, soft	60	155
Shells, dark, soft	5	160
Shale, hard, dark, soft	50	210
Limestone, light, hard	90	300
Sand, light, hard	150	450
Shale, hard, dark, soft, (water 480)	30	480
Sand, white, soft, (water 510)	30	510
Shale, hard, white, soft	10	520
Shale, dark, soft	16	536

Mississippian System.

	Thickness	Depth
Sandstone, light, hard (Big Lime in part)	345	881
Shale, light, soft	204	1,085
Sandstone, soft (Sunbury)	20	1,105
Sandstone, light, hard	115	1,220

Devonian System.

Shale, brown, shelly (Chattanooga)	500	1,720
Limestone, white, gritty	15	1,735
Shale, hard, white, soft	110	1,845
Limestone, hard, dark	17	1,862
Total depth		1,862

Log No. 418

J. C. Holbrook, No. 1, lessor. Union Oil & Gas Co., lessee. Location: Blaine Creek.

Strata.

Pennsylvanian System

	Thickness	Depth
Soil	25	25
Quicksand	30	55
Water sand	205	260
Shale	75	335

Mississippian System.

Limestone (Big Lime)	180	515
Sandstone	5	520
Shale, sandy	270	790
Sandstone (Wier)	45	835
Shale, blue	35	870
Shale, black (Sunbury)	30	900
Sandstone (Berea)	40	940
Shale, sandy	31	971
Total depth		971

Log No. 419

Jos. A. Hutchison, No. 1, lessor. Ohio Fuel Oil & Gas Co., lessee. Location: Busseyville District. Commenced: May 19, 1913. Completed: June 25, 1913. Shot July 4, 1913, 125 quarts. Production: 2½ bbls. oil.

Strata.

Pennsylvanian System

	Thickness	Depth
Gravel	11	11
Sandstone	25	36
Shale	14	50

Pennsylvanian System.	Thickness	Depth
Sandstone	25	75
Shale	40	115
Sandstone	80	195
Shale	35	230
Sandstone	20	250
Shale	20	270
Sandstone	15	285
Shale	165	450
Sandstone	100	550
Shale	40	590
Sand (salt), (water)	45	635
Shale	90	725
Sandstone	111	836
Shale	9	845

Mississippian System.		
Sand	10	855
Shale	5	860
Limestone (Big Lime)	170	1,030
Shale	5	1,035
Sandstone (Big Injun)	131	1,166
Limestone, shell, shale	374	1,540
Shale, brown (Sunbury)	21½	1,561½
Sandstone (Berea)	65	1,626½
Shale	3½	1,630
Total depth		1,630

Log No. 420

Jos. A. Hutchison, No. 2, lessor. Ohio Fuel Oil & Gas Co. lessee. Commenced: April 18, 1918. Completed: June 1, 1918. Shot June 7, 1918, 60 quarts. Production: 3 bbls. oil per day.

Strata.

Pennsylvanian System.	Thickness	Depth
Soil	5	5
Sandstone	85	90
Shale	15	105
Sandstone	85	190
Coal	3	193
Shale	17	210
Sandstone	110	320
Coal	4	324
Shale	11	335

Pennsylvanian System.

	Thickness	Depth
Sandstone	80	415
Shale	135	550
Sandstone	10	560
Coal	2	562
Shale	73	635
Limestone	55	690
Shale	24	714
Limestone	46	760
Sand (salt)	120	880
Shale and shells	30	910

Mississippian System.

	Thickness	Depth
Sand (Maxon)	40	950
Shale	8	958
Shale, red, sandy	10	968
Shale	32	1,000
Limestone (Little Lime)	28	1,028
Shale	2	1,030
Limestone (Big Lime)	130	1,160
Sandstone (Big Injun)	85	1,245
Shale	5	1,250
Shale and shells	426	1,676
Shale, brown (Sunbury)	24	1,700
Sand, (oil)	13	1,713
Shale, (oil 1713, gas 1716)	5	1,718
Shale and sand	7	1,725
Total depth		1,725

Log No. 421

L. N. Hutchison, No. 3, lessor. Ohio Fuel Oil & Gas Co., lessee. Commenced: April 9, 1917. Completed: May 7, 1917. Shot May 9, 1917, 100 quarts. Production: 3 bbls. per day.

Strata.

Pennsylvanian System

	Thickness	Depth
Soil	16	16
Shale	14	30
Sandstone	40	70
Shale	80	150
Sand and shale alternating	600	750
Sand	55	805
Sandstone (salt), (water)	100	905
Shale	95	1,000

Pennsylvanian System. Thickness Depth
Sand	50	1,050
Shale	10	1,060
Sand	26	1,086

Mississippian System.
Limestone (Big Lime)	94	1,180
Sand	183	1,363
Shale and shells	392	1,755
Shale, brown (Sunbury)	19	1,774
Sand (Berea)	42	1,816
Total depth		1,816

Log No. 422

L. N. Hutchison, No. 5, lessor. Ohio Fuel Oil & Gas Co., lessee. Commenced: February 26, 1920. Completed: March 12, 1920. Shot May 13, 1920, 60 quarts. Production: 3 bbls. per day.

Strata.

Pennsylvanian System Thickness Depth
Soil	12	12
Sandstone	108	120
Shale	20	140
Shale, red, sandy	15	155
Shale	75	230
Sandstone	15	245
Shale	55	300
Sandstone	25	325
Shale	75	400
Sandstone	250	650
Shale	50	700
Sandstone	75	775
Shale	25	800
Sandstone	30	830
Shale	60	890

Mississippian System.
Sandstone (Maxon), (gas 890-900)	10	900
Limestone (Big Lime)	130	1,030
Shale	5	1,035
Sandstone (Big Injun)	105	1,140
Shale	410	1,550
Sandstone, brown (Berea)	57	1,607
Sandstone (Berea)	24	1,631
Total depth		1,631

NOTE—The Sunbury shale was not noted by the driller, it occurring in the base of the 410 feet of shale above 1550. The Maxon sand above the Big Lime is very thin.

Log No. 423

L. N. Hutchison, No. 6, lessor.   Ohio Fuel Oil & Gas Co., lessee.
Commenced: March 31, 1920.   Completed: May 4, 1920.   Shot May
5, 1920, 110 quarts.   Production: 3 bbls. oil.

Strata.

Pennsylvanian System	Thickness	Depth
Soil, sandy	30	30
Sandstone	95	125
Shale	75	200
Sandstone	25	225
Shale	75	300
Sandstone	15	315
Shale	45	360
Sandstone	240	600
Shale	60	660
Sandstone	40	700
Shale	25	725
Sandstone	85	810
Shale	35	845
Sandstone	5	850
Shale	10	860
Mississippian System.		
Limestone (Big Lime)	120	980
Sandstone (Big Injun)	203	1,183
Shale	347	1,530
Shale, brown (Sunbury)	32	1,562
Sand (Berea), (oil show)	44	1,606
Total depth		1,606

Log No. 424

D. C. Hughes, No. 2, lessor.   Ohio Fuel Oil & Gas Co., lessee.
Commenced: September 17, 1917.   Completed: October 15, 1917.   Shot
October 18, 1917, 60 quarts.   Production: 2 bbls. oil.

Strata.

Pennsylvanian System	Thickness	Depth
Shale	9	9
Sandstone	11	20
Shale, soft	25	45
Sandstone	125	170
Shale	30	200

Pennsylvanian System.

	Thickness	Depth
Sandstone	12	212·
Shale and shells	14	226
Sandstone	22	248
Shale	42	290
Sandstone	10	300
Shale and shell	15	315
Coal	4	319
Shale and sand	31	350
Sand	15	365
Coal ·	4	369
Shale and shells	156	525
Shale, soft	20	545
Shale and shell·	55	600
Sand	6	606
Shale	74	680
Sand (salt)	345	1,025

Mississippian System.

	Thickness	Depth
Shale, soft	33	1,058
Limestone (Little Lime)	13	1,071
Shale (pencil cave)	2	1,073
Limestone (Big Lime)	122	1,195
Sandstone (Big Injun)	55	1,250
Shale	5	1,255
Limestone	20	1,275
Shale and shells	270	1,545
Limestone	13	1,558
Shale and shells	113	1,671
Shale, brown (Sunbury)	25	1,696
Sandstone (Berea)	21½	1,717½
Total depth		1,717½

Log No. 425

M. H. Johns, No. 2, lessor. New Domain Oil & Gas Co., lessee. Location: Near Louisa. Shot January 30, 1920, 80 quarts. Production: 3 bbls. oil per day.

Strata.

Pennsylvanian System

	Thickness	Depth
Gravel	40	40
Sandstone	22·	62
Coal	4	66
Sandstone	14	80

Pennsylvanian System.

	Thickness	Depth
Sandstone	150	230
Sandstone (cow run)	30	260
Shale	50	310
Sandstone	100	410
Shale	225	635
Sand (salt)	173	808
Shale	15	823
Sandstone	68	891

Mississippian System.

	Thickness	Depth
Shale (pencil cave)	22	913
Limestone (Big Lime)	120	1,033
Sandstone (Big Injun)	55	1,088
Shale and shells	566	1,654 .
Shale, black (Sunbury)	20	1,674
Sandstone (Berea)	27½	1,701½
Total depth		1,701½

First oil, 1674-1684.
Second oil, 1692-1696.

Log No. 426

Wm. Justice, No. 1, lessor. Ohio Fuel Oil & Gas Co., lessee. Locotion: Near Louisa. Commenced: May 26, 1920. Completed: June 28, 1920. Shot June 29, 1920, 90 quarts. Production: 4 bbls. oil per day.

Strata.

Pennsylvanian System

	Thickness	Depth
Soil	14	14
Sandstone	16	30
Shale	30	60
Sandstone	/ 40	100
Shale, blue	40	140
Sandstone	70	210
Shale, blue	495	705
Sand (salt)	185	890
Shale, blue	10	900
Sandstone (salt)	130	1,030

Mississippian System.

	Thickness	Depth
Sandstone (Maxon)	35	1,065
Shale (pencil cave)	15	1,080
Limestone (Big Lime)	160	1,240

Mississippian System.

	Thickness	Depth
Sandstone (Big Injun)	80	1,320
Shale, blue	340	1,660
Limestone and shells	25	1,685
Shale, blue	115	1,800
Shale, brown (Sunbury)	15½	1,815½
Sandstone (Berea), (pay oil 1817-1842)	29½	1,845
Total depth		1,845

Log No. 427

Hannah Lackey, No. 1, lessor. Ohio Fuel Oil & Gas Co., lessee. Location: Near Louisa. Commenced: January 12, 1914. Completed: February 17, 1914. Shot February 18, 1914, 120 quarts. Production: 4 or 5 bbls. oil when shot.

Strata.

Pennsylvanian System

	Thickness	Depth
Sand, gravel	19	19
Shale and sand	406	425
Sandstone (Little Dunkard)	35	460
Shale and sand	470	930

Mississippian System.

	Thickness	Depth
Limestone (Little Lime)	30	960
Limestone (Big Lime)	175	1,135
Shale	25	1,160
Sandstone (Big Injun)	82	1,242
Shale and shells	353	1,595
Shale, brown (Sunbury)	21	1,616
Sandstone (Berea)	42½	1,658½
Total depth		1,658½

Log No. 428

Hannah Lackey, No. 4, lessor. Ohio Fuel Oil & Gas Co., lessee. Commenced: Nov. 20, 1919. Completed: Jan. 9, 1919. Shot Jan. 10, 1919, 40 qts. Production: 5 bbls. oil.

Strata.

Pennsylvanian System.

	Thickness	Depth
Soil	11	11
Sandstone	59	70
Shale, shelly	5	75

Pennsylvanian System.

	Thickness	Depth
Shale, soft	80	155
Sandstone, buff	45	200
Shale, soft	20	220
Sandstone	10	230
Shale, soft	15	245
Sandstone	50	295
Shale, soft	15	310
Shale	50	360
Sandstone	20	380
Shale	30	410
Sandstone	30	440
Shale	120	560
Shale, hard	20	580
Shale	20	600
Sandstone	55	655
Shale, hard	35	690
Shale	10	700
Sandstone	90	790
Shale	20	810
Sand (salt)	90	900
Shale	10	910

Mississippian System.

	Thickness	Depth
Sand (Maxon)	85	995
Shale (pencil cave)	5	1,000
Limestone, (Big Lime)	150	1,150
Sandstone (Big Injun)	119	1,269
Shale, sandy, fine	31	1,300
Shale	5	1,305
Shale, sandy, fine	45	1,350
Shale	100	1,450
Shale, sandy, fine	25	1,475
Shale and shell	85	1,560
Sandstone, fine	10	1,570
Shale and shell	152	1,722
Shale, brown (Sunbury)	28	1,750
Sandstone (Berea), (oil 1,750-1,759)	21	1771
Total depth		1,771

Log No. 429

Hannah Lackey, No. 5, lessor. Ohio Fuel Oil & Gas Co., lessee. Commenced: Sept. 27, 1919. Completed: Oct. 2, 1919. Shot Oct. 23, 1919, 40 qts. Production: 9 bbls. oil per day.

Strata.

Pennsylvanian System.	Thickness	Depth
Soil	16	16
Sandstone	64	80
Shale	70	150
Sandstone	100	250
Sha'e	70	320
Coal, (little water 320)	3	323
Shale	77	400
Shale, hard	40	440
Shale	35	475
Sandstone	70	545
Shale, (water)	55	600
Shale, hard	30	630
Shale and shells	130	760
Sandstone (salt), (water, hole flooded)	180	940

Mississippian System.		
Shale and shells	75	1,015
Sandstone (Maxon)	30	1,045
Shale (pencil cave)	21	1,066
Limestone (Big Lime)	152	1,218
Sandstone (Big Injun)	117	1,335
Shale and shell	428	1,763
Shale, brown (Sunbury)	24½	1,787½
Sandstone (Berea), (oil pay 1,789-1,799)	19	1,806½
Total depth		1,806½

Log No. 430

Hannah Lackey, No. 6, lessor. Ohio Fuel Oil & Gas Co., lessee. Commenced: July 18, 1918. Completed: Aug. 20, 1918. Shot 60 qts. Production: 7 bbls. oil per day.

Strata.

Pennsylvanian System.	Thickness	Depth
Soil (41)	16	16
Sandstone	64	80
Shale	70	150
Sandstone	100	250
Shale	70	320
Coal (water)	3	323
Shale	77	400
Shale, hard	40	440
Shale	35	475

Pennsylvanian System.	Thickness	Depth
Sandstone, (water 500)	70	545
Shale ..:	55	600
Shale, hard	30	630
Shale and shells	75	705
Sand (salt)	180	885
Shale and shells	75	960

Mississippian System.		
Sandstone (Maxon)	30	990
Shale	18	1,008
Shale (pencil cave)	3	1,011
Limestone (Big Lime)	152	1,163
Sandstone (Big Injun)	89	1,252
Shale and shells	450	1,702
Shale, brown (Sunbury)	23	1,725
Sandstone (Berea), (pay oil 1,725-1,736)	24½	1,749½
Total depth		1,749½

Log No. 431

Floyd McCown, No. 1, lessor. Reuben Fork Oil Co., lessee. Location: near Busseyville, on Reuben Creek.

Strata.

Pennsylvanian System.	Thickness	Depth
Soil	13	13
Sand, shale, etc., ...	187	200
Sandstone	100	300
Sandstone (Cow Run)	105	405
Sandstone (salt)	205	610
Coal	2	612
Sandstone	13	625
Shale	60	685
Sandstone (second salt)	110	795
Shale	90	885
Sandstone (third salt)	65	950
Shale	10	960

Mississippian System.		
Sand (Maxon)	80	1,040
Shale	12	1,052
Limestone (Little Lime)	5	1,057
Shale	5	1,062
Limestone (Big Lime)	188	1,250

Mississippian System.	Thickness	Depth
Shale	5	1,255
Sandstone (Big Injun)	100	1,355
Shale and shells	345	1,700
Shale, coffee (Sunbury)	22	1,722
Sandstone (Berea)	54	1,776
Total depth		1,776

Log No. 432

James McGlinn, No. 1, lessor. Location: Louisa Precinct. Completed: July 16, 1920. Production: 3 bbls. oil. Authority: New Domain Oil & Gas Co.

Strata.

Pennsylvanian System.	Thickness	Depth
Gravel	16	16
Sandstone	40	56
Shale, hard	80	136
Sandstone	35	171
Shale, hard	60	231
Sandstone	50	281
Shale .....:.............................	150	431
Sandstone	79	510
Shale, hard	95	605
Sandstone	34	639
Shale, hard	100	739
Sand (salt)	85	824
Shale, hard ....:......................	60	884

Mississippian System.	Thickness	Depth
Sand (Maxon)	100	984
Shale, hard	80	1,064
Limestone (Big Lime)	170	1,234
Sandstone (Big Injun)	110	1,344
Shale, hard, and shells	460	1,804
Shale, brown (Sunbury)	20	1,824
Sandstone (Berea), (oil)	27	1,851
Total depth		1,851

Log No. 433

E. G. McKinster, No. 1, lessor. Little Blaine Oil & Gas Co., lessee. Location: Right fork of Little Blaine's. Commenced: June, 1912. Completed: July 13, 1912.

Strata.

Pennsylvanian System.	Thickness	Depth
Soil	15	15
Shale	9	24
Coal	4	28
Sandstone	4	32
Shale	193	225
Sandstone	60	285
Coal	3	288
Sandstone	17	305
Shale	30	335
Shale, fine, hard	55	390
Sandstone (salt)	158	548
Shale	42	590
Sandstone	70	660
Shale	10	670
Sandstone	12	682
Shale	36	718
Coal	3	721
Shale	6	727

Mississippian System.		
Limestone (Big Lime)	158	885
Sandstone (Big Injun)	105	990
Shale	260	1,250
Sandstone, fine, hard	15	1,265
Shale	62	1,327
Shale, coffee (Sunbury)	20	1,347
Sandstone (Berea)	65	1,412
Shale	23	1,435
Total depth		1,435

Log No. 434

Sophia Moffett, No. 2, lessor. Ohio Fuel Oil & Gas Co., lessee. Location: near Busseyville. Commenced: Apr. 7, 1920. Completed: May 24, 1920. Shot May, 25, 1920. 60 qts. Production: 2 bbls. oil per day.

Strata.

Pennsylvanian System.	Thickness	Depth
Clay	8	8
Shale, blue	177	185
Sand, mountain	60	245
Shale, blue	5	250

Pennsylvanian System.   Thickness Depth
    Sandstone, (gas show) 130   380
    Shale, blue, shells 450   830
    Sand (salt) 105   935
    Shale, blue 15   950

Mississippian System.
    Sandstone (Maxon) and shale 130   1,080
    Shale, blue 15   1,095
    Limestone (Little Lime) 10   1,105
    Shale, blue, shells 5   1,110
    Limestone (Big Lime) 140   1,250
    Sandstone (Big Injun) 40   1,290
    Shale, blue 20   1,310
    Limestone shell 50   1,360
    Shale, blue, shells 430   1,790
    Shale, brown (Sunbury) 17   1,807
    Sandstone (Berea) 35   1,842
        Total depth 1,842

Log No. 435

A. L. Moore, No. 2, lessor. New Domain Oil & Gas Co., lessee. Location: near Louisa. Production: 2 bbls. oil.

    Strata.
Pennsylvanian System.   Thickness Depth
    Gravel 21   21
    Shale 80   101
    Sandstone 40   141
    Shale 60   201
    Sandstone 16   217
    Shale 83   300
    Sandstone 65   365
    Shale 145   510
    Sandstone 90   600
    Shale 150   750
    Sand (salt) 100   850
    Shale 70   920
    Sandstone 35   955
    Shale 40   995

Mississippian System.
    Sandstone (Maxon) 20   1,015
    Limestone (Big Lime) 165   1,180
    Sandstone (Big Injun) 75   1,255

Mississippian System.   Thickness Depth

    Shale, shelly   485   1,740
    Shale, brown (Sunbury)   20   1,760
    Sandstone (Berea)   30   1,790

        Total depth   1,790

Log No. 436

A. L. Moore, No. 4, lessor. New Domain Oil & Gas Co., lessee.
Shot Jan. 20, 1920, 80 qts. Production: 2 bbls. oil.

        Strata.

Pennsylvanian System.   Thickness Depth

    Gravel   16   16
    Sandstone   30   46
    Shale   85   131
    Sandstone   60   191
    Shale   45   236
    Sandstone   50   286
    Shale   120   406
    Sandstone   35   441
    Shale   60   501
    Sandstone   30   531
    Shale   170   701
    Sandstone (1st salt)   90   791
    Shale   35   826
    Sandstone (2nd salt)   115   941
    Shale   45   986

Mississippian System.

    Sandstone (Maxon)   30   1,016
    Limestone (Big Lime)   145   1,161
    Sandstone (Big Injun)   90   1,251
    Shale and shells   485   1,736
    Shale, brown (Sunbury)   20   1,756
    Sandstone (Berea)   27   1,783

        Total depth   1,783

      First oil, 1,757-1,767.

      Gas, 1,773-1,777.

Log No. 437

A. L. Moore No. 5, lessor.   New Domain Oil & Gas Co., lessee.
Shot April 23, 1920, 80 qts.   Production: Dry hole.

Strata.

Pennsylvanian System.	Thickness	Depth
Gravel	14	14
Sandstone	26	40
Shale	60	100
Sandstone	80	180
Shale	200	380
Sandstone	150	530
Shale	75	605
Sand (salt)	320	925
Shale	90	1,015
Mississippian System.		
Limestone (Big Lime)	155	1,170
Sandstone (Big Injun)	45	1,215
Shale and shells	463	1,678
Shale, brown (Sunbury)	20	1,698
Sandstone (Berea)	30½	1,728½
Total depth		1,728½

Log No. 438

W. D. O'Neal, No. 2, lessor.   Venora Oil & Gas Co., Huntington,
W. Va., lessee.   Location: Busseyville.

Strata.

Pennsylvanian System.	Thickness	Depth
Clay, yellow	12	12
Sandstone, white	28	40
Shale, black	140	180
Sandstone, white	20	200
Shale, black	400	600
Sand (salt), white, (water 615)	390	990
Shale, blue	10	1,000
Mississippian System.		
Limestone (Little Lime), black	30	1,030
Limestone (Big Lime), white	120	1,150
Sandstone (Big Injun), brown	15	1,165
Shale, white	10	1,175

Mississippian System.

	Thickness	Depth
Sandstone, white	25	1,200
Shale and shells	300	1,500
Shale, white	133	1,633
Shale, brown (Sunbury)	20	1,653
Sandstone (Berea)	61	1,714
Total depth		1,714

Log No. 439

R. J. Peters, No. 1, lessor. New Domain Oil & Gas Co., lessee. Location: near Louisa. Shot 80 qts. Production: 2 bbls. oil.

Strata.

Pennsylvanian System.

	Thickness	Depth
Clay, blue	40	40
Quicksand	10	50
Sandstone	10	60
Coal	2	62
Clay	25	87
Sandstone	33	120
Shale	60	180
Sandstone (cow run)	20	200
Shale	120	320
Sandstone	30	350
Shale	70	420
Sandstone	40	460
Shale	115	575
Sandstone (first salt)	40	615
Shale	30	645
Sandstone (second salt)	155	800
Shale	20	820
Sandstone	30	850
Shale	15	865

Mississippian System.

	Thickness	Depth
Sand (Maxon)	25	890
Shale	30	920
Limestone (Big Lime)	150	1,070
Sandstone (Big Injun)	55	1,125
Shale and shells	500	1,625
Shale, brown (Sunbury)	20	1,645
Sandstone (Berea)	28	1,673
Total depth		1,673

Log No. 440

R. J. Peters, No. 4, lessor.   New Domain Oil & Gas Co., lessee.
Shot 80 qts. Production: 3 bbls. oil.

Strata.

Pennsylvanian System.	Thickness	Depth
Gravel	22	22
Shale	27	49
Sandstone	30	79
Shale	40	119
Sandstone	50	169
Shale	131	300
Sandstone	45	335
Shale	115	450
Sand (salt)	60	510
Shale	290	800
Sand (salt)	40	840
Shale	35	875

Mississippian System.		
Sand (Maxon)	50	925
Shale	25	950
Limestone (Big Lime)	150	1,100
Sandstone (Big Injun)	67	1,167
Shale and shell	467	1,634
Shale, brown (Sunbury)	20½	1,654½
Sandstone (Berea), (pay oil and gas)	28½	1,683
Total depth		1,683

Pay sand, 1,667-1,678.
Oil and gas, 1,674-1,676.

Log No. 441

R. J. Peters, No. 5, lessor.   New Domain Oil & Gas Co., lessee.
Shot 80 qts. Production: 3 bbls. oil.

Strata.

Pennsylvanian System	Thickness	Depth
Gravel	21	21
Shale	25	46
Sandstone	100	146
Shale	50	196
Sandstone	200	396

Pennsylvanian System.

	Thickness	Depth
Shale	154	550
Sandstone	70	620
Shale	28	648
Sand (salt)	130.	778
Shale	130	908

Mississippian System.

Sand (Maxon)	45	953
Shale	40	993
Limestone (Big Lime)	150	1,143
Sandstone (Big Injun)	70	1,213
Limestone and shale	478	1,691
Shale, black (Sunbury)	20	1,711
Sandstone (Berea), (oil 1,712-1,726, 1,731-1,734)	29	1,740
Total depth		1,740

Log No. 442

R. J. Peters, No. 7, lessor. New Domain Oil & Gas Co., lessee. Shot 80 qts. Production: 3 bbls. oil per day.

Strata.

Pennsylvanian System

	Thickness	Depth
Clay, blue	40	40
Quicksand	10	50
Sandstone	10	60
Coal	2	62
Clay	25	87
Sandstone	33	120
Shale	60	180
Sandstone (cow run)	20	200
Shale	120	320
Sandstone	30	350
Shale	70	420
Sandstone	40	460
Shale	115	575
Sandstone (1st salt)	40	615
Shale	30	645
Sandstone (2nd salt)	155	800
Shale	20	820
Sandstone	30	850
Shale	15	865

Mississippian System.

	Thickness	Depth
Sand (Maxon)	25	890
Shale	25	915
Limestone (Big Lime)	150	1,065
Sandstone (Big Injun)	55	1,120
Shale and shells	500	1,620
Shale, brown, (Sunbury)	20	1,640
Sandstone (Berea)	28	1,668
Total depth		1,668

Log No. 443

R. J. Peters, No. 8, lessor. New Domain Oil & Gas Co., lessee. Shot Dec. 20, 1919, 60 qts. Production: 1 bbl. oil per day.

Strata.

Pennsylvanian System

	Thickness	Depth
Gravel	19	19
Sandstone	29	48
Shale	72	120
Sandstone	40	160
Shale	35	195
Sandstone	40	235
Shale	125	360
Sandstone	75	435
Shale	110	545
Sandstone	60	605
Shale	80	685
Sand (salt)	125	810
Shale	40	850
Sandstone	35	885

Mississippian System.

	Thickness	Depth
Shale	105	990
Limestone (Big Lime)	160	1,150
Sandstone (Big Injun)	60	1,210
Shale and shell	488	1,698
Shale, brown (Sunbury)	20	1,718
Sandstone (Berea)	30½	1,748½
Total depth		1,748½

NOTE—Although not recognized by the driller, the 105 feet of shale above 990 feet probably contains the Maxon sand.

Log No. 444

R. J. Peters, No. 9, lessor. New Domain Oil & Gas Co., lessee.
Shot 80 quarts. Production: 6 barrels oil per day.

Strata.

Pennsylvanian System	Thickness	Depth
Gravel	16	16
Sandstone	40	56
Shale	80	136
Sandstone	35	171
Shale	60	231
Sandstone	50	281
Shale	150	431
Sandstone	80	511
Shale	95	606
Sandstone	34	640
Shale	100	740
Sand (salt)	85	825
Shale	60	885

Mississippian System.		
Sandstone	100	985
Shale	80	1,065
Limestone (Big Lime)	170	1,235
Sandstone (Big Injun)	110	1,345
Limestone, shale and shell	460	1,805
Shale, brown· (Sunbury)	20	1,825
Sandstone (Berea)	27½	1,852½
Total depth		1,852½

Log No. 445

R. J. Peter, No. 1, lessor. New Domain Oil & Gas Co., lessee.
Shot Feb. 20, 1920, 80 qts. Production: 6 bbls. oil.

Strata.

Pennsylvanian System	Thickness	Depth
Gravel	22	22
Sandstone	50	72
Shale	40	112
Sandstone	85	197
Shale	35	232
Sandstone	45	277
Shale	150	427

Pennsylvanian System.	Thickness	Depth
Sandstone	80	507
Shale	65	572
Sandstone	90	662
Shale	100	762
Sand (salt)	80	842
Shale	25	867
Sandstone	125	992
Shale	40	1,032
Mississippian System.		
Limestone (Big Lime)	145	1,177
Sandstone (Big Injun)	65	1,242
Shale and shells	531	1,773
Shale, brown (Sunbury)	20	1,793
Sandstone (Berea)	25	1,818
Total depth		1818

1st oil, 1,793-1,808.

Oil and gas, 1,808-1,814.

Log No. 446

W. B. Pfost, No. 2, lessor. Ohio Fuel Oil & Gas Co., lessee. Commenced: November 19, 1919. Completed: May 3, 1920. Shot March 4, 1920, 40 quarts. Production: 5 bbls. oil.

Strata.

Pennsylvanian System.	Thickness	Depth
Soil	12	12
Sandstone	14	26
Shale (red rock)	54	80
Sandstone	40	120
Shale	30	150
Sandstone	10	160
Shale	80	240
Sandstone	45	285
Shale and shells	65	350
Sandstone	20	370
Shale and shells	365	735
Shale, shelly	102	837
Sand (salt)	80	917
Shale and sand	83	1,000
Sandstone	30	1,030

A CLIFF OF BEREA SANDSTONE.

The Berea Sandstone, productive of both oil and gas in Lawrence, Johnson and other counties, is a prominent rather evenly bedded formation on outcrop. Photo near Vanceburg by Charles Butts.

Mississippian System.

	Thickness	Depth
Shale	30	1,060
Sandstone (Maxon)	55	1,115
Shale (pencil cave)	3	1,118
Shale, white	39	1,157
Limestone (Big Lime)	158	1,315
Sandstone (Big Injun)	80	1,395
Shale and shell	429	1,824
Shale, brown (Sunbury)	25	1,849
Sandstone (Berea)	20½	1,869½
Total depth		1,869½

Log No. 477

Thad Ranson, No. 1, lessor. Ohio Fuel Oil & Gas Co., lessee. Commenced: December 15, 1919. Completed: January 22, 1920. Shot January 24, 1920, 60 quarts. Production: 4 bbls. oil per day.

Strata.

Pennsylvanian System.

	Thickness	Depth
Soil	2	2
Sandstone bluff	28	30
Shale	100	130
Sand, mountain	105	235
Shale and shells	435	670
Sand (salt)	150	820
Shale	20	840
Sandstone	135	975
Shale	5	980

Mississippian System.

	Thickness	Depth
Sandstone (Maxon), (gas at 985)	20	1,000
Shale	15	1,015
Limestone (Big Lime)	160	1,175
Sandstone (Big Injun)	77	1,252
Shale and shells	473	1,725
Shale, brown (Sunbury)	24	1,749
Sandstone (Berea), (oil pay 1750-1765)	23	1,772
Total depth		1,772

Log No. 448

Thad Ranson, No. 2, lessor. Ohio Fuel Oil & Gas Co., lessee. Lo_cation: Near Louisa. Commenced: February 27, 1920. Completed: March 31, 1920. Production: Well dry.

Strata.

Pennsylvanian System.	Thickness	Depth
Soil	20	20
Shale, blue	80	100
Sand, mountain	50	150
Shale, blue	315	465
Sandstone, (oil show 500)	85	550
Shale, blue	25	575
Sand (salt)	215	790
Shale, blue	15	805
Mississippian System.		
Sandstone (Maxon)	70	875
Shale, blue	20	895
Limestone (Little Lime)	20	915
Shale, blue	10	925
Limestone (Big Lime)	145	1,070
Shale, blue	15	1,085
Sandstone (Big Injun)	65	1,150
Shale, blue	5	1,155
Limestone	34	1,189
Shale and shells	441	1,630
Shale, brown (Sunbury)	22½	1,652½
Sandstone (Berea)	55	1,707½
Shale, shelly, (dry)	11½	1,719
Total depth		1,719

Log No. 449

Thad Ranson, No. 3, lessor. Ohio Fuel Oil & Gas Co., lessee. Commenced: April 19, 1920. Completed: May 15, 1920. Shot May 15, 1920, 60 quarts. Production: 4 bbls. oil.

Strata.

Pennsylvanian System.	Thickness	Depth
Soil	16	16
Shale	89	105
Coal (water)	2	107
Shale	3	110

Pennsylvanian System.	Thickness	Depth
Shale, hard	35	145
Shale	10	155
Sandstone, yellow, (water)	70	225
Shale	75	300
Sandstone	62	362
Shale	45	407
Limestone	31	438
Shale	62	500
Sandstone	27	527
Shale and shells	233	760
Sand (salt), (water flooded)	172	932
Mississippian System.		
Shale and shells	28	960
Shale and shells	128	1,088
Shale (pencil cave)	3	1,091
Limestone (Big Lime)	165	1,256
Sandstone (Big Injun)	85	1,341
Shale and shells	452	1,793
Shale, brown (Sunbury)	21	1,814
Sandstone (Berea), (1st 12 feet pay oil)	28	1,842
Total depth		1,842

Log No. 450

J. N. Roberts, No. 1, lessor. Ohio Fuel Oil & Gas Co., lessee. Location: Busseyville District. Commenced: January 15, 1919. Completed: February 14, 1919. Shot February 19, 1919, 60 quarts. Production: 3 bbls. oil.

Strata.

Pennsylvanian System.	Thickness	Depth
Soil	20	20
Sandstone	30	50
Shale	70	120
Sandstone, yellow	50	170
Shale	50	220
Sandstone	60	280
Shale	70	350
Limestone	25	375
Shale	35	410
Sandstone, (gas 420)	68	478
Shale	52	530
Shale and shells	70	600

Pennsylvanian System.

	Thickness	Depth
Limestone	18	618
Shale and shells	112	730
Sandstone (salt), (gas 735) (water 750-810)	260	990

Mississippian System.

Sandstone (Maxon)	20	1,010
Limestone (Little Lime)	25	1,035
Shale (pencil cave)	5	1,040
Limestone (Big Lime)	140	1,180
Shale	5	1,185
Sandstone (Big Injun)	83	1,268
Shale	2	1,270
Sandstone, fine, hard	30	1,300
Shale and shells	417	1,717
Shale, brown (Sunbury)	26	1,743
Sandstone (Berea)	24½	1,767½
Total depth		1,767½

Log No. 451

J. N. Roberts, No. 2, lessor. Ohio Fuel Oil & Gas Co., lessee. Commenced: April 28, 1920. Completed: May 29, 1920. Shot May 31, 1920, 60 quarts. Production: 6 bbls. oil.

Strata.

Pennsylvanian System.

	Thickness	Depth
Soil	16	16
Shale and shells	134	150
Sandstone	75	225
Shale	125	350
Shale, hard	50	400
Shale and shells	380	780
Sandstone (salt)	320	1,100
Shale	30	1,130

Mississippian System.

Limestone (Little Lime)	5	1,135
Shale (pencil cave)	3	1,138
Limestone (Big Lime)	170	1,308
Sandstone (Big Injun)	62	1,370
Shale and shells	452	1,822
Shale, black (Sunbury)	20	1,842
Sandstone (Berea)	30½	1,872½
Total depth		1,872½

Log No. 452

H. B. Salters, No. 1, lessor. Ohio Fuel Oil & Gas Co., lessee. Location: Twin Branch District. Commenced: August 26, 1919. Completed: September 23, 1919. Shot September 24, 1919, 60 quarts. Production: Gas, 150,000 cubic feet.

Strata.

Pennsylvanian System.	Thickness	Depth
Soil	16	16
Shale	120	136
Coal	2	138
Shale	92	230
Sandstone	22	252
Shale	128	380
Sand and shale	230	610
Shale	46	656
Sand (salt)	239	895
Mississippian System.		
Sandstone (Maxon)	35	930
Shale (pencil cave)	15	945
Limestone (Big Lime)	175	1,120
Sandstone (Big Injun)	80	1,200
Shale and shells	477	1,677
Shale, brown (Sunbury)	20	1,697
Sandstone (Berea)	27	1,724
Total depth		1,724

Log No. 453

E. E. Shannon, No. 1, lessor. Ohio Fuel Oil & Gas Co., lessee. Location: Louisa District. Commenced: January 30, 1920. Completed: February 27, 1920. Shot February 28, 1920, 58 quarts. Production: 4 bbls. oil when pumped.

Strata.

Pennsylvanian System.	Thickness	Depth
Soil	20	20
Shale	80	100
Sandstone, yellow	45	145
Shale	105	250
Sandstone	75	325
Shale, hard	75	400
Shale (water)	10	410

Pennsylvanian System.   Thickness Depth

Sandstone (Cow Run)	30	40
Shale	28	468
Sandstone	12	480
Shale and shells	120	600
Shale, hard	65	665
Shale, (water)	10	675
Sandstone (salt)	175	850
Shale and shells	100	950

Mississippian System.

Sandstone (Maxon)	10	960
Shale (pencil cave)	5	965
Limestone (Big Lime)	160	1,125
Sandstone (Big Injun)	91	1,216
Shale and shells	476	1,692
Shale, brown (Sunbury)	22	1,714
Sandstone (Berea), (oil pay 1715-1730)	23	1,737
Total depth		1,737

Log No. 454

E. E. Shannon, No. 1, lessor. New Domain Oil & Gas Co., lessee. Location: Lower Louisa Precinct. Completed: June 2, 1920. Shot June 3, 1920, 60 quarts. Production: 2½ bbls. oil per day.

Strata.

Pennsylvanian System.   Thickness Depth

Clay	18	18
Sandstone	16	34
Shale, hard	50	84
Sandstone	40	124
Shale, hard	25	149
Sandstone	35	184
Shale, hard	140	324
Sandstone	60	384
Shale, hard	95	479
Sandstone	80	559
Shale, hard	125	684
Sand (salt)	65	749
Sandstone	25	774
Shale, hard	200	974

Mississippian System.	Thickness	Depth
Sandstone (Maxon)	20	994
Limestone (Big Lime)	20	1,014
Sandstone (Big Injun)	155	1,169
Shale, shelly and sandstone	552	1,721
Shale, brown (Sunbury)	20	1,741
Sandstone (Berea)	26	1,767
Total depth		1,767

Log No. 455

Martha Taylor, No. 2, lessor. Ohio Fuel Oil & Gas Co., lessee.
Commenced: October 8, 1918. Completed: October 28, 1918.   Shot
November 1, 1918, 60 quarts. Production: 2 bbls. oil daily.

Strata.

Pennsylvanian System.	Thickness	Depth
Soil	8	8
Sandstone	6	14
Shale and mud	46	60
Sandstone	20	80
Coal	2	82
Shale, black	23	105
Sandstone	43	148
Shale, soft	6	154
Sandstone	61	215
Shale and shells	20	235
Coal	3	238
Shale and shells	97	335
Sandstone	17	352
Coal	3	355
Shale	35	390
Sandstone	10	400
Shale	220	620
Shale, hard	20	640
Shale, white	10	650
Sandstone	66	716
Shale	8	724
Sandstone (salt)	116	840
Shale	6	846
Sandstone	14	860
Limestone, black	30	890
Shale	6	896

Mississippian System.   Thickness Depth

  Sand (Maxon) 29 925
  Shale, hard 25 950
  Shale, soft 15 965
  Limestone (Little Lime) 20 985
  Shale (Pencil Cave) 3 988
  Limestone (Big Lime) 162 1,150
  Sandstone (Big Injun) 62 1,212
  Sandstone, fine, hard 48 1,260
  Shale and shells 400 1,660
  Shale, brown (Sunbury) 28 1,688
  Sandstone (Berea) 24½ 1,712½

    Total depth 1,712½

Log No. 456

  T. W. Taylor, No. 5, lessor. Location: Lower Louisa Precinct. Completed: April 22, 1920. Production: The well was abandoned. Authority: The New Domain Oil & Gas Co.

    Strata.

Pennsylvanian System.   Thickness Depth

  Gravel 14 14
  Sandstone .... : 26 40
  Shale, hard 60 100
  Sandstone 80 180
  Shale, hard 200 380
  Sandstone 150 530
  Shale, hard 75 605
  Sand (salt), (salt water) 320 925
  Shale, hard 90 1,015

Mississippian System.

  Limestone (Big Lime) 155 1,170
  Sandstone (Big Injun) 45 1,215
  Shale, hard, and limestone 463 1,678
  Shale, brown (Sunbury) 20 1,698
  Sandstone (Berea) 30½ 1,728½

    Total depth 1,728½

Log No. 457

  John B. Thompson, No. 1, lessor. New Domain Oil & Gas Co., lessee. Shot Nov. 11, 1919, 80 quarts. Production: ¼ bbl. oil.

Strata.

Pennsylvanian System.	Thickness	Depth
Clay	20	20
Sandstone	175	195
Shale	330	525
Shale, white	90	615
Sandstone	200	815
Shale, hard	79	894

Mississippian System.		
Limestone (Big Lime)	145	1,039
Sandstone (Big Injun)	60	1,099
Shale and shell	387	1,486
Shale, brown (Sunbury)	20	1,506
Sandstone (Berea)	68	1,574
Total depth		1,574

Log No. 458

John B. Thompson, No. 2, lessor. New Domain Oil & Gas Co., lessee. Shot January 6, 1920, 70 quarts. Production: 1 bbl. oil.

Strata.

Pennsylvanian System.	Thickness	Depth
Clay	20	20
Sandstone	180	200
Shale	110	310
Shale	340	650
Sandstone	350	1,000
Shale	51	1,051

Mississippian System.		
Limestone (Big Lime)	135	1,186
Shale	20	1,206
Sandstone (Big Injun)	25	1,231
Shale and shells	417	1,648
Shale, brown (Sunbury)	20	1,668
Sandstone (Berea)	57½	1,725½
Total depth		1,725½

Log No. 459

John B. Thompson, No. 3, lessor. New Domain Oil & Gas Co., lessee. Location: Busseyville Precinct. Completed: July 19, 1920. Production: 2 or 3 bbls. oil.

Strata.

Pennsylvanian System.	Thickness	Depth
Gravel	14	14
Sandstone	150	164
Shale, hard	200	364
Sandstone	300	664
Sand (salt)	360	1,024
Shale, hard	52	1,076
Mississippian System.		
Limestone (Big Lime)	150	1,226
Sandstone (Big Injun)	35	1,261
Limestone and shells	125	1,386
Shale, hard	261	1,647
Shale, brown (Sunbury)	20	1,667
Sand (Berea)	66½	1,733½
Total depth		1,733½

Log No. 460

C. M. Waller, No. 1, lessor. Ohio Fuel Oil & Gas Co., lessee. Location: Near Potters. Commenced: November 2, 1918. Completed December 18, 1918. Shot December 21, 1918, 60 quarts. Production: 5 or 6 bbls. oil.

Strata.

Pennsylvanian System.	Thickness	Depth
Clay	14	14
Sandstone	18	32
Wood	48	80
Sandstone, yellow	70	150
Shale, (water)	25	175
Shale, hard	51	226
Shale	29	255
Sandstone	60	315
Shale	45	360
Sandstone	90	450
Shale and shells	50	500
Shale, hard	70	570

Pennsylvanian System.	Thickness	Depth
Shale	30	600
Sandstone, (oil show 610)	30	630
Shale and shells	100	730
Shale	20	750
Sandstone (salt), (water 780)	80	830
Shale	5	835
Shale, hard·	35	870·
Mississippian System.		
Shale, broken, and shells	60	930
Sandstone (Maxon)	20	950·
Shale (pencil cave)	5	955
Limestone (Big Lime)	160	1,115
Sandstone (Big Injun)	85	1,200·
Shale	10	1,210
Sand, hard, fine	30	1,240
Shale and shells	444	1,684
Shale, brown (Sunbury)	25	1,709
Sandstone (Berea), (gas and oil)	27½	1,736½
Total depth		1,736½

Log No. 461

C. M. Waller, No. 2, lessor. Ohio Fuel Oil & Gas Co., lessee. Commenced: May 5, 1919. Completed: July 7, 1919. Shot July 8, 1919, 60 quarts. Production: 3 bbls. oil.

Strata.

Pennsylvanian System.	Thickness	Depth
Soil	16	16
Sandstone	1,49	165
Shale	50	215
Sandstone	75	290
Shale	10	300
Sandstone	20	320
Coal	3	323
Sandstone, limy	77	400·
Shale	80	480
Sandstone	75	555·
Shale	55	610
Sandstone	20	630
Shale and shells	160·	790
Sandstone	19	809
Shale	60	869·

Pennsylvanian System.

	Thickness	Depth
Shale, hard, and sand	115	984
Sandstone	95	1,079
Shale	80	1,159

Mississippian System.

	Thickness	Depth
Sand (Maxon)	25	1,184
Shale	50	1,234
Shale (Pencil Cave)	3	1,237
Limestone (Big Lime)	152	1,389
Sandstone (Big Injun)	90	1,479
Shale	26	1,505
Sand	40	1,545
Shale and shells	356	1,901
Shale, brown (Sunbury)	24	1,925
Sandstone (Berea), (pay oil 1926-1941)	25	1,950
Total depth		1,950

Log No. 462

Laura Webb, No. 1, lessor. Vanora Oil & Gas Co., Huntington, W. Va., lessee. Commenced: January 25, 1921. Completed: February 26, 1912.

Strata.

Pennsylvanian System.

	Thickness	Depth
Gravel, brown	30	30
Shale, white	10	40
Coal, black	3	43
Shale, black	17	60
Sandstone, white	20	80
Shale, white	15	95
Sandstone, white	25	120
Shale, black	180	300
Sandstone, white	25	325
Shale, brown	50	375
Shale, white	75	450
Shale, black	30	480
Sandstone, white	405	885

Mississippian System.

	Thickness	Depth
Limestone (Big Lime), white	130	1,015
Sandstone (Big Injun), white	10	1,025
Shale and shells	453	1,478
Shale, gray	21	1,499

Mississippian System.	Thickness	Depth
Sand	35	1,534
Shale, black (Sunbury)	3	1,537
Sandstone (Berea), white	21	1,558
Shale, black	26	1,584 ·
Total depth		1,584

NOTE—This record is irregular in the last 26 feet. Black shale does not occur as a parting in the Berea sandstone.

Log No. 463

•

F. H. Yates, No. 1, lessor. Ohio Fuel Oil & Gas Co., lessee. Location: Near Louisa. Commenced: April 30, 1912. Completed: June 4, 1912. Production: 3 bbls. oil. Authority: Wayne Oil Co.

Strata.

Pennsylvanian System.	Thickness	Depth
Sandstone, gray	20	20
Shale	40	60
Sandstone, (2 bailers at 65)	72	132
Shale	8	140
Sandstone	70	210
Shale	120	330
Sandstone, (show oil)	30	360
Shale and sand	100	460
Sandstone	10	470
Shale	90	560
Sandstone	40	600
Shale	47	647
Sand (salt), (water flood 705)	163	810
Shale	5	815
Shale and sandstone	145	960

Mississippian System.	Thickness	Depth
Shale	10	970
Limestone (Little Lime)	20	990
Shale (Pencil Cave)	5	995
Limestone (Big Lime)	165	1,160
Shale	5	1,165
Sandstone (Big Injun)	60	1,225
Shale and shells	480	1,705
Sandstone (Berea), (pay oil)	48	1,753
Total depth		1,753

Abbreviated Logs and "Sand" Records of Lawrence County.

Log No. 464

John D. Adkins, No. 1, lesor. Big Blaine Oil & Gas Co., lessee. Shot 60 quarts. Well abandoned.

Top of Berea sand	1,564
Pay sand	12
Bottom hole	1,621

Log No. 465

H. C. Austin, No. 1, lessor. Big Blaine Oil & Gas Co., lessee. Shot 65 quarts.

Top of Berea sand	1,825
Pay	1,825-1,860
Total depth	1,868

Log No. 466

H. C. Austin, No. 2, lessor. Big Blaine Oil & Gas Co., lessee. Shot 70 quarts.

Top of Berea sand	1,847½
Pay sand	1,847½-1,852
Break, 3 feet	
Total depth	1,878½

Log No. 467

Tom Ball, lessor. Frank Yates, lessee. Location: Mattie. Production: Gas and oil.

Top of limestone (Big Lime)	540
Top of sandstone (Berea)	1,090

Log No. 468

F. R. Bussey, No. 1, lessor. New Domain Oil & Gas Co., lessee. Commenced: June 5, 1917.

	Thicknes	Depth
Limestone (Big Lime)	147	971
Sandstone (Big Injun) and (Squaw)	50	1,021
Shale, brown (Sunbury)	25	1,475
Sandstone (Berea)		1,475
Total depth		1,475

Log No. 469

F. R. Bussey, No. 3, lessor. New Domain Oil & Gas Co., lessee.
Shot 60 quarts.

Top of limestone (Big Lime)	815
Sandstone (Berea)	1,450-1,516

Log No. 470

Hester Carter, No 1, lessor. Ohio Fuel Oil & Gas Co., lessee. Location: Near Fallsburg. Commenced: December 27, 1915. Completed: January 27, 1916. Production: ½ bbl. oil, 150,000 cubic feet gas. Shot January 29, 1916, 120 quarts, 1 bbl. oil after shot.

Strata.

	Thickness	Depth
Pennsylvanian System.		
Sandstone (salt)	258	858
Mississippian System.		
Limestone (Big Lime)	183	1,083
Sandstone (Big Injun)	40	1,140
Sandstone (Berea)	40	1,694
Total depth		1,694

Log No. 471

Hester Carter, No. 2, lessor. Ohio Fuel Oil & Gas Co., lessee. Commenced: December 22, 1916. Completed: January 29, 1917. Shot January 31, 1917, 110 quarts. Production: 5 bbls. oil per day.

	Thickness	Depth
Salt, sand and water	...	...
Big Lime	125	1,230
Berea sand	6	1,835
Shale	6	1,842
Sand	22	1,865
Gas sand, 1,843		
Total depth		1,865

Log No. 472

J. W. Carter, No. 1, lessor. Ophir Oil Co., lessee. Location: Near Fullers, on Buck Branch, and Big Blaine Creek.

Strata.

	Thickness	Depth
Pennsylvanian System.		
Sandstone, (fresh water)	600	600
Sandstone (salt)	110	710

Mississippian System.   Thickness Depth

	Thickness	Depth
Sandstone (Maxton), (½ million cubic feet gas)	140	850
Limestone (Big Lime)	150	1,000
Sandstone (Big Injun)	106	1,106
Sandstone (Berea)	449	1,555
Total depth		1,555

14 feet in.

Log No. 473

Joseph Carter, No. 1, lessor. Ohio Fuel Oil & Gas Co., lessee. Commenced: July 1, 1914. Completed: July 22, 1914. Shot July 30, 1914, 120 quarts. Production: 1½ bbls. oil per day.

Strata.

Pennsylvanian System.   Thickness Depth

	Thickness	Depth
Sandstone (cow run), (show oil)	25	495
Sandstone (salt)	220	890

Mississippian System.

	Thickness	Depth
Limestone (Big Lime)	165	1,115
Sandstone (Big Injun)	90	1,205
Sandstone (Berea)	46½	1,707
Total depth		1,707

Log No. 474

John R. Chapman, No. 1, lessor. Dameron Oil. Co., lessee. Location: On Lick Creek. Commenced: May 3, 1910. Production: 3 bbls. oil.

Strata.

Pennsylvanian System.   Thickness Depth

	Thickness	Depth
Soil	16	16
Sandstone (salt)	928	944

Mississippian System.

	Thickness	Depth
Limestone (Big Lime)	164	1,092
Shale	557	1,649
Sand, coffee	42	1,691
Sandstone (Berea), (oil)	13	1,704
Total depth		1,704

Pay 1651-1670.

Log No. 475

James L. Clark, No. 1, lessor. Ohio Fuel Oil & Gas Co., lessee. Location: Near Busseyville. Commenced: July 28, 1915. Completed: August 20, 1915. Shot August 23, 1915. Production: Gas, 200,000 cubic feet.

Strata.

Pennsylvanian System.	Thickness	Depth
Coal	5	270
Sandstone (salt)	165	890
Mississippian System.		
Limestone (Big Lime)	155	1,185
Sandstone (Big Injun)	108	1,293
Sandstone (Berea)	52	1,805
Total depth		1,805

Log No. 476

A. Collinsworth, No. 1, lessor. Ohio Fuel Oil & Gas Co., lessee. Location: Near Louisa. Commenced: May 8, 1915. Completed: June 22, 1915. Shot June 28, 1915, 140 quarts. Production: 2 bbls. oil per day.

· Strata.

Pennsylvanian System.	Thickness	Depth
Sandstone (salt)	207	1,000
Mississippian System.		
Limestone (Big Lime)	165	1,285
Sandstone (Big Injun)	45	1,330
Sandstone (Berea)	49	1,893
Total depth		1,893

Log No. 477

A. Collinsworth, No. 2, lessor. Ohio Fuel Oil & Gas Co., lessee. Commenced: June 28, 1915. Completed: July 22, 1915. Shot July 24, 1915, 175 quarts. Production: ½ bbl. oil per day.

Strata.

Pennsylvanian System.	Thickness	Depth
Coal	2	132
Coal	4	234
Sandstone (salt)	180	750
Mississippian System.		
Limestone (Big Lime)	130	1,198
Sandstone (Berea)	55	1,712
Total depth		1,712

Log No. 478

A. Collinsworth, No. 3, lessor. Ohio Fuel Oil & Gas Co., lessee. Commenced: August 27, 1915. Completed: November 16, 1915. Shot November 22, 1915, 115 quarts, 3 bbls. after shot. Production: 4 bbls. oil per day.

Strata.

Pennsylvanian System.	Thickness	Depth
Coal	3	198
Coal	2	242
Sandstone (salt)	150	850
Mississippian System.		
Limestone (Big Lime)	140	1,245
Sandstone (Big Injun)	25	1,300
Sandstone (Berea)	48	1,808
Total depth		1,808

Log No. 479

Malinda Dameron, No. 1, lessor. New Domain Oil & Gas Co., lessee. Completed: March 8, 1912. Shot 90 quarts.

Red Rock (Mauch Chunk)		835
Limestone (Big Lime)		885
Sandstone (Berea)		1,492 - 1,522
Total depth		1,522

Log No. 480

Aleck. Dial, No. 1, lessor. Location: Upper Laurel. Commenced: August 20, 1919. Completed: September, 1919. Shot 60 quarts. Production: 8 bbls. oil.

Top of Berea sand	740
Feet of sand	14
Total depth	764

Log No. 481

D. W. Diamond, No. 1, lessor. Ohio Fuel Oil & Gas Co., lessee. Location: Near Busseyville. Commenced: February 23, 1915. Completed: March 19, 1915. Shot March 19, 1915, 115 quarts. Production: 3 bbls. oil per day.

Strata.

Pennsylvanian System.	Thickness	Depth
Sandstone (salt)	160	900

Mississippian System.

Limestone (Big Lime)	175	1,235
Sandstone (Big Injun)	90	1,350
Shale	35	1,385
Sandstone (Berea)	56	1,834
Total depth		1,834

Log No. 482

Minerva Diamond, No. 1, lessor. Ohio Fuel Oil & Gas Co., lessee. Location: Near Busseyville. Commenced: October 1, 1915. Completed: October 27, 1915. Shot November 5, 1915, 90 quarts. Production: 3 bbls. oil.

Strata.

Pennsylvanian System.	Thickness	Depth
Coal	5	135
Coal	5	270
Sandstone (salt)	165	860

Mississippian System.		
Limestone (Big Lime)	150	1,150
Sandstone (Big Injun)	117	1,267
Sandstone (Berea)	52½	1,757½
Total depth		1,757½

Log No. 483

J. J. Gambill, No. 1, lessor. Union Oil & Gas Co., Mr. A. B. Ayres, Pres., Indianapolis, Ind., lessee. Location: Spring Branch. Commenced: Nov., 1917. Completed: Feb., 1918. Not shot. Production: 18 bbls. oil per day.

Top of (Berea) sand	650
Feet of sand	40
Total depth of well	690

Log No. 484

J. J. Gambill, No. 2, lessor. Location: Spring Branch. Commenced: Oct., 1919. Completed: Nov., 1919. Well shot 60 qts., pumping water July 5, 1920. Production: 35 bbls. oil.

Top of (Berea) sand	711
Feet of sand	40
Total depth	758

Log No. 485

J. J. Gambill, No. 3, lessor. Location: Spring Branch. Commenced: March 31, 1920. Completed: June 12, 1920. Production: 20 bbls. oil.

Top of Berea sand	982
Feet of sand	32
Total depth	1,022

Log No. 486

Lafe Hayes, No. 1, lessor. Cambrian Oil Co., lessee. Location: near Charles P. O. Drilled 1917. Shot and pumped. Production: Small oil and gas.

Top of Big Lime	600
Sandstone (Berea)	1,911-1,271

Log No. 487

John Hayes, No. 1, lesor. Cumberland Petroleum Co., lessee.

Top of Limestone (Big Lime)	847
Top of Sandstone (Berea)	1,447

Log No. 488

John C. Holbrook, No. 1, lessor. Location: Blaine Creek. Commenced: Jan., 1920. Completed: Feb. 7, 1920. Not shot. Production: 750,000 cu. ft. gas.

Top of Berea sand	714
Feet of sand	40
Total depth	754

Log No. 489

D. C. Hughes, No. 1, lessor. Ohio Fuel Oil & Gas Co., lessee. Location: Busseyville Precinct. Purchased by lessee from Wayne Oil Co. Completed: May 19, 1913. Shot Oct. 18, 1917, 60 qts. Production: 2 bbls. oil per day.

Strata	Thickness	Depth
Pennsylvanian System.		
Sandstone (salt)	430	975
Mississippian System.		
Limestone (Big Lime)	165	1,160
Sandstone (Berea)	60½	1,718½
Total depth		1,719½

Hole full of water 660.
Break, 1,670-1,688.

Log No. 490

L. N. Hutchison, No. 1, lessor. Ohio Fuel Oil & Gas Co., lessee. Location: 2 miles northeast of Yatesville Purchased by lessee from Wayne Oil Co. Completed: Feb. 4, 1914. Production: 3 bbls. oil per day.

Strata	Thickness	Depth
Pennsylvanian System.		
Sandstone (salt)	80	720
Mississippian System.		
Limestone (Big Lime)	175	1,095
Sandstone (Big Injun)	85	1,180
Sandstone (Berea)	50	1,651
Total depth		1,651½

Hole full of water, 665.
Break in Injun, 1,145-1,155.
Oil, 1,601-1,621.
Show salt water in bottom of Berea.

Log No. 491

L. N. Hutchison No. 2, lessor. Ohio Fuel Oil & Gas Co., lessee. Commenced: Feb. 14, 1917. Completed: March 14, 1917. Shot: March 19, 1917, 100 qts. Production: 3 bbls. oil per day.

Strata	Thickness	Depth
Mississippian System.		
Limestone (Big Lime)	125	1,025
Sandstone (Big Injun)	45	1,170
Sandstone (Berea)	37	1,647
Total depth		1,647

Log No. 492

Kane, No. 1, lessor. Big Blaine Oil & Gas Co., lessee. Shot: 70 qts. Production: 2 bbls. oil.

Top of Berea sand	1,635
Pay sand	1,635-1,655
Gas pay	1,663
Total depth	1,675

Log No. 493

Mary Kelley, No. 1, lessor. Big Blaine Oil & Gas Co., lessee: Shot: 60 qts.

Top of Berea sand	1,872½
Pay sand	1,872½-1,878
Two ft. break	
Total depth	1,895

Log No. 494

Roscoe C. Miller, No. 1, lessor. Location: Blaine Creek. Commenced: Feb. 14, 1920. Completed: Feb. 26, 1920. Shot: 100 qts. Production: 8 bbls. oil.

Top of Berea sand	689
Feet of sand	41
Total depth	730

Log No. 495

Roscoe C. Miller, No. 2, lessor. Location: Blaine Creek. Commenced: March 22, 1920. Completed: April 9, 1920. Shot 100 qts. Production: 8 bbls. oil.

Top of Berea sand	688
Feet of sand	28
Total depth	716

Log No. 496

John Moore, No. 1, lessor. Location: Tarkin Branch. Commenced: June, 1918. Completed: June, 1919. Not shot. Production: 60,000 cu. ft. gas.

Top of Berea sand	900
Feet of sand	6
Total depth	1,860

Log No. 497

L. B. Mullen, No. 1, lessor. Kentucky & Oklahoma Oil Co., lessee. Location: On Brushy Creek, near Cordell.

Limestone (Big Lime)	403-563
Sandstone (Berea), (oil and gas show)	952-1,057

Log No. 498

W. D. O'Neal, No. 1, lessor. Venora Oil & Gas Co., Huntington, W. Va., lessee. Location: Busseyville. Commenced: Nov. 8, 1911. Completed: Nov. 28, 1911. Production: One million ft. gas in Big Injun exhausted in 1,085.

Strata

	Thickness	Depth
Pennsylvanian System.		
Sandstone, (gas)	35	515
Sandstone (salt), (water flood 540)	385	900
Mississippian System.		
Limestone (Little Lime)	15	915
Shale (pencil cave)	10	925
Limestone (Big Lime)	160	1,085
Sandstone (Big Injun)	25	1,110
Sandstone (Berea)	23	1,580
Total depth		1,580

Log No. 499

W. B. Pfost, No. 1, lessor. Ohio Fuel Oil & Gas Co., lessee. Location: Louisa District. Commenced: April 15, 1915. Completed: May 8, 1915. Shot May 8, 1915, 120 qts. Production: 3 or 4 bbls. daily.

	Strata	Thickness	Depth
Pennsylvanian System.			
Sandstone (salt)		180	940
Mississippian System.			
Limestone (Big Lime)		160	1,260
Sandstone (Big Injun)		75	1,335
Shale		35	1,370
Shale, brown (Sunbury)		415	1,785
Shale, hard		36	1,821
Sandstone (Berea)		49	1,870
Total depth			1,870

Log No. 500

Harry Phillips, No. 1, lessor. Location: Upper Laurel. Commenced: July, 1919. Completed: Aug. 1, 1919. Not shot. Production: gas, 100,000 cu. ft.

Top of Berea sand	837
Feet of sand	10
Total depth	936

Log No. 501

C. A. Rice, No. 1, lessor. Location: Blaine Creek. Commenced: July 8, 1918. Completed: Aug. 11, 1918. Shot 20 qts. Production: oil; not pumping now.

Top of Berea sand	814
Feet of sand	19
Total depth	833

Log No. 502

Savage, No. 4, lessor. Big Blaine Oil & Gas Co., lessee. Shot 70 qts.

Top of Berea sand	1,588
Pay	1,588-1,600
Gas sand	1,618-1,623
Total depth	1,629

Log No. 503

.Savage, No. 5, lessor. Big Blaine Oil & Gas Co., lessee. Shot 75 qts.

Top of Berea sand	1,605½
Bottom hole	1,643
Total depth	1,643

Log No. 504

Savage, No. 6, lessor. Big Blaine Oil & Gas Co., lessee. Shot 70 qts.

Top of Berea sand	1,819½
Bottom sand	1,857
Total depth	1,857

Log No. 505

Savage, No. 7, lessor. Big Blaine Oil & Gas Co., lessee. Shot 70 qts.

Top of Berea sand	1,608
Bottom hole	1,647
Total depth	1,647

Log No. 506

Savage, No. 8, lessor. Big Blaine Oil & Gas Co., lessee. Shot 70 qts.

Top of Berea sand	1,573
Pay sand	1,573-1,585
Total depth	1,603

Log No. 507

Savage No. 9, lessor. Big Blaine Oil & Gas Co., lessee. Shot 70 qts.

Top of Berea sand	1,598
Pay sand	1,598-1,628
Total depth	1,628

Log No. 508

D. W. Skaggs, No. 1, lessor. Location: Blaine Creek. Commenced: May, 1918. Completed: June, 1918. Shot 60 qts. Production: oil; not pumping now.

Top of Berea sand	800
Feet of sand	40
Total depth of well	840

Log No. 509

D. W. Skaggs, No. 2, lessor. Location: Blaine Creek. Commenced: March, 1919. Completed: April 3, 1919. Shot: 60 qts. Production: 20 bbls. oil

Top of Berea sand	762
Feet of sand	36
Total depth	808

Log No. 510

D. W. Skaggs, No. 3, lessor. Location: Blaine Creek. Commenced: May, 1919. Completed: June, 1919. Shot: 60 qts. Production: 24 bbls. oil.

Top of Berea sand	748
' Feet of sand	30
Total depth	790

Log No. 511

D. W. Skaggs, No. 4, lessor. Location: Blaine Creek. Commenced: Oct., 1919. Completed: Dec. 13, 1919. Shot 60 qts. Production: 24 bbls. oil.

Top of sand	1,019
Feet of sand	34
Total depth	1,053

Log No. 512

Daniel Skaggs, No. 2, lessor. Location: Blaine Creek. Commenced: Dec. 28, 1919. Completed: Feb. 25, 1920. Shot 80 qts. Production: 20 bbls. oil.

Top of Berea sand	948
Feet of sand	35
Total depth	983

Log No. 513

M. L. Skaggs, No. 1, lessor. Location: Barn Rock Branch. Commenced: Feb. 24, 1920. Completed: March 24, 1920. Not shot. Production: Oil; not pumping.

Top of Berea sand	625
Feet of sand	40
Total depth	665

Log No. 514

Oscar Skaggs, No. 1, lessor. Location: Big Lick Branch. Commenced: April, 1918. Completed: May, 1918. Shot 60 qts. Production: 12 bbls. oil.

Top of Berea sand	730
Feet of sand	25
Total depth	755

Log No. 515

Lafayette Wellman, No. 1, lessor. Ohio Fuel Oil & Gas Co., lessee. Purchased by lessee from Wayne Oil Co., June 1, 1914. Location: Busseyville District. Completed: July 18, 1913. Production: 1 bbl. oil per day.

Strata	Thickness	Depth
Pennsylvanian System.		
Sandstone (salt)	308	938
Mississippian System.		
Limestone (Big Lime)	175	1,155
Sandstone (Berea)	52	1,707
Total depth		1,707

Water at 665.
Oil 1,647-1,667.
Oil 1,680-1,700.

Log No. 516

John Yates, No. 1, lessor. Big Blaine Oil & Gas Co., lessee. Shot 70 qts.

Top of Berea sand		1,553
Pay		1,553-1,568
Shale		1,570
Sand	22	1,592
Total depth		1,592

Log No. 517

John Yates, No. 2, lessor. Big Blaine Oil & Gas Co., lessee. Shot 70 qts.

Top of Berea sand		$1,602\frac{1}{2}$
Pay		$1,602\frac{1}{2}$-1,615
Sandstone	28	1,643
Total depth		1,643

CHAPTER VI.

LEE COUNTY.

Production: Oil and Gas. Producing Sands: Corniferous (Devonian).
Niagaran (Silurian).

Log No. 518

G. G. Adams, No. 1, lessor. Irvine Development Co., lessee. Location: Hell Creek section. Commenced: May 22, 1918. Completed: June 28, 1918. Production: 25 bbls. oil per day. Authority: Irvine Development Co.

Strata.

	Thickness	Depth
Pennsylvanian System.		
Soil	10	10
Sandstone and shale	320	330
Mississippian System.		
Limestone	145	475
Shale, gray	508	983
Devonian System.		
Shale, brown	152	1,135
Shale (fire clay)	12	1,147
Shale, hard and black	8	1,155
Limestone "sand"	10	1,165
Total depth		1,165

Drilled 3 inches into salt water. Well filled 325 feet while being drilled. Pumped off at 25 barrels in 4 hours.

Log No. 519

G. G. Adams, No. 2, lessor. Location: Hell Creek section. Commenced: July 24, 1918. Completed: August 9, 1918. Authority: Irvine Development Co.

Strata.

	Thickness	Depth
Pennsylvanian System.		
Soil	15	15
Sandstone and shale	355	370
Mississippian System.		
Limestone	145	515
Shale, gray	495	1,010
Devonian System.		
Shale, brown	148	1,158
Shale (fire clay)	14	1,172
Shale, hard, black	10	1,182
Limestone "sand," (oil)	8	1,190
Total depth		1,190

Log No. 520

G. G. Adams, No. 3, lessor. Irvine Development Co., lessee. Location: Hell Creek section. Production: Salt water was pumped from the well for 2 days; then 7 bbls. oil, then salt water again. Authority: Irvine Development Co.

Strata.

Pennsylvanian System.	Thickness	Depth
Soil	35	35
Sandstone and shale	345	380
Mississippian & Devonian Systems.		
Limestone	170	550
Shale, brown, and other strata	665	1,215
Sand, (oil)	6	1,221
Total depth		1,221

The sand was very hard and fine.

Log No. 521

G. G. Adams, No. 4, lessor. Irvine Development Co., lessee. Location: Hell Creek section. Commenced: March 7, 1918. Completed: March 7, 1918. Shot 30 qts. between 1,200 and 1,205 ft. Nov. 17, 1918. Authority: Irvine Development Co.

Strata.

Pennsylvanian System.	Thickness	Depth
Soil ....'................	30	30
Sandstone and shale	360	390
Mississippian & Devonian Systems.		
Limestone	135	525
Shale, green and brown	600	1,125
Fire clay and shale, brown	34	1,159
Shale, hard, black	40	1,199
Limestone "sand," (oil)	7	1,206
Total depth		1,206

Showing for a 15 barrel well. A large amount of gas with heavy pressure.

Log No. 522

G. G. Adams, No. 6, lessor. Atlantic Oil Producing Co., lessee.
Location: Hell Creek Section. Completed: Sept. 9, 1918. Shot: 20
qts. Sept. 11, 1919, between 1,137 and 1,142 ft. Production: light
oil show oil. Authority: Atlantic Oil Producing Co.

Strata.

Pennsylvanian System.	Thickness	Depth
Soil	5	5
Sandstone (mountain)	110	115
Shale, hard	25	140
Mississippian & Devonian Systems.		
Shale, shelly	10	150
Shale, hard, and shell	100	250
Limestone (Little Lime)	50	300
Shale	25	325
Limestone (Big Lime)	115	440
Shale, green and brown (lower part Chattanooga)	660	1,100
Shale (fire clay)	20	1,120
Shale, hard, black	17	1,137
Limestone ''sand,'' hard	9½	1,146½
Total depth		1,146½

Casing pulled, and well plugged and abandoned.

Log No. 523

G. G. Adams, No. 7, lessor. Irvine Development Co, lessee. Location: Hell Creek section. Commenced: Aug. 29, 1919. Completed:
Sept. 18, 1919. Shot: 20 qts. Sept. 19, 1919, between 1,212 and
1,216 feet. Pumped production after shot, 10 bbls per day. Authority:
Atlantic Oil Producing Co.

Strata.

Pennsylvanian System.	Thickness	Depth
Soil	65	65
Sandstone (mountain)	105	170
Shale, shelly	205	375
Mississippian System.		
Limestone (Little Lime)	50	425
Shale, hard	5	430
Limestone (Big Lime)	95	525
Limestone, shelly, and shale, hard	475	1,000

Devonian System.	Thickness	Depth
Shale, brown (Chattanooga)	185	1,185
Shale (fire clay)	15	1,200
Shale, hard, black	10	1,210
Limestone "sand," (oil)	6½	1,216½
Total depth		1,216½

Log No. 524

G. G. Adams, No. 8, lessor. Irvine Development Co., lessee. Location: Hell Creek section. Commenced: Sept. 20, 1919. Completed: Oct. 15, 1919. Shot: 20 qts. Oct. 16, 1919, between 1,227 and 1,232 feet. Production: 20 bbls. oil after shot. Authority: Atlantic Oil Producing Co.

Strata.

Pennsylvanian System.	Thickness	Depth
Soil	50	50
Sandstone	10	60
Shale, hard	20	80
Sandstone, yellow	120	200
Sandstone, white	30	230
Shale, hard	10	240
Sandstone, (fresh water)	20	260
Shale, hard	75	335

Mississippian System.		
Limestone, sandy	45	380
Shale, hard	45	425
Limestone (Big Lime)	140	565
Shell and shale, hard	515	1,080

Devonian System.		
Shale, brown (Chattanooga)	125	1,205
Shale (fire clay)	11	1,216
Shale, hard, black	2	1,218
Limestone (pay), hard	2	1,220
Limestone (cap)	6¼	1,226¼
Limestone "sand,"	7-1/10	1,234⅓
Total depth		1,234⅓

Well showed strong for 4 inches into sand.

Log No. 526

G. G. Adams, No. 9, lessor. Irvine Development Co., lessee. Location: Hell Creek section. Shot 20 qts. Dec. 15, 1919, between 1,283 and 1,288 ft. Average daily production: 3 bbls. oil. Authority: Irvine Development Co.

Strata.

Pennsylvanian System.	Thickness	Depth
Soil	130	130
Sand, hard, gray	35	165
Shale, hard, black	10	175
Sand, white, soft	80	255
Shale, hard, black	70	325
Sand, white, hard	15	340
Shale, hard, black	20	360
Mississippian System.		
Shale, hard, blue	20	380
Limestone (Little Lime), hard	20	400
Shale, blue, soft	8	408
Limestone (Big Lime), white, hard	115	523
Shale, blue, soft	15	538
Shale and shells	457	995
Devonian System.		
Shale, brown (Chattanooga)	170	1,160
Shale (fire clay)	15	1,175
Shale, black	12	1,187
Sand, gray, medium	6	1,193
Total depth		1,193

There was a light show of oil and gas.

Log No. 527

G. G. Adams, No. 10, lessor. Irvine Development Co., lessee. Location: Hell Creek section. Commenced: Oct. 13, 1919. Completed: Nov. 7, 1919. Shot: 20 qts. Nov. 8, 1919, between 1,162 and 1,167 feet. Production: beginning Nov. 10, 1919, 30 bbls. per 24 hrs. Authority and Contractor: Atlantic Oil Producing Co.

Strata.

Pennsylvanian System.	Thickness	Depth
Sandstone (mountain), yellow, medium	137	137
Shale and sand	40	177

Mississippian System.   Thickness Depth

	Thickness	Depth
Shale, blue	88	265
Limestone, blue, sandy	20	285
Shale, blue	22	307
Limestone, blue, sandy	8	315
Shale, gray	6	321
Limestone, blue	87	408
Limestone, white	96	504
Shale, green	31	535
Shale, gray	442	977

Devonian System.

	Thickness	Depth
Shale, brown (Chattanooga)	158	1,135
Shale (fire clay)	15	1,150
Shale, black	10	1,160
Sand	10½	1,170½
Total depth		1,170½

There was a good showing of gas and oil.

Log No. 528

G. G. Adams, No. 11, lessor. Irvine Development Co., lessee. Location: Hell Creek Section. Commenced: November 1, 1919. Completed: November 12, 1919. Shot 30 quarts. Nov. 13, 1919, between 1211 and 1218 feet. Production: 15 bbls. per 24 hours. Authority: Atlantic Oil Producing Co.

Strata.

Pennsylvanian System.   Thickness Depth

	Thickness	Depth
Soil, yellow, soft	15	15
Shale, hard, black	40	55
Sandstone (mountain), gray, hard	170	225
Shale, gray, soft, (water)	85	310

Mississippian System.

	Thickness	Depth
Limestone (Little Lime), gray, hard	40	350
Sand, white, soft	30	380
Limestone (Big Lime)	55	435
Shale (break), hard	5	440
Limestone, gray	10	450
Sand, white	30	480
Limestone, white, hard	70	550
Shale, blue, hard	485	1,035
Limestone, shelly	5	1,040

Devonian System.	Thickness	Depth
Shale, brown (Chattanooga)	150	1,190
Shale (fire clay), white, soft	12	1,202
Shale, black	9	1,211
Sand	10½	1,221½
Total depth		1,221½

There was a good showing of oil and gas.

Log No. 529

G. G. Adams, No. 12, lessor. Irvine Development Co., lessee. Location: Fincastle Section. Commenced: November 25, 1919. Completed: December 23, 1919. Shot 30 quarts December 30, 1919. Production: Beginning January 2, 1920, 25 bbls. per 24 hours; 40 bbls. were pumped after the shot. Authority and Contractor: Atlantic Oil Producing Co.

Strata.

Pennsylvanian System.	Thickness	Depth
Soil	6	6
Sandstone (mountain)	114	120
Sand, red	55	175
Sand and shale, blue, (water)	96	271
Shale	24	295
Sand, white	12	307
Shale, blue, (water)	18	325
Sand, white	5	330
Mississippian System.		
Shale, blue	64	394
Limestone, blue (Big Lime)	98	492
Limestone, white (Big Lime)	26	518
Shale, green ....:..	430	948
Shale, gray	5	953
Pink rock	18	971
Devonian System.		
Shale, brown, hard (Chattanooga)	170	1,141
Shale (fire clay)	9	1,150
Shale, black	15⅓	1,165⅓
Sand	6	1,171⅓
Total depth		1,171⅓

The well filled up 300 feet.

Log No. 530

G. G. Adams, No. 13, lessor. Irvine Development Co., lessee. Location: Fincastle Section. Authority: Atlantic Oil Producing Co.

Strata.

Pennsylvanian System.	Thickness	Depth
Soil, yellow, soft	14	14
Shale, black, hard	126	140
Sandstone (mountain), white	125	265
Shale, blue, hard	50	315
Limestone, white	10	325
Shale, hard	15	340
Sand	25	365
Shale, hard	25	390
Sand, white, (water)	30	420
Mississippian System.		
Shale, hard	40	460
Limestone (Little Lime)	10	470
Shale, hard	5	475
Limestone (Big Lime)	25	500
Shale, blue, hard	290	790
Shale, gray, hard	155	945
Devonian System.		
Shale, brown (Chattanooga)	160	1,105
Shale (fire clay)	7	1,112
Shale, black	18	1,130
Limestone (cap), gray	3	1,133
Limestone "sand," (dry)	8	1,141
Total depth		1,141

Log No. 531

G. G. Adams, No. 14, lessor. Irvine Development Co., lessee. Location: Fincastle Section, southwest on Cliff. Commenced: January 23, 1920. Completed: March 2, 1920. Shot 30 quarts March 12, 1920, between 1175 and 1182 feet. Production: Oil, best well on lease. Authority and contractor: Atlantic Oil Producing Co.

Strata.

Pennsylvanian System.	Thickness	Depth
Soil	9	9
Sandstone (mountain)	127	136

Pennsylvanian System. Thickness Depth

 Sand, red 12 148
 Sand and shale 47 195
 Shale 85 280
 Sand, (water 290) 25 305

Mississippian System.

 Shale 30 335
 Limestone (Little Lime) 25 360
 Shale 10 370
 Shale (water 380) 42 412
 Limestone (Big Lime) 90 502
 Shale 5 507
 Limestone 5 512
 Shale 18 530
 Limestone 5 535
 Shale 447 982

Devonian System.

 Shale, brown (Chattanooga) 163 1,145
 Shale (fire clay) 15 1,160
 Shale, black 10 1,170
 Limestone "sand" 7 1,177
  Total depth 1,177

Log No. 532

G. G. Adams, No. 15, lessor. Irvine Development Co., lessee. Location: West side of center; Fincastle Section. Commenced: February 7, 1920. Completed: March 12, 1920. Shot 30 quarts March 19, 1920, between 1198 and 1207 feet. Production: 20 bbls. per 24 hours, after shot. Authority: Irvine Development Co.

  Strata.

Pennsylvanian System. Thickness Depth

 Soil 40 40
 Sand, (water) 130 170
 Shale, hard 110 280
 Shale, hard (water) 10 290
 Sand 20 310
 Shale, hard 25 335

Mississippian System.

 Limestone 15 350
 Shale, hard 10 360

Mississippian System.	Thickness	Depth
Limestone (Little Lime)	20	380
Shale, hard	10	390
Limestone (Big Lime)	145	535
Shale, hard	460	995

Devonian System.		
Shale, brown (Chattanooga)	179	1,174
Shale (fire clay)	12	1,186
Shale, black, hard	12	1,198
Sand	7	1,205
Total depth		1,205

There was a fair show of oil and a little gas.

Log No. 533

Frailey, No. 1, lessor. Atlantic Oil Producing Co., lessee. Location: Airdale Section. Commenced: June 14, 1919. Completed: July 19, 1919. Authority: Atlantic Oil Producing Co.

Strata.

Pennsylvanian System.	Thickness	Depth
Soil	20	20
Sandstone (mountain), (fresh water)	80	100
Shell and shale, hard	180	280

Mississippian System.		
Limestone (Little Lime)	35	315
Shale	5	320
Limestone (Big Lime)	120	440
Limestone, shell and shale, hard	475	915

Devonian System.		
Shale, brown (Chattanooga)	160	1,075
Shale (fire clay)	15	1,090
Shale, black, hard	13	1,103
Sand, (salt water), dry	12	1,115
Total depth		1,115

Casing was pulled and well abandoned. Casing record:

Length 17', 460'. Size 8¼", 6¼".

Log No. 534

Frailey, No. 2, lessor. Atlantic Oil Producing Co., lessee. Location: Hell Creek Section. Commenced: September 13, 1919. Completed: October 23, 1919. Authority: Atlantic Oil Producing Co.

Strata.

Pennsylvanian System.	Thickness	Depth
Soil	90	90
Shale, blue	30	120
Shale, blue, hard-pan	20	140
Shale, hard	15	155
Sandstone (mountain), white	69	224
Shale, blue, hard	20	244
Shale, hard, dark	72	316

Mississippian System.		
Limestone (Big Lime)	140	456
Shale, blue, hard	34	490
Shell and shale, hard	60	550
Shale, blue	389	939

Devonian System.		
Shale, brown (Chattanooga)	160	1,099
Shale (fire clay)	6	1,105
Shale, black, hard	12½	1,117½
Limestone "sand," (show of oil with salt water)	10	1,127½
Total depth		1,127½

Casing record:

Length, 256', 460'. Size, 8¼", 6¼".

The casing was pulled and the well plugged and abandoned.

Log No. 535

Dan Frailey, No. 2, lessor. Commenced: September 17, 1918. Authority: The Ohio Oil Co.

Strata.

Pennsylvanian System.	Thickness	Depth
Soil, brown, soft	20	20
Shale, shelly, brown, hard	20	40
Shale blue, soft	100	140

Mississippian System.		
Limestone (Big Lime), hard, white	135	275

Mississippian System.	Thickness	Depth
Shale, green, soft	15	290
Shale and shells, hard and blue	80	370
Shale, hard, blue, soft	330	700

Devonian System.		
Shale brown, soft	175	875
Shale (fire clay), gray, soft	25	900
Limestone (cap rock), hard, brown	27	927
Limestone ''sand,'' brown, soft	5	932
Limestone, hard, white	58	990
Total depth		990

Log No. 536

Taylor Gilbert, No. 1, lessor. Location: Southwest of and near Fincastle. Commenced: August 3, 1919. Completed: August 22, 1919. Production: 2 bbls. oil per day. Shot with 20 quarts August 25, 1919, between 1246 and 1254 feet. Authority: Empire Oil & Gas Co.

Strata.

Pennsylvanian System.	Thickness	Depth
Clay, yellow, soft	30	30
Shale, black, soft	30	60
Shell, sandy, hard	20	80
Shale, dark, soft	20	100
Sandstone (mountain), light, soft	85	185
Shale, black, hard	165	350
Sand, white, soft	35	385
Shale, black, soft	60	445

Mississippian System.		
Limestone (Big Lime), white, hard	120	565
Shale, sandy, green, hard	125	690
Shale and shells, blue, hard, soft	60	750
Shale, gray, soft	305	1,055

Devonian System.		
Shale, brown, soft	167	1,222
Shale (fire clay), white, soft	12	1,234
Shale, brown, soft	11	1,245
Limestone (cap), dark gray, hard	2	1,247
Limestone "sand," gray, hard	9	1,256
Total depth		1,256

Log No. 537

Hopewell, No. 5. (Shearer Tract.) Authority: W. E. Thompson.

Strata.

Pennsylvanian System.	Thickness	Depth
Soil	15	15
Sandstone (mountain)	95	110
Sandstone, (water)	15	125
Shale, soft, mud and lime	195	320

Mississippian System.		
Limestone (Big Lime), (6½ in. casing 520) ..	200	520
Shale, soft	427	947
Shale, red soft	8	955
Shale (fire clay)	10	965

Devonian System.		
Shale (Chattanooga)	135	1,100
Shale, soft, red	5	1,105
Shale, black	17	1,122
Fire clay (cap)	8	1,130
Limestone	38	1,168
Total depth		1,168

NOTE—While the top of the Mississippian System is placed just above the "Big Lime" in this and many succeeding records, it is done so simply because the driller did not differentiate the several separate formations immediately above. In this and similar cases it is altogether probable that the base of the Pottsville would come somewhat above the top of the "Big Lime."

Log No. 538

Kincaid, No. 1, lessor. Atlantic Oil Producing Co., lessee. Location: Hell Creek Section. Commenced: June 19, 1918. Completed: July 5, 1918. Production: Dry. Authority: Atlantic Oil Producing Co.

Strata.

Pennsylvanian System.	Thickness	Depth
Soil	10	10
Sandstone (mountain)	70	80
Shale	25	105
Sand and shale, hard	105	210

Pennsylvanian System.	Thickness	Depth
Sand	140	350
Shale, hard	25	375
Mississippian and Devonian Systems.		
Limestone (Little Lime)	5	380
Shale, hard	15	395
Limestone (Big Lime)	115	510
Shale, brown, hard	685½	1,195½
Limestone (cap)	2	1,197½
Limestone "sand"	5½	1,203
Limestone "sand," (show at 1211)	10	1,213
Limestone "sand" and lime, (water)	19	1,232
Total depth		1,232

Pulled, plugged below fresh water, and abandoned.

Log No. 539

Kincaid, No. 2, lessor. Atlantic Oil Producing Co., lessee. Location: Hell Creek Section. Commenced: June 29, 1918. Completed: September 30, 1918. Authority: Atlantic Oil Producing Co.

Strata.

Pennsylvanian System.	Thickness	Depth
Soil	12	12
Gravel	6	18
Sandstone (mountain)	57	75
Shale, hard	135	210
Mississippian System.		
Limestone (Big Lime), white	110	320
Shale, hard	20	340
Limestone	40	380
Limestone and shale, hard	200	580
Shale, hard •..............................	260	840
Devonian System.		
Shale, brown (Chattanooga)	140	980
Shale (fire clay)	14	994
Shale, black, hard	11	1,005
Limestone "sand," (oil show 1008)	6	1,011
Limestone "sand," (salt and water)	11½	1,022½
Limestone "sand," (salty)	24½	1,047

Devonian System.   Thickness Depth

    Limestone "sand," dark 20   1,067

    Limestone "sand," light 24   1,091

    Shale 20   1,111

        Total depth   1,111

Casing record: Length, 19', 4", 340'. Size 8¼", 6¼".

Log No. 540

Kincaid, No. 3, lessor. Atlantic Oil Producing Co., lessee. Location: Hell Creek Section. Commenced: February 13, 1919. Completed: March 19, 1919. Shot with 20 quarts February 19th, 1919. Production: 42 bbls. oil per 24 hours.

    Strata.

Pennsylvanian System.   Thickness Depth

    Clay and gravel 12   12

    Sandstone (mountain) 20   32

    Shale, black 40   72

    Shale and sand 26   98

    Shale, black 36   134

    Shale, hard 12   146

    Shale, sandy 35   181

    Shale, black, hard 122   303

    Sand, hard, white 31   334

Mississippian System.

    Limestone (Little Lime) 25   359

    Shale (fire clay)   7   366

    Limestone (Big Lime) 134   500

    Shale, hard 10   510

    Shale, green, hard, (cased 520' 2") 18   528

    Shale, gray 450   978

Devonian System.

    Shale brown, soft (Chattanooga) 150 1,128

    Shale (fire clay) 15 1,143

    Shale black, hard 15 1,158

    Limestone (cap), (pay 10" in cap) 2 1,160

    Limestone "sand," (pay) 2 1,162

    Pocket .... 3'9" 1,165'9"

        Total depth ....   1,165'9"

Casing record: Length, 17', 520', 2". Size, 6¼".

Log No. 541

D. B. Kincaid, No. 4, lessor.   Atlantic Oil Production Co., lessee.
Location: Hell Creek Section. Shot with 20 quarts May 6, 1919. Production, beginning May 8, 1919, 20 bbls. oil per 24 hours. Authority:
Atlantic Oil Production Co.

Strata.

Pennsylvanian System.	Thickness	Depth
Clay, gravel and sandstone	75	75
Shale, sandy	60	135
Shale and limestone	187	322
Mississippian System.		
Limestone (Big Lime)	132	454
Shale, green, hard	116	570
Shale	170	740
Shale, blue, hard	130	870
Shale black, hard	50	920
Devonian System.		
Shale, brown	111	1,031
Shale, hard and mixed	64	1,095
Shale (fire clay)	15	1,110
Shale black, hard	7	1,117
Limestone "sand"	6	1,123
Total depth		1,123

Casing record: Length 17', 451' 10", 1120'.   Size 8¼",
6¼" 2".

Log No. 542

D. B. Kincaid, No. 5, lessor.   Location: Airdale Section, at Squires
Branch. Commenced: June 4, 1919. Completed: July 12, 1919. Shot
with 20 quarts, July 14, 1919. Production: After shot, 9 bbls. pumped
Authority: Atlantic Oil Producing Co.

Strata.

Pennsylvanian System.	Thickness	Depth
Soil, yellow	10	10
Shale and sand	75	85
Shale, hard	30	115
Sand	25	140
Sand	110	250

Mississippian System.

	Thickness	Depth
Limestone (Big Lime)	195	445
Shale, green	12	457
Shale, gray	47	504
Limestone and shale	60	564
Shale	56	620
Limestone, dark	10	630
Shale, hard	48	678
Shale	237	915
Shale (red rock)	5	920
Shale, hard	20	940

Devonian System.

Shale, brown (Chattanooga)	164	1,104
Shale (fire clay)	7	1,111
Shale, black, hard	4	1,115
Sand	11	1,126
Total depth		1,126

Casing record: Length 462', 18'. Size 6¼", 8¼".

Log No. 543

D. B. Kincaid, No. 6, lessor. Atlantic Oil Producing Co., lessee. Location: Hell Creek Section. Commenced: August 2, 1919. Completed: September 1, 1919. Shot with 20 quarts, September 2, 1919. Production, beginning September 2, 1919, 2 bbls. oil per 24 hours. Average daily production after 7 days 10 bbls. Authority: Atlantic Oil Producing Co.

Strata.

Pennsylvanian System.

	Thickness	Depth
Soil	11	11
Sandstone (mountain)	94	105
Coal, (good vein)	5	110
Sandstone	40	150
Shale	75	225
Sandstone	72	297
Shale	45	342

Mississippian System.

Limestone (Big Lime)	177	519
Shale, green	51	570
Shale, gray	447	1,017

Devonian System.   Thickness Depth
    Shale, brown   142   1,159
    Shale (fire clay)   10   1,169
    Shale, black hard ....   14'8"   1,183'8"
    Limestone "sand," (oil and gas) ....   7'8"   1,191'4"
       Total depth ....   1,191'4"

    Casing record: Length 21', 525'. Size 8¼", 6¼".

Log No. 544

D. B. Kincaid, No. 7, lesor. Atlantic Oil Producing Co., lessee.
Location: Hell Creek Section. Commenced: September 27, 1919. Completed: October 13, 1919. Production: Dry; plugged with lead plug and abandoned. Authority: Atlantic Oil Producing Co.

    Strata.
Pennsylvanian System.   Thickness Depth
    Sandstone   20   20
    Shale, hard   200   220
    Shale, soft and hard ....   30`   250
    Shale, hard, shelly   45   295
    Shale, hard   20   315
Mississippian and Devonian Systems.
    Limestone (Big Lime)   150   465
    Shell and shale, hard   658   1,123
    Limestone (cap)   2   1,125
    Limestone "sand," hard (salt water 1123)   10   1,135
       Total depth   1,135

    Casing record: Length 11', 470'. Size, 8¼", 6¼".

Log No. 545

D. B. Kincaid, No. 8, lessor. Atlantic Oil Producing Co., lessee.
Location: Airdale Section. Commenced: October 31, 1919. Completed: November 26, 1919. Authority: Atlantic Oil Producing Co.

    Strata.
Pennsylvanian System.   Thickness Depth
    Soil   5   5
    Shale, black   45   50
    Shale, broken, white   50   100

Pennsylvanian System.

	Thickness	Depth
Sandstone (mountain)	85	185
Sand and shell, (fresh water)	40	225
Shale, black, hard	25	250

Mississippian System.

Limestone, white	25	275
Shale, black, hard	5	280
Limestone (Big Lime)	130	410
Shale, hard	5	415
Shale, hard, green	15	430
Shells	80	510
Shale, hard	40	550
Shale, hard	350	900

Devonian System.

Shale, brown	160	1,060
Shale (fire clay)	10	1,070
Shale, black, hard	5 1/2	1,075 1/2
Limestone "sand," (dry)	8 1/2	1,084
Total depth		1,084

Casing record: Length, 430', 11'. Size, 6 1/4", 8 1/4. Pulled.

Log No. 546

Shoemaker, No. 1, lessor. Atlantic Oil Producing Co., lessee. Location: Fincastle Section. Commenced: June 12, 1918. Completed: August 7, 1918. Shot 20 quarts. Production: Commenced producing 2 bbls. per 24 hours. Average daily production after 3 months was 7 bbls. per day. Average daily production after 6 months was 7 bbls. per day. Authority: Atlantic Oil Producing Co.

Strata.

Pennsylvanian System.

	Thickness	Depth
Soil	20	20
Sandstone (mountain)	60	80
Shale	180	260

Mississippian System.

Limestone (Little Lime)	20	280
Shale	70	350
Limestone (Big Lime)	102	452
Shale, hard	508	960

Devonian System.	Thickness	Depth
Shale, brown (Chattanooga)	155	1,115
Shale (fire clay)	10	1,125
Shale, black, hard	8½	1,133½
Limestone "sand" (oil)	10	1,143½
Total depth		1,143½

Casing record: Length, 20', 460". Size 8¼", 6¼".

Log No. 547

Shoemaker, No. 2, lessor. Atlantic Oil Producing Co., lessee. Location: Fincastle Section. Commenced: Sept. 25, 1918. Completed: Nov. 26, 1918. Shot 20 qts. between 1,191 and 1,187 feet. Production: 16 bbls. oil per 24 hours. Authority: Atlantic Oil Producing Co.

Strata.

Pennsylvanian System.	Thickness	Depth
Soil	7	7
Shale, sandy, soft	11	18
Sand, gray, hard, (water 23)	14	32
Shale, hard, dark	28	60
Sandstone (mountain), medium and hard	95	155
Shale, sandy	153	308
Mississippian System.		
Limestone (Little Lime)	40	348
Shale	42	390
Limestone (Big Lime)	92	482
Limestone, sandy	70	552
Limestone, green and medium	173	725
Shale, hard, black	132	857
Shale, gray	109	966
Pink rock	7	973
Shale, hard, green	25	998
Limestone, slag	9	1,007
Devonian System.		
Shale, brown, soft (Chattanooga)	148	1,155
Shale (fire clay), soft	10	1,165
Shale, hard, black	18	1,183
Shale, hard, brown	4½	1,187½
Limestone "sand," (oil)	5	1,192½
Total depth		1,192½

Casing record:

Length	Size
53'4"	8¼"
487'1"	6¼"

Log No. 548

Shoemaker, No. 3, lessor. Atlantic Oil Producing Co., lessor. Location: Fincastle Section. Commenced: Oct. 23, 1918. Completed: Nov. 14, 1918. Shot Nov. 16, 1918, 30 qts. Authority: Atlantic Oil Producing Co.

Strata.

Pennsylvanian System.	Thickness	Depth
Shale, soil and water	105	105
Sandstone (mountain)	70	175
Shale, hard	163	358
Sandstone	60	398
Mississippian System.		
Shale, hard	39	437
Limestone (Big Lime)	113	550
Shale, hard	507	1,057
Devonian System.		
Shale, brown (Chattanooga)	141	1,198
Shale, (fire clay)	15	1,213
Shale, hard, black	7'6"	1,220'6"
Limestone (cap rock)	9'6"	1,230
Total depth		1,230

Shot into salt water.

Casing was pulled and well plugged and abandoned April 28 1919.

Log No. 549

Shoemaker, No. 4, lessor. Atlantic Oil Producing Co., lessee. Location: Fincastle Section. Commenced: Jan. 23, 1919. Completed: Feb. 19, 1919. Shot 20 qts. Feb. 21, 1919, between 1,196 and 1,200 feet. Authority: Atlantic Oil Producing Co.

Strata.

Pennsylvanian System.	Thickness	Depth
Sand, gravel and clay	85	85
Sandstone (mountain)	68	153
Shale	162	315
Sandstone	30	345
Shale	92	437
Coal	3	440

Mississippian System.	Thickness	Depth
Limestone (Big Lime)	93½	533½
Shale, blue, soft and muddy	481½	1,015
Devonian System.		
Shale, brown, soft (Chattanooga)	148	1,163
Shale, white, and fire clay	10	1,173
Shale, black	15	1,188
Limestone ''sand,'' soft, (oil)	12½	1,200½
Total depth		1,200½

NOTE—The occurrence of the 3 feet of coal just above and in contact to the ''Big Lime'' is very unusual. The fact that ''coal'' is not reported in the other Shoemaker wells is also significant, and points toward a probable error of identification of the cuttings on the part of the driller.

Log No. 550

Shoemaker, No. 5, lessor. Atlantic Oil Producing Co., lessee. Location: Fincastle Section. Commenced: Jan. 8, 1919. Completed: Feb. 24, 1919. Shot: 20 qts. Feb. 24, 1919, between 1,205 and 1,210 feet.

Strata.

Pennsylvanian System.	Thickness	Depth
Soil	32	32
Sand, blue	19	51
Shale, hard	16	67
Shale, sandy	18	85
Sandstone (mountain)	90	175
Shale, sandy	71	246
Limestone	16	262
Shale, dark	68	330
Mississippian System.		
Limestone (Little Lime)	41	371
Shale, sandy, white	4	375
Sand, white, (water)	13	388
Limestone, sandy	12	400
Shale, sandy, dark brown	16	416
Limestone (Big Lime)	121	537
Shale, shelly	24	561
Shale, green	174	735
Shale, hard, black	138	873
Shale	112	985
Sandstone, flinty	6	991
Shale	15	1,006
Shale, shelly, brown	9	1,015

Devonian System.

	Thickness	Depth
Shale, brown (Chattanooga) ....,	150	1,165
Shale (fire clay)	15	1,180
Shale, hard, black	20	1,200
Shale	5½	1,205½
Limestone ''sand,'' (6" pay)	6	1,211½
Total depth		1,211½

Log No. 551

Shoemaker, No. 6, lessor. Atlantic Oil Producing Co., lessee. Location: Fincastle Section. Commenced: March 10, 1919. Completed: April 29, 1919. Shot with 20 qts. between 1,515 and 1,567 feet. Shot with 20 qts. between 1,205 and 1,211 feet. Production: beginning April 29, 1919, 3 bbls. oil per day. Authority: Atlantic Oil Producing Co.

Strata.

Pennsylvanian System.

	Thickness	Depth
Soil	25	25
Sand	5	30
Shale	50	80
Sand, brown, hard	90	170
Shale	152	322
Sand, hard	20	342

Mississippian System.

Shale, blue, hard	68	410
Limestone (Big Lime), hard	120	530
Shale, green	30	560
Shale, gray	70	630
Shale, red, hard	15	645
Shale, gray	380	1,025

Devonian System.

Shale, brown (Chattanooga)	150	1,175
Shale (fire clay), white	15	1,190
Shale, hard, black	10	1,200
Limestone ''sand'' (show of oil)	20	1,220
Limestone ''sand,'' white, hard, (salt water)	8	1,228
Limestone ''sand,'' gray, hard (oil at 1278)	50	1278

Silurian System.

Limestone ''sand,'' gray, hard	27	1,305
Shale, blue, hard	15	1,320
Limestone, red, shaly	5	1,325

THE IRVINE-PAINT CREEK FAULT.

This section occurs a short distance above Glencarin on the Middle Fork of the Red River in Wolfe County, Kentucky, in a cut of the L. & N. R. R. The "Big Lime" (Ste. Genevieve-St. Louis) (right) is here opposite the Cuyahoga group (left) and the displacement is about 140 ft. The downthrow is on the right.

Silurian System.

	Thickness	Depth
Shale, blue, hard	43	1,368
Limestone, red, shaly	22	1,390
Shale, black, hard	35	1,425
Limestone, red, shaly	5	1,430
Shale, black, hard	5	1,435
Limestone, red, shaly	5	1,440

Ordovician System.

	Thickness	Depth
Shale, black, hard	5	1,445
Sand, gray	35	1,480
Shale, blue, hard	5	1,485
Limestone, gray, medium hard	310	1,795
Total depth		1,795

NOTE—The Devonian-Silurian contact is placed just below 1,278 feet—a driller's division. It is probable however that it occurs in the lower part of the 50 feet of limestone showing oil at 1,278, and in such a case the oil would be of Silurian origin.

Log No. 552

Shoemaker, No. 7, lessor. Atlantic Oil Producing Co., lessee. Location: Fincastle Section. Commenced: March 10, 1919. Authority: Atlantic Oil Producing Co.

Strata.

Pennsylvanian System.

	Thickness	Depth
Soil	25	25
Sandstone (mountain)	89	114
Shale, blue	11	125
Shale, sandy, gray	131	256

Mississippian System.

	Thickness	Depth
Limestone (Little Lime), gray, sandy (water)	24	280
Shale, blue	79	359
Limestone (Big Lime), hard	116	475
Shale, green	22	497
Shale, green	118	615
Shale, dark blue	135	750

Mississippian System.

	Thickness	Depth
Shale, light blue	125	875
Shale, green	34	909
Pink rock	14	923
Shale, hard, dark	37	960

Devonian System.

	Thickness	Depth
Shale, brown (Chattanooga)	150	1,110
Shale (fire clay), white	16	1,126
Shale, black, hard	12	1,138
Limestone "sand," (dry)	10	1,148
Total depth		1,148

Log No. 553

Shoemaker, No. 8, lessor. Atlantic Oil Producing Co., lessee. Location: Fincastle Section. Commenced: May 29, 1919. Completed: June 20, 1919. Authority: Atlantic Oil Producing Co.

Strata.

Pennsylvanian System.

	Thickness	Depth
Soil	14	14
Sandstone (mountain), (fresh water 33)	126	140
Shale, hard	15	155
Shale, shelly	45	200
Shale	56	256
Shale, blue	44	300
Sand, gray, (fresh water)	22	322
Shale, gray	15	337

Mississippian System.

	Thickness	Depth
Limestone (Little Lime)	46	383
Shale, blue	28	411
Limestone (Big Lime)	149	560
Shale, hard, green	20	580
Shale, shelly	20	600
Shell and shale, hard	100	700
Shale hard	275	975
Shale (red rock)	10	985
Shale, hard	5	990
Limestone, white	5	995

Devonian System.

	Thickness	Depth
Shale, brown (Chattanooga)	165	1,160
Shale (fire clay)	10	1,170
Shale, hard, black, (dry)	14	1,184
Total depth		1,184

Log No. 554

Shoemaker, No. 9, lessor. Atlantic Oil Producing Co., lessee. Location: Fincastle Section. Commenced: May 24, 1919. Completed: June 16, 1919. Shot with 30 quarts. June 17, 1919, between 1159 and 1166 feet. Production: Commencing June 19, 1919, 7 bbls. per day. Average daily production after 1 day, 3½ bbls. Average daily production after shot, 3½ bbls. Authority: Atlantic Oil Producing Co.

Strata.

Pennsylvanian System.	Thickness	Depth
Soil	17	17
Sandstone (mountain)	93	110
Shale, sandy	25	135
Shale, blue	120	255
Mississippian System.		
Limestone (Little Lime)	37	292
Sand	53	345
Shale, blue	10	355
Limestone (Big Lime)	130	485
Shale, green	45	530
Shale, gray	400	930
Shale, red, sandy	10	940
Shale, gray	32	972
Devonian System.		
Shale, brown (Chattanooga)	154	1,126
Shale (fire clay)	14	1,140
Shale, black, hard	15	1,155
Limestone "sand"	15	1,170
Total depth		1,170

The heaviest volume of gas on this lease was in this well.

Log No. 555

Rhodes Hall, No. 1, lessor. Interstate Petroleum Co., lessee. Commenced: July 10, 1918. Shot with 50 quarts, September 16, 1918. Well cleaned and fully completed September 21, 1918. Authority: L. Beckner, and approved by Geo. Ogden.

Strata.

Pennsylvanian System.	Thickness	Depth
Soil	25	25
Shale, hard	25	50
Sand, (water)	25	75

Pennsylvania'n System.

	Thickness	Depth
Shale, hard, black	30	105
Shale sandy	30	135
Shale, gray, hard	40	175
Shale, dark, hard	15	190
Shale, limy	15	205
Shale, white, hard	10	215

Mississippian System.

Limestone (Little Lime)	10	225
Sand, hard	5	230
Shale, hard, black	10	240
Limestone (Big Lime), (6¼" casing)	150	390
Shale, hard, white, green	414	804

Devonian System.

Shale, brown (Chattanooga)	140	944
Shale (fire clay), (top of Irvine sand)	22	966
Limestone (Irvine sand), (first pay)	12	978
Limestone (Irvine sand), (second pay)	18	996
Limestone "sand," (pay and pocket)	4	1,000
Total depth		1,000

Log No. 556

Rhodes Hall, No. 2, lessor. Lantz & Ogden, drilling contractors. Commenced: September 27, 1918. Completed: November 14, 1918. Shot with 40 quarts, November 22, 1918. Well cleaned and fully completed November 28, 1918. Authority: L. Beckner, and approved by George Ogden.

Strata.

Pennsylvanian System.

	Thickness	Depth
Soil and shelly rock	15	15
Shale, hard, white	45	60
Sand, watery	20	80
Shale, hard, black	30	110
Limestone, sandy	30	140
Shale, hard, gray	40	180
Shale, hard, dark	15	195
Shell, limy	15	210
Shale, hard, white	10	220

Mississippian System.

	Thickness	Depth
Limestone (Little Lime)	12	232
Sandstone, hard	5	237
Shale, hard, black, (6¼" casing 247)	10	247
Limestone (Big Lime)	123	370
Shale, green, hard	17	387
Sandstone, red, shaly (Big Injun)	9	396
Shale, hard, white	426	822

Devonian System.

	Thickness	Depth
Shale, hard, chocolate	110	932
Shale (fire clay)	12	944
Shale, hard, black	5	949
Limestone "sand"	9	958
Limestone "sand," (first pay)	4	962
Limestone, sandy	8	970
Limestone "sand," (second pay)	4	974
Total depth		974

Log No. 557

Richardson, No. 1, lessor. Lantz & Ogden, drilling contractors. Commenced: December 4, 1918. Completed: January 18, 1919. Shot, 2 shots of 20 quarts each, January 23, 1919. Authority: L. Beckner, and approved by George Ogden.

Strata.

Pennsylvanian System.

	Thickness	Depth
Soil	10	10
Sand, shelly, (8¼" casing at 23)	15	25
Sand, hard	15	40
Sand, soft and yellow	15	55
Shale, hard, dark	80	135
Shell, limy	10	145
Shale hard, black	50	195

Mississippian System.

	Thickness	Depth
Limestone (Little Lime)	40	235
Shale, hard, white	15	250
Limestone (Big Lime)	125	375
Shale, hard, green, (6¼" casing at 380)	20	395
Red rock or Pink (Big Injun)	5	400
Shale, hard, white	435	835

Devonian System.	Thickness	Depth
Shale, hard chocolate	109	944
Shale (fire clay), (top of cap rock)	12	956
Limestone (cap rock)	9	965
Limestone, (pay)	2	967
Limestone	15	982
Total depth		982

Log No. 558

J. D. Smyth, No. 1, lessor. Ohio Fuel Oil & Gas Co., lessee. Location: Bulen Springs, Lee County. Commenced: May 26, 1920. Completed: June 23, 1920. Shot June 23, 1920, 10 quarts. Production: 5 bbls. oil naturally.

Strata.

Pennsylvanian System.	Thickness	Depth
Sandstone	50	50
Shale and shells	90	140
Mississippian System.		
Limestone (Little Lime)	25	165
Shale	10	175
Limestone (Big Lime)	80	255
Shale and shells	522	777
Devonian System.		
Shale, brown (Chattanooga)	128	905
Shale (fire clay)	11	916
Limestone "sand," (oil)	61	977
Total depth		977

First oil, 975.

Best pay, 965-970.

Log No. 559

J. D. Smyth, No. 2, lessor. Ohio Fuel Oil & Gas Co., lessee. Location: Bulen Springs, Lee County. Commenced: June 26, 1920. Completed: July 9, 1920. Shot July 9, 1920, 10 quarts. Production: 10 bbls. naturally.

Strata.

Pennsylvanian System.	Thickness	Depth
Soil	16	16
Sandstone	44	60
Shale	50	110
Sandstone (Little water 120)	15	125
Shale	25	150

Mississippian System.		
Limestone (Little Lime)	20	170
Shale	10	180
Limestone (Big Lime)	90	270
Shale	40	310
Shale, shelly	25	335
Shale and shells	430	765
Limestone	5	770

Devonian System.		
Shale, brown (Chattanooga)	130	900
Shale (fire clay)	20	920
Limestone "sand," (oil)	62	982
Total depth		982

First oil, 963.

Best pay, 972-982.

Log No. 560

Flahaven Logs (3-109 following).

Flahaven Land Co., No. 3, lessor. National Refining Co., Beattyville, Ky., lessee. Location: This tract consists of the eastern 1000 acres of the Eveleth Heirs farm of 2490 acres, which is situated at and above the juncture of Little Sinking and Big Sinking Creeks in Lee County, Ky. This particular 1000 acre lease lies on the waters of Big Sinking Creek, and was leased by the Flahaven Land Co.—Charles Eveleth, Pres. (Eveleth Heirs), to the National Refining Co., et. al. Another block of 1000 acres partitioned off of this same farm and located south of the mouth of Little Sinking Creek, was operated by the Ohio Oil Co. Representative logs of this latter tract are given on another page. Commenced: August 27, 1918. Completed: September 16, 1918. Production: Commenced producing September 20, 1918; production 41 hours after shot was 120 bbls. oil. Drilling contractor: McKay Bros., Fixer, Ky. Authority: National Refining Co., Beattyville, Ky., for this and the immediately following Flahaven Land Co. records (3 to 109).

Strata.

Mississippian System.	Thickness	Depth
Soil	28	28
Limestone (Big Lime), hard, gray	124	152
Shale brown, soft	458	610

Devonian System.		
Shale, black, soft (Chattanooga)	140	750
Shale (fire clay), gray, soft	27	777
Limestone, gray, medium	71	848
Total depth		848

NOTE—The Silurian-Devonian contact is toward the bottom of the last 71 feet of limestone. The driller missed the "break."

Log No. 561

Flahaven, No. 4, lessor. National Refining Co. and Le Roy Adams, lessees. Completed: October 18, 1918. Production: After shot was 250 bbls. oil. Drilling contractor: John Cain, Fixer, Ky.

Strata.

Pennsylvanian System.	Thickness	Depth
Limestone (Big Lime), hard, gray	115	115
Shale, brown, soft	510	625

Devonian System.		
Shale, black, soft (Chattanooga)	135	760
Shale (fire clay), gray, soft	18	778
Shale, gray, soft	4	782
Limestone, brownish gray, medium	62	844
Total depth		844

Log No. 562

Flahaven, No. 5, lessor. Commenced: September 28, 1918. Completed: October 18, 1918. Drilling contractor: J. A. Ross. Production: 48 hours after shot, 170 bbls. oil.

Strata.

Mississippian System.	Thickness	Depth
Soil and other strata	·46	46
Limestone (Big Lime), hard, gray	114	160
Shale, brown, soft	525	685

Devonian System.   Thickness Depth
   Shale, black, soft (Chattanooga) 135   820
   Shale (fire clay), gray, soft 15   835
   Limestone, brown, medium 44   879
      Total depth   879

Log No. 564

Flahaven, No. 8. Commenced: October 12, 1918. Completed: November 15, 1918. Production: Commenced producing October 31, 1918; production 48 hours after shot, 480 bbls. Drilling contractor: McKay Bros., Fixer, Ky.

   Strata.

Mississippian System.   Thickness Depth
   Shale, hard, and shells, soft and gray 85   85
   Limestone (Big Lime), hard, gray 112   197
   Shale, hard, gray 20   217
   Limestone, shelly, and shale, hard, gray 100   317
   Shale, hard, gray 373   690

Devonian System.
   Shale, black, gray, soft (Chattanooga) 145   835
   Shale (fire clay), gray, soft 18   853
   Limestone, brown, gray, soft 35   888
      Total depth   888

Log No. 565

Flahaven No. 9. Commenced: October 18, 1918. Completed: November 9, 1918. Production: Commenced producing November 9, 1918; production 24 hours after shot, 175 bbls.

   Strata.

Mississippian System.   Thickness Depth
   Soil, gray 38   38
   Limestone, hard, broken (Big Lime in part) .. 227   265
   Shale, soft 515   780

Devonian System.
   Shale, gray, soft (Chattanooga) 120   900
   Shale (fire clay), brown, soft 12   912
   Limestone, medium 55   967
      Total depth   967

Log No. 566

Flahaven, No. 10. Commenced: Oct. 31, 1918. Completed: Nov. 27, 1918. Production: pumped 48 hours, made 300 bbls per day.

Strata.

Mississippian System.	Thickness	Depth
Soil	15	15
Limestone (Big Lime), hard, gray	185	200
Shale, hard, gray, sandy	150	350
Shale, hard, blue, soft	310	660
Shale, hard, red, soft	15	675
Shale, hard, blue, soft	15	690
Devonian System.		
Shale, black, soft (Chattanooga)	145	835
Shale (fire clay), gray, soft	18	853
Limestone, black, soft	6	859
Limestone, brown, medium	37	896
Total depth		896

Log No. 567

Flahaven, No. 11. Commenced: November 10, 1918. Completed: December 5, 1918.

Strata.

Mississippian System.	Thickness	Depth
Soil, hard, dark	16	16
Limestone (Big Lime), gray, soft	86	102
Shale, hard and soft, gray	503	605
Devonian System.		
Shale, brown, soft (Chattanooga)	127	732
Shale (fire clay), gray, soft	20	752
Shale, black, soft	5	757
Limestone, brown, medium	32	789
Total depth		789

Log No. 568

Flahaven, No. 12. Commenced: November 16, 1918. Completed: December 7, 1918. Production: Well flowed at the rate of 350 bbls. until shut in.

Strata.

Mississippian System.	Thickness	Depth
Soil	10	10
Limestone (Big Lime), hard, gray	75	85
Shale, hard, and shells, soft, gray	505	590
Devonian System.		
Shale, brown, soft (Chattanooga)	140	730
Shale (fire clay), gray, soft	15	745
Shale, black, soft	7	752
Limestone, brown, medium	31	783
Total depth		783

Log No. 569

Flahaven, No. 13. Commenced: November 20, 1918. Completed: December 5, 1918. Production: Commenced producing December 11, 1918; production 24 hours after shot, 100 bbls.

Strata.

Mississippian System.	Thickness	Depth
Soil and sandrock, soft	42	42
Limestone, hard, white, (Big Lime)	172	214
Shale, white, soft	459	673
Devonian System.		
Shale black (Chattanooga)	165	838
Shale (fire clay) and sand	18	856
Sand	37	893
Total depth		893

Log No. 570

Flahaven, No. 17. Commenced: December 16, 1918. Completed: January 20, 1919. Production: Commenced producing January 30, 1919; production 24 hours after shot, 100 bbls.

Strata.

Pennsylvanian System.	Thickness	Depth
Soil	48	48
Sand	172	220

Mississippian System.		
Limestone (Big Lime)	140	360
Shale	115	475
Shale, hard, shelly	330	805
Shale, sandy, red	35	840
Shale, hard	20	860

Devonian System.		
Shale, black (Chattanooga)	140	1,000
Shale (fire clay.)	18	1,018
Limestone	41	1,059
Total depth		1,059

Log No. 571

Flahaven, No. 19. Completed: March 20, 1919. Production: 24 hours after shot, 175 bbls.

Strata.

Mississippian System.	Thickness	Depth
Soil, hard, brown, sandy	5	5
Limestone (Big Lime), gray, soft	100	105
Shale, green, hard	95	200
Shale, white hard	30	230
Shale, hard and soft brown	330	560
Shale, red, hard	10	570
Shale, hard, blue, soft	30	600

Devonian System.		
Shale, brown, soft (Chattanooga)	140	740
Shale (fire clay), brown, soft	15	755
Shale, brown, soft	10	765
Limestone "sand," hard, gray	35	800
Total depth		800

Log No. 572

Flahaven, No. 21. Commenced: January 3, 1919. Completed: February 6, 1919.

Strata.

	Thickness	Depth
Pennsylvanian System.		
Sandrock and shale	160	160
Mississippian System.		
Limestone (Big Lime)	160	320
Shale	443	763
Devonian System.		
Shale, black (Chattanooga)	180	943
Shale (fire clay)	22	965
Limestone "sand"	38	1,003
Total depth		1,003

Log No. 573

Flahaven, No. 22. Commenced: March 6, 1919. Completed: March 24, 1919.

Strata.

	Thickness	Depth
Pennsylvanian System.		
Soil and sandrock, hard, dark	156	156
Mississippian System.		
Limestone (Big Lime), hard, gray	166	322
Shale, brown, soft	460	782
Devonian System.		
Shale, black, soft (Chattanooga)	180	962
Shale (fire clay), gray, soft	22	984
Limestone "sand," hard, gray	33	1,017
Total depth		1,017

Log No. 574

Flahaven, No. 24.   Commenced: January 12, 1919.   Completed: January 29, 1919.

Strata.

Pennsylvanian System.	Thickness	Depth
Soil	18	18
Shale, hard	142	160
Mississippian System.		
Limestone (Big Lime)	110	270
Shale, brown	433	703
Devonian System.		
Shale, black (Chattanooga)	180	883
Shale (fire clay)	25	908
Limestone	37	945
Total depth		945

Log No. 575

Flahaven, No. 25.   Commenced: January 13, 1919.   Completed: February 14, 1919.

Strata.

Pennsylvanian System.	Thickness	Depth
Soil	17	17
Shale, hard	108	125
Mississippian System.		
Limestone (Big Lime)	200	325
Shale, brown	310	635
Devonian System.		
Shale, black (Chattanooga)	315	950
Shale (fire clay)	25	975
Limestone	$40\frac{1}{2}$	$1,015\frac{1}{2}$
Total depth		$1,015\frac{1}{2}$

Log No. 576

Flahaven, No. 28. Commenced: May 7, 1919. Completed: June 13, 1919.

Strata.

Pennsylvanian System.	Thickness	Depth
Sand and soil, dark, hard	162	162

Mississippian System.		
Limestone, gray, hard (Big Lime in part) .... _	203	365
Shale, soft	451	816

Devonian System.		
Shale, black, soft (Chattanooga)	160	976
Shale (fire clay), gray, soft	10	986
Shale, hard, gray	14	1,000
Limestone ''sand,'' hard, brown	66	1,066
Total depth		1,066

NOTE—The Silurian-Devonian contact is toward the bottom of the last 66 feet of limestone.

Log No. 577

Flahaven, No. 29. Commenced: April 24, 1919. Completed: May 5, 1919.

Strata.

Pennsylvanian System.	Thickness	Depth
Soil, dark, soft	15	15
Shale, hard and soft, dark	140	155

Mississippian System.		
Limestone (Big Lime), hard, gray	150	·305
Shale, hard, dark, (set casing)	10	315
Shale, red, sandy, hard	450	765
Shale hard, dark	10	775
Shale, soft	15	790

Devonian System.		
Shale, gray, soft (Chattanooga)	150	940
Shale, hard, dark	10	950
Limestone, hard, dark	15	965
Limestone ''sand,'' gray, hard	33	998
Total depth		998

Log No. 578

Flahaven, No. 30. Commenced: March 15, 1919. Completed: April 2, 1919.

Strata.

Pennsylvanian System.	Thickness	Depth
Soil and sandrock	235	235

Mississippian System.

	Thickness	Depth
Limestone (Big Lime), hard, gray	100	335
Shale, hard and soft, brown	509	844

Devonian System.

	Thickness	Depth
Shale, black, soft (Chattanooga)	135	979
Shale (fire clay), gray, soft	25	1,004
Shale, brown, soft	10	1,014
Limestone, brown, hard, (pay sand)	36	1,050
Total depth		1,050

Log No. 579

Flahaven, No. 31. Commenced: April 26, 1919. Completed: May 19, 1919.

Strata.

Pennsylvanian System.	Thickness	Depth
Sand rock, gray, soft	220	220

Mississippian System.

	Thickness	Depth
Limestone (Big Lime), hard, gray	125	345
Shale, hard, and shells, brown, soft	511	856

Devonian System.

	Thickness	Depth
Shale, black, soft (Chattanooga)	120	976
Shale (fire clay), gray, soft	25	1,001
Shale, brown, soft	5	1,006
Limestone "sand," brown, hard	36	1,042
Total depth		1,042

Log No. 580

Flahaven, No. 32. Commenced: February 13, 191.9. Completed: March 7, 1919.

Strata.

	Thickness	Depth
Pennsylvanian System.		
Soil, dark, soft	190	190
Mississippian System.		
Limestone (Big Lime), hard, gray	95	285
Shale, sandy	495	780
Devonian System.		
Shale, brown, soft (Chattanooga)	140	920
Shale (fire clay), gray, soft	20	940
Limestone "sand," gray, hard	43	983
Total depth		983

Log No. 581

Flahaven, No. 33. Commenced: June 9, 1919. Completed: June 24, 1919.

Strata.

	Thickness	Depth
Pennsylvanian System.		
Soil and sand, hard, dark	201	201
Mississippian System.		
Limestone (Big Lime), hard, gray	145	346
Shale, brown, soft	491	837
Devonian System.		
Shale, black, soft (Chattanooga)	125	962
Shale (fire clay), brown, soft	25	987
Shale, hard, black	4	991
Limestone "sand," hard, brown	36	1,027
Total depth		1,027

Log No. 582

Flahaven, No. 34. Commenced: July 11, 1919. Completed: July 24, 1919.

Strata.

	Thickness	Depth
Pennsylvanian System.		
Soil and sandstone, hard, dark	115	115
Mississippian System.		
Limestone (Big Lime), gray, soft	134	249
Shale, hard, gray	496	745
Devonian System.		
Shale, black, soft (Chattanooga)	140	885
Shale, brown, soft	15	900
Limestone "sand," hard, brown	39	939
Total depth		939

Log No. 583

Flahaven, No. 35.   Commenced: March 30, 1919.   Completed: May 8, 1919.

Strata.

Mississippian System.	Thickness	Depth
Soil, hard, dark	10	10
Sand, hard, gray	10	20
Limestone (Big Lime), hard, gray	180	200
Shale, shelly, gray, hard	40	240
Shale, hard, white	110	350
Shells, gray, hard, limy	10	360
Shale, hard, white	296	656
Shale (red rock), hard	20	676
Devonian System.		
Shale, hard, black (Chattanooga)	94	770
Shale, brown, soft (Chattanooga)	64	834
Shale (fire clay), gray, soft	20	854
Limestone, gray, hard	5	859
Limestone (pay ''sand'' gray, hard	20	879
Limestone "sand"	18	897
Total depth		897

Log No. 584

Flahaven, No. 36.   Commenced: May 24, 1919.   Completed: June 10, 1919.

Strata.

Pennsylvanian System.	Thickness	Depth
Soil, brown, hard	5	5
Limestone, hard, gray	8	13
Shale, hard, gray	17	30
Sand, hard, gray	60	90
Shale, hard, gray	30	120
Shale, hard, white	55	175
Mississippian System.		
Limestone, sandy, brown, soft	50	225
Shale, hard, white	5	230
Limestone (Big Lime), brown, hard	118	348
Shale, hard, brown	22	370
Shale, hard, white	440	810

Mississippian System.	Thickness	Depth
Shale (red rock), soft	20	830
Shale, hard, white	15	845
Limestone, shelly, brown, soft	5	850

Devonian System.		
Shale, brown, soft (Chattanooga)	135	985
Shale (fire clay), gray, soft	22	1,007
Limestone "sand," brown, hard	38	1,045
Total depth		1,045

Log No. 585

Flahaven, No. 37. Commenced: February 25, 1919. Completed: March 10, 1919.

Strata.

Pennsylvanian System.	Thickness	Depth
Soil and sand, hard, dark	190	190

Mississippian System.		
Limestone (Big Lime), hard, gray	175	365
Shale, brown, soft	445	810

Devonian System.		
Shale, black, soft (Chattanooga)	178	988
Shale (fire clay), gray, soft	18	1,006
Sand, gray, hard	32	1,038
Total depth		1,038

Log No. 586

Flahaven, No. 38. Commenced: April 15, 1919. Completed: April 22, 1919.

Strata.

Pennsylvanian System.	Thickness	Depth
Sand and soil, hard, dark	170	170

Mississippian System.		
Limestone (Big Lime), hard, gray	168	338
Shale, brown, soft	255	593

Devonian System.	Thickness	Depth
Shale, black, soft (Chattanooga)	170	763
Shale (fire clay), gray, soft	22	785
Limestone ''sand,'' hard, gray	30	815
Total depth		815

Log No. 587

Flahaven, No. 39. Commenced: May 14, 1919. Completed: May 25, 1919.

Strata.

Pennsylvanian System.	Thickness	Depth
Soil and sand, gray, soft	36	36
Shale, brown, soft	134	170
Mississippian System.		
Limestone (Big Lime), hard, gray	160	330
Shale, brown, soft	455	785
Devonian System.		
Shale, black, soft (Chattanooga)	165	950
Shale (fire clay), gray, soft	22	972
Limestone ''sand,'' hard, brown	29	1,001
Total depth		1,001

Log No. 588

Flahaven, No. 40. Commenced: June 11, 1919. Completed: June 25, 1919.

Strata.

Pennsylvanian System.	Thickness	Depth
Soil, hard, dark	20	20
Shale, brown, soft	145	165
Mississippian System.		
Limestone (Big Lime), gray, hard	175	340
Shale, brown, soft	452	792
Devonian System.		
Shale, black, soft (Chattanooga)	170	962
Shale (fire clay), gray, hard	22	984
Limestone "sand," hard, gray	36½	1,020½
Total depth		1,020½

Log No. 589

Flahaven, No. 41. Commenced: July 21, 1919. Completed: August 8, 1919.

Strata.

Pennsylvanian System.	Thickness	Depth
Soil, brown, soft	180	180
Mississippian System.		
Limestone (Big Lime), hard, white	165	345
Shale, gray, soft	473	818
Devonian System.		
Shale black, soft (Chattanooga)	155	973
Shale (fire clay), white, soft	18	991
Limestone "sand," brown, hard	36	1,027
Total depth		1,027

Log No. 590

Flahaven, No. 42. Commenced: August 16, 1919. Completed: September 1, 1919.

Strata.

Pennsylvanian System.	Thickness	Depth
Soil and sand, gray, soft	168	168
Mississippian System.		
Limestone (Big Lime), hard, white	197	365
Shale, brown, soft	422	787
Devonian System.		
Shale, black, soft (Chattanooga)	166	953
Shale (fire clay), gray, soft	18	971
Limestone "sand," brown, hard	36	1,007
Total depth		1,007

Log No. 591

Flahaven, No. 43. Commenced: September 15, 1919. Completed: October 2, 1919.

Strata.

Pennsylvanian System.	Thickness	Depth
Soil, brown, soft	12	12
Shale, hard, blue, shelly	198	210

Mississippian System.		
Limestone (Big Lime), hard, white	125	335
Shale, hard, blue	20	355
Shale, white, soft	315	670
Shale, hard, dark	4	674
Shale, hard, blue	126	800
Shale, red, soft	20	820
Shale, hard, white	40	860

Devonian System.		
Shale, black, soft (Chattanooga)	149	1,009
Limestone (cap rock), hard, gray	3	1,012
Limestone ''sand,'' coarse, hard, gray	15	1,027
Limestone "sand," fine, gray, soft	41	1,068

Silurian System.		
Limestone ''sand,'' brown sugar sand, soft ..	15	1,083
Limestone "shale," hard, blue	8	1,091
Total depth		1,091

Log No. 592

Flahaven, No. 44.   Commenced: September 2, 1919.   Completed: September 19, 1919.

Strata.

Pennsylvanian System.	Thickness	Depth
Soil, brown, soft	18	18
Shale, hard, blue	190	208

Mississippian System.		
Limestone (Big Lime), hard, white	120	328
Shale and sand, hard, blue	522	850

Devonian System.		
Shale, black, soft (Chattanooga)	122	972
Shale (fire clay), white, soft	18	990
Shale, hard, black	2	992
Limestone ''sand,'' gray, soft	62	1,054

Devonian System.   Thickness Depth
    Limestone "sand," gray and brown, hard   5   1,059
    Limestone "sand," hard   10   1,069
    Shale, hard, blue   10   1,079
        Total depth   1,079

NOTE—The Silurian-Devonian contact is toward the bottom of the
62 feet of limestone above 1054.

Log No. 593

Flahaven, No. 45. Commenced: August 4, 1919. Completed
August 23, 1919.

    Strata.
Pennsylvanian System.   Thickness Depth
    Soil and sand, brown, soft   190   190
Mississippian System.
    Limestone (Big Lime), hard, white   175   365
    Shale, gray, soft   473   838
Devonian System.
    Shale, black, soft (Chattanooga)   170   1,008
    Shale (fire clay), white, soft   18   1,026
    Limestone "sand," hard, brown   36   1,062
        Total depth   1,062

Log No. 594

Flahaven, No. 47. Commenced: October 6, 1919. Completed:
October 15, 1919.

    Strata.
Pennsylvanian System.   Thickness Depth
    Soil, brown, soft   20   20
    Limestone (Big Lime), hard, white   70   90
    Shale, hard, green   20   110
    Shale, hard, blue   435   545
    Shale, hard, red   10   555
    Shale, hard, gray   30   585

Devonian System.   Thickness Depth

    Shale, black, hard (Chattanooga)   150   735
    Shale (fire clay), white, soft   10   745
    Limestone "sand," gray, soft   10   755
    Limestone "sand," hard, dark   10   765
    Limestone "sand," hard, gray   10   775
    Limestone "sand," hard, white   10   785
    Limestone "sand," hard, gray   12   797

Silurian System.

    Limestone "sand," gray, soft, (pay)   10   807
    Limestone "sand," hard, white   10   817
    Shale, hard, blue ....   9·   826
        Total depth   826

Log No. 595

    Flahaven No. 48.   Commenced: July 4, 1919.   Completed: July 21, 1919.

        Strata.

Pennsylvanian System.   Thickness Depth

    Soil, hard, dark   20   20
    Shale, hard, gray   45   65

Mississippian System.

    Limestone, gray, soft, sandy   25   90
    Shale, hard, gray   5   95
    Limestone (Big Lime) gray, soft   100   195
    Shale, green, soft   30   225
    Shale, hard, gray   425   650
    Shale, red, hard   20   670
    Shale, hard, black   25   695

Devonian System.

    Shale, brown, soft (Chattanooga)   135   830
    Shale (fire clay), brown, soft   21   851
    Limestone "sand," brown, hard   40   891
        Total depth   891

Log No. 596

Flahaven, No. 51. Commenced: March 15, 1919. Completed:
April 3, 1919.

       Strata.

Mississippian System.	Thickness	Depth
Soil, dark, soft	62	62
Limestone (Big Lime), hard, gray	105	167
Shale, brown, soft	10	177
Shale brown, soft	483	660

Devonian System.		
Shale, black, soft (Chattanooga)	145	805
Shale (fire clay), gray, soft	18	823
Limestone ''sand,'' brown, hard, (oil)	54	877
Total depth		877

Log No. 597

Flahaven, No. 53.

       Strata.

Pennsylvanian System.	Thickness	Depth
Soil and shale, hard, dark	100	100

Mississippian System.		
Shale, hard, gray	40	140
Limestone (Big Lime) hard, gray	120	260
Shale, hard and gray, and lime shells	505	765

Devonian System.		
Shale, brown, soft (Chattanooga)	135	900
Shale (fire clay), gray, soft	10	910
Lime shells, hard, gray	5½	915½
Limestone ''sand,'' hard, gray, (oil)	32	947½
Total depth		947½

Log No. 598

Flahaven, No. 54. Commenced: April 16, 1919. Completed:
April 29, 1919.

       Strata.

Pennsylvanian System.	Thickness	Depth
Soil, dark, soft	24	24
Shale, hard	111	135

Mississippian System.	Thickness	Depth
Limestone (Big Lime), hard, gray	100	235
Shale, hard, and shells	495	730
Devonian System.		
Shale, brown, soft (Chattanooga)	135	865
Shale (fire clay), gray, soft	15	880
Limestone (cap rock), gray, hard	6	886
Limestone ''sand,'' gray, hard	32	918
Total depth		918

Log No. 599

Flahaven, No. 55.   Commenced: April 30, 1919.   Completed: May 23, 1919.

Strata.

Mississippian System.	Thickness	Depth
Sand and shale, hard and gray	62	62
Limestone (Big Lime), gray, soft	44	106
Shale, hard, and shells, gray soft	549	655
Devonian System.		
Shale brown, soft (Chattanooga)	145	800
Shale (fire clay), gray, soft	25	825
Limestone ''sand,'' brown, hard, (oil)	36	861
Total depth		861

Log No. 600

Flahaven, No. 56.   Commenced: May 29, 1919.   Completed: June 16, 1919.

Strata.

Mississippian System.	Thickness	Depth
Clay, gray, hard	15	15
Shale, hard, brown	65	80
Limestone (Big Lime), gray, soft	115	195
Shale, hard, brown	460	655
Shale (red rock), hard	10	665
Shale, hard, brown	15	680
Devonian System.		
Shale, black, soft (Chattanooga)	150	830
Shale (fire clay), gray, soft	10	840
Shale, hard, black	11	851
Limestone ''sand,'' hard, brown, (oil)	37	888
Total depth		888

Log No. 601

Flahaven, No. 57.   Commenced: May 22, 1919.   Completed: June 5, 1919.

Strata.

Pennsylvanian System.	Thickness	Depth
Shale, brown, soft	170	170
Mississippian System.		
Limestone (Big Lime), hard, gray	100	270
Shale and limestone, gray and hard	505	775
Devonian System.		
Shale, brown, soft (Chattanooga)	130	905
Shale (fire clay) and limestone (cap rock), gray, soft	21	926
Limestone "sand," brown, hard, (oil)	37½	963½
Total depth		963½

Log No. 602

Flahaven, No. 58.   Commenced: June 4, 1919.   Completed: June 20, 1919.

Strata.

Pennsylvanian System.	Thickness	Depth
Sand and shale, hard and gray	125	125
Mississippian System.		
Limestone (Big Lime), gray, soft	105	230
Shale and shells, hard and gray	490	720
Devonian System.		
Shale, brown, soft (Chattanooga)	140	860
Shale (fire clay), gray, soft	22	882
Limestone "sand," brown, hard	69	951
Total depth		951

NOTE—The Silurian-Devonian contact is toward the bottom of the last 69 feet of this record.

Log No. 603

Flahaven, No. 61.   Commenced: July 23, 1919.   Completed:
September 11, 1919.

Strata.

Pennsylvanian System.	Thickness	Depth
Soil, brown, soft	28	28
Sand and shale, hard and blue	112	140

Mississippian System.		
Limestone (Big Lime), hard, white	100	240
Shale, hard, blue	514	754

Devonian System.		
Shale, black, soft (Chattanooga)	140	894
Shale (fire clay), white, soft	20	914
Shale, black, hard	5	919
Limestone ''sand,'' gray, soft, (pay sand)	8	927
Limestone "sand," hard, dark	20	947
Limestone "sand," light, dark	10	957
Limestone ''sand,'' hard, dark	15	972

Silurian System.		
Limestone "sand," light, soft, (pay)	18	990
Limestone ''sand,'' blue, soft	10	1,000
Total depth		1,000

Log No. 604

Flahaven, No. 62. ·Commenced: November 28, 1919.   Completed:
January 12, 1920.

Strata.

Pennsylvanian System.	Thickness	Depth
Sand and rock, brown, hard	13	13
Sand and shale, hard, gray	77	90

Mississippian System.		
Limestone (Big Lime), white, soft	90	180
Shale, hard, and shells, gray, soft	490	670

Devonian System.		
Shale brown, soft (Chattanooga)	150	820
Shale (fire clay)	16	836
Limestone "sand," (oil) and shale, hard, brown	78	914
Total depth		914

NOTE—The Silurian-Devonian contact is within the last 78 feet of the record.

NOTE—Beginning with this lease, Flahaven No. 62, and continuing through No. 109, LeRoy Adams is given as joint lessee with the National Refining Co.

Log No. 605

Flahaven, No. 63. Commenced: July 24, 1919. Completed August 16, 1919.

Strata.

	Thickness	Depth
Pennsylvanian System.		
Soil, gray, soft	6	6
Shale, hard, blue	30	36
Mississippian System.		
Limestone (Big Lime), hard, white	100	136
Shale, hard, blue	14	150
Shale, hard, light, gritty	125	275
Shale, hard, dark	311	586
Shale, red, soft	20	606
Shale, hard, white	20	626
Devonian System.		
Shale, black (Chattanooga)	139	765
Shale, white, soft	20	785
Limestone "sand," black, hard	3	788
Limestone "sand," gray, hard	37	825
Total depth		825

Log No. 606

Flahaven, No. 64. Commenced: August 1, 1919. Completed: August 16, 1919.

Strata.

	Thickness	Depth
Pennsylvanian System.		
Soil, sandy, brown, hard	18	18
Shale, hard, gray	44	62
Sand, hard, gray	19	81
Shale, hard, gray	21	102
Sand, hard, white	18	120

Mississippian System.	Thickness	Depth
Shale (red rock), hard	20	140
Limestone (Big Lime), gray, soft	111	251
Shale, hard, green	14	265
Shale, hard, gray	465	730

Devonian System.		
Shale, brown, soft (Chattanooga)	130	860
Shale (fire clay), brown, soft	20	880
Limestone "sand," brown, hard, (oil)	36	916
Total depth		916

Log No. 607

Flahaven, No. 66.   Commenced: August 2, 1919.   Completed: August 21, 1919.

Strata.

Mississippian System.	Thickness	Depth
Clay, brown, soft	16	16
Limestone (Big Lime), gray, soft	89	105
Shale, hard, green	15	120
Shale, hard, brown	10	130
Limestone shells, gray, hard	55	185
Shale, hard, green	15	200
Shale and shells, hard, gray	10	210
Shale and shells, hard, black	220	430
Shale, hard, red	130	560
Shale (red rock), brown, hard	20	580
Shale, brown, soft	30	610

Devonian System.		
Shale, gray, soft (Chattanooga)	133	743
Shale (fire clay), gray, soft	17	760
Limestone (cap rock), brown, hard	3½	763½
Limestone "sand," brown, hard, (oil)	36½	800
Total depth		800

Log No. 608

Flahaven, No. 67.   Commenced: September 1, 1919.   Completed: September 13, 1919.

Strata.

Mississippian System.	Thickness	Depth
Soil, brown, soft	14	14
Sand, brown, soft	16	30
Limestone (Big Lime), hard, white	120	150
Shale, green, soft	250	400
Shale, gray, soft	210	610
Shale (red rock), soft	25	635
Shale, hard, white	25	660

Devonian System.		
Shale, black, soft (Chattanooga)	133	793
Shale (fire clay), white, soft	15	808
Limestone "sand," hard, gray	3	811
Limestone "sand," gray, soft	9	820
Limestone "sand," hard, dark	10	830
Limestone "sand," light, hard	5	835
Limestone "sand," gray, soft	8	843
Limestone "sand," hard, dark	20	863
Limestone "sand," light, soft	14	877
Shale, hard, blue	10	887
Total depth		887

NOTE—The Silurian-Devonian contact is within the 20 feet above 863.

Log No. 609

Flahaven, No. 69. Commenced: September 1, 1919. Completed: September 14, 1919.

Strata.

Pennsylvanian System.	Thickness	Depth
Soil, gray, soft	18	18
Shale, hard, gray, shelly	62	80

Mississippian System.		
Limestone (Big Lime), hard, white	140	220
Shale, green, soft	480	700

Devonian System.		
Shale, black, soft (Chattanooga)	154	854
Shale (fire clay), white, soft	15	869
Limestone "sand," gray, soft	8	877

Devonian System.	Thickness	Depth
Limestone "sand," hard, dark	20	897
Limestone "sand," light, hard	8	905
Limestone "sand," dark, hard	10	915
Silurian System.		
Limestone "sand," light, hard	26½	941½
Shale, hard, blue	10	951½
Total depth		951½

Log No. 610

Flahaven, No. 70.   Commenced: October 8, 1919.   Completed: October 19, 1919.

Strata.

Pennsylvanian System.	Thickness	Depth
Soil, yellow, soft	20	20
Sand, white, soft	30	50
Shale, hard, blue	43	93
Mississippian System.		
Limestone, hard, white, sandy	42	135
Limestone (Big Lime), hard, white	105	240
Shale, hard, black	126	366
Shale, hard, gray	329	695
Shale, red, soft, sandy	15	710
Shale, hard, blue	20	730
Devonian System.		
Shale, brown, soft (Chattanooga)	150	880
Shale (fire clay), white, soft	17	897
Limestone "sand," hard, dark	5	902
Limestone "sand," hard, gray	21	923
Limestone "sand," brown, hard	17	940
Limestone "sand," gray, hard	21	961
Silurian System.		
Shale, blue, soft	11	972
Total depth		972

Log No. 611

Flahaven, No. 71.   Commenced: September 16, 1919.   Completed: September 30, 1919.

Strata.

Pennsylvanian System.	Thickness	Depth
Soil, hard, brown, sandy	18	18
Shale, hard, gray	42	60

Pennsylvanian. System.   Thickness Depth
    Sand, hard, white 32   92
    Shale, hard, gray 130   222
Mississippian System. ....
    Limestone (Big Lime), hard, gray 100   322
    Shale, green, soft 18   340
    Shale, gray, soft 427   767
Devonian System.
    Shale, brown, soft (Chattanooga) 176   943
    Shale (fire clay), gray, soft 23   966
    Limestone "sand," hard, dark, (oil) (1st pay
        981, 2d pay 1022) 71   1,037
    Shale, hard, gray 12   1,049
        Total depth 1,049

NOTE—The Devonian-Silurian contact is within the 71 feet above 1037.

Log No. 612

Flahaven, No. 72. Commenced: July 24, 1919. Completed: August 7, 1919.
        · Strata.
Mississippian System.   Thickness Depth
    Soil, dark, hard 30   30
    Limestone (Big Lime), hard, gray 100   130
    Shale, hard, brown 20   150
    Shale, brown, soft 480   630
Devonian System.
    Shale, black, soft (Chattanooga) 93   723
    Shale (fire clay); gray, soft 20   743
    Limestone "sand," gray, hard 38   781
        Total depth 781

Log No. 613

Flahaven, No. 73. Commenced: August 1, 1919. Completed: August 9, 1919.
        Strata.
Pennsylvanian & Mississippian Systems.   Thickness Depth
    Soil and blue mud, dark, soft 80   80
    Limestone (Big Lime), hard, gray 120   200
    Shale, hard, gray, and lime shells 477   677

Devonian System.	Thickness	Depth
Shale, brown, soft (Chattanooga)	150	827
Shale (fire clay), gray, soft	21	848
Limestone "sand," brown, hard	36	884
Total depth		884

Log No. 614

Flahaven, No. 74. Commenced: August 26, 1919. Completed: October 1, 1919.

Strata.

Pennsylvanian System.	Thickness	Depth
Soil, brown, soft	20	20
Shale, hard, blue	70	90

Mississippian System.		
Limestone (Big Lime), hard, white	95	185
Shale, hard, green	15	200
Shale, hard, blue	100	300
Shale, hard, light, sandy	100	400
Shale, hard, blue	245	645
Shale, red, soft, sandy	15	660
Shale, hard, white	25	685

Devonian System.		
Shale, black, soft (Chattanooga)	139	824
Shale (fire clay), white, soft	20	844
Limestone "sand," gray, soft	12	856
Limestone "sand," hard, dark	39	895

Silurian System.		
Limestone "sand," light, soft	16	911
Shale, hard, blue	11⅓	922⅓
Total depth		922⅓

Log No. 615

Flahaven, No. 75. Commenced: October 15, 1919. Completed: October 30, 1919.

Strata.

Mississippian System.	Thickness	Depth
Soil, brown, soft	15	15
Shale, hard, blue	8	23
Limestone (Big Lime), hard, white	103	126
Shale, hard, blue	504	630

Devonian System.

	Thickness	Depth
Shale, black, soft (Chattanooga)	136	766
Shale (fire clay), white, soft	20	786
Limestone (cap rock), hard, black	2	788
Limestone "sand," soft, gray	8	796
Limestone "sand," hard, dark	15	811
Limestone "sand," light, hard	15	826
Limestone "sand," hard, dark	18	844

Silurian System.

Limestone "sand," light, soft	16	860
Shale, hard, blue	7	867
Total depth		867

Log No. 616

Flahaven, No. 77. Commenced: November 17, 1919. Completed:
November 29, 1919.

Strata.

Pennsylvanian System.

	Thickness	Depth
Soil, brown, soft	12	12
Soil, sandy, light, hard	4	16
Soil, sandy, light, soft	61	77

Mississippian System.

Limestone (Big Lime), white, hard	108	185
Shale, hard, white	35	220
Shale, hard, gray	195	415
Shale, hard, dark	235	650
Shale (red rock), soft	15	665
Shale, hard, gray	25	690

Devonian System.

Shale, black, soft (Chattanooga)	140	830
Shale (fire clay), white, soft	17	847
Limestone (cap rock), hard, black	2	849
Limestone "sand," gray, soft	6	855
Limestone "sand," hard, dark	22	877
Limestone "sand," gray, medium	5	882
Limestone "sand," gray, medium	25	907

Silurian System.

Limestone "sand," light, medium	9	916
Shale, blue, soft	11	927
Total depth		927

Log No. 617

Flahaven, No. 78. Commenced: December 12, 1919. Completed: January 16, 1920.

Strata.

Pennsylvanian System.	Thickness	Depth
Soil, yellow, soft	20	20
Sand, yellow, soft	40	60
Shale, hard, gray	130	190
Mississippian System.		
Limestone (Big Lime), hard, white	115	305
Shale, hard, gray	305	610
Shale, hard, black, soft	136	746
Shale (red rock), soft	15	761
Shale, hard, black, soft	30	791
Devonian System.		
Shale, brown, soft (Chattanooga)	140	931
Shale (fire clay), white, soft	15	946
Limestone "sand," hard, dark	8	954
Limestone "sand," light, hard	14	968
Limestone "sand," hard, dark	32	1,000
Silurian System.		
Limestone "sand," light, hard	22	1,022
Shale, blue, soft	10	1,032
Total depth		1,032

Log No. 618

Flahaven, No. 80. Commenced: November 8, 1919. Completed: November 29, 1919.

Strata.

Pennsylvanian System.	Thickness	Depth
Soil, brown, soft	18	18
Sand, light, white, soft	172	190
Mississippian System.		
Limestone (Big Lime), hard, white	143	333
Shale, hard, blue	63	396
Shale, hard, gray	420	816

Devonian System.   Thickness Depth
    Shale, black, soft (Chattanooga)   145   961
    Shale (fire clay), white, soft   25   986
    Limestone "sand," gray, soft   8   994
    Limestone "sand," hard, dark   15   1,009
    Limestone "sand," light, soft   7   1,016
    Limestone "sand," hard, dark   30   1,046

Silurian System.
    Limestone "sand," light, soft   18   1,064
    Limestone "sand," gray, soft   5   1,069
    Shale, hard, blue, soft   12   1,081
        Total depth   1,081

Log No. 619

Flahaven, No. 81. Commenced: October, 1919. Completed: October 20, 1919.

        Strata.

Pennsylvanian System.   Thickness Depth
    Soil, brown, soft   5   5
    Sand, yellow, soft   35   40
    Shale, hard, blue, soft   150   190

Mississippian System.
    Limestone (Big Lime), hard, white   140   330
    Shale, hard, blue, soft   440   770
    Shale, hard, light, soft   30   800

Devonian System.
    Shale, black, soft (Chattanooga)   149   949
    Shale, red, soft   20   969
    Shale (fire clay), white, soft   15   984
    Limestone "sand," gray, soft, (pay)   8   992
    Limestone "sand," hard, dark, (no good)   17   1,009
    Limestone "sand," gray, soft, (some pay)   15   1,024

Silurian System.
    Limestone "sand," light, soft, (watery)   10   1,034
    Limestone "sand," hard, dark   10   1,044
    Limestone "sand," light, soft   10   1,054
    Shale, hard, blue, soft .... $9\frac{1}{4}$ $1,063\frac{1}{4}$
        Total depth .... $1,063\frac{1}{4}$

Log No. 620

Flahaven, No. 82.   Commenced: October 30, 1919.   Completed: November 11, 1919.

Strata.

Pennsylvanian System.	Thickness	Depth
Soil, brown, soft	3	3
Soil, sandy, light, soft	40	43
Shale, hard, blue, soft	157	200

Mississippian System.		
Limestone (Big Lime), hard, white	136	336
Shale, hard, blue, soft	430	766
Shale (red rock), soft	20	786
Shale, hard, blue, soft	15	801

Devonian System.		
Shale, black, soft (Chattanooga)	155	956
Shale (fire clay), white, soft	20	976
Limestone (cap rock), hard, black	4	980
Limestone "sand," gray, soft, (good pay)	8	988
Limestone "sand," hard, dark	8	996
Limestone "sand," hard, light	18	1,014
Limestone "sand," hard, dark	19	1,033

Silurian System.		
Limestone "sand," light, soft	12	1,045
Shale, hard, blue, soft	9½	1,054½
Total depth		1,054½

Log No. 621

Flahaven No. 83.   Commenced: Dec. 29, 1919.   Completed: Jan. 13, 1920.

Strata.

Pennsylvanian System.	Thickness	Depth
Sand and soil, gray and soft	27	27
Shale, hard, gray, soft	43	70
Sand, gray, soft	10	80
Shale hard, gray, soft	45	125

Mississippian System.   Thickness Depth

    Limestone (Big Lime), white, hard 115 240
    Shale, hard, green, soft 5 245
    Shale hard, gray, soft 20 265
    Shale hard, green, soft 6 271
    Shale, hard, gray, soft 481 752

Devonian System.

    Shale brown, soft (Chattanooga) 123 875
    Shale (fire clay), soft 18 893
    Limestone "sand," brown, hard, (oil) (1st
       pay 896-911, 2nd pay 939-951) 69 962
    Shale, hard, blue, soft 12 974

        Total depth 974

Log No. 622

    Flahaven, No. 88.   Commenced: Oct. 11, 1919.   Completed: Nov. 21, 1919.

        Strata.

Pennsylvanian System.   Thickness Depth

    Soil, brown, soft 15 15
    Shale hard, blue, soft 100 115

Mississippian System.

    Limestone (Big Lime), hard, white 110 225
    Shale, hard, yellow, soft 20 245
    Shale hard, blue, soft 425 670
    Shale (red rock) soft 25 695
    Shale, hard, blue, soft 20 715

Devonian System.

    Shale, black, soft (Chattanooga) 140 855
    Shale (fire clay), white, soft 15 870
    Limestone "sand," gray, soft 57 927

Silurian System.

    Limestone "sand," hard, dark 11 938
        Total depth 938

Log No. 623

Flahaven, No. 89. Commenced: May 4, 1920. Completed: May 18, 1920.

Strata.

Pennsylvanian System.	Thickness	Depth
Soil	7	7
Shale, hard	153	160
Mississippian System.		
Limestone (Big Lime)	105	265
Shale, hard	490	755
Shale (red rock)	10	765
Shale, hard	5	770
Devonian System.		
Shale, black (Chattanooga)	140	910
Shale (fire clay)	10	920
Shale, black	8	928
Limestone "sand"	68	996
Shale, hard	42½	1,038½
Total depth		1,038½

NOTE—The Silurian-Devonian contact occurs within the 68 feet above 996 feet in depth.

Log No. 624

Flahaven, No. 91. Commenced: Aug. 8, 1919. Completed: Sept. 16, 1919.

Strata.

Pennsylvanian System.	Thickness	Depth
Soil, gray, soft	30	30
Shale, hard, gray, soft, shelly	70	100
Shale, hard, white, soft	40	140
Mississippian System.		
Sand and limestone, gray, soft	53	193
Limestone (Big Lime), hard, white	102	295
Shale, green, soft	30	325
Shale, gray, soft	25	350
Grit, white, soft	35	385
Shale, gray, soft	45	430
Shells, gray, soft	5	435

Mississippian System.

	Thickness	Depth
Shale, gray, soft	265	700
Shale, hard, black	55	755
Shale (red rock), soft	15	770
Shale, hard, white, soft	20	790
Shells, hard, dark	2	792

Devonian System.

Shale brown, soft, (Chattanooga)	83	875
Shells, brown, hard, (Chattanooga)	10	885
Shale brown, soft, (Chattanooga)	47	932
Shale (fire clay) white, hard	20	952
Shale, black, hard	5	957
Limestone ''sand,'' gray, soft	5	962
Limestone ''sand,'' hard, dark	10	972
Limestone ''sand,'' light, hard	8	980
Limestone ''sand,'' dark, hard	5	985
Limestone ''sand,'' dark, hard	5	990
Limestone ''sand,'' light, hard	5	995
Limestone ''sand,'' dark, hard	10	1,005

Silurian System.

Limestone ''sand,'' light, hard	20	1,025
Shale, hard, gray soft	10½	1,035½
Total depth		1,035½

Log No. 625

Flahaven, No. 93. Commenced: Sept. 1, 1919. Completed: Sept. 13, 1919.

Strata.

Mississippian System.

	Thickness	Depth
Soil and sand, gray, soft	20	20
Limestone (Big Lime), white, hard	115	135
Shale and shells, hard and soft, light	467	602
Shale (red rock), soft	20	622

Devonian System.

Shale, black, soft (Chattanooga)	130	752
Shale (fire clay), white, soft	16	768
Shale, black, soft	5	773
Limestone ''sand,'' gray, soft, (pay sand) ...	10	783
Limestone ''sand,'' dark, hard	14	797
Limestone ''sand,'' white, hard	13½	810½

Silurian System.	Thickness	Depth
Limestone ''sand,'' hard, dark	28½	839
Limestone ''sand,'' light, soft	8	847
Shale, hard, blue, soft	8⅔	855⅔
Total depth		855⅔

Log No. 626

Flahaven, No. 94.   Commenced: Sept. 11, 1919.   Completed: Sept. 23, 1919.

Strata.

Mississippian System.	Thickness	Depth
Soil, brown, soft	19	19
Limestone (Big Lime), hard, white	78	97
Shale, hard, blue, soft	458	555
Shale (red rock), soft	20	575
Devonian System.		
Shale, black, soft (Chattanooga)	135	710
Shale (fire clay), white, soft	15	725
Limestone ''sand,'' gray, soft	8	733
Limestone ''sand,'' light, soft	4	737
Limestone ''sand,'' hard, dark, (fine stuff) ..	8	745
Limestone ''sand,'' hard, dark, (coarse)	8	753
Limestone ''sand,'' hard, gray	4	757
Limestone ''sand,'' and shale, hard, dark, soft, (break)	12	769
Silurian System.		
Limestone ''sand,'' brown sugar, gritty, brown	12	781
Shale, blue, soft	8	789
Total depth		789

Log No. 627

Flahaven, No. 95.   Commenced: Aug. 20, 1919.   Completed: Aug. 29, 1919.

Strata.

Pennsylvanian System.	Thickness	Depth
Soil, gray, soft	25	25
Shale, hard, blue, soft	105	130

Mississippian System.

	Thickness	Depth
Limestone (Big Lime) white, hard	112	242
Shale, hard, blue, soft	20	262
Shale, hard, light, hard	125	387
Shale, hard, soft	333	720
Shale (red rock), soft	20	740
Shale, hard, light, soft	20	760

Devonian System.

Shale, black, soft (Chattanooga)	119	879
Shale (fire clay), white, soft	20	899
Shale, black, and limestone (cap rock), hard..	3	902
Limestone ''sand,'' gray, hard	36	938
Total depth		938

Log No. 628

Flahaven, No. 96.   Commenced: Sept. 16, 1919.   Completed: Sept. 30, 1919.

Strata.

Pennsylvanian System.

	Thickness	Depth
Soil, brown, soft	14	14
Shale, hard, blue, soft	125	139
Shell, dark, hard	11	150

Mississippian System.

Shale, hard, blue, soft	20	170
Shale, hard, white, soft	15	185
Limestone (Big Lime), hard, white	138	323
Shale, hard, blue, soft	100	423
Shale, hard, light, soft, sandy	100	523
Shale, hard, blue, soft.....................	267	790
Shale (red rock), soft	15	805
Shale, hard, white, soft	20	825

Devonian System.

Shale, black, soft (Chattanooga)	142	967
Shale (fire clay), white, soft	18	985
Shale, hard, black, (cap rock)	2	987
Limestone ''sand,'' gray, soft	10	997
Limestone "sand," hard, dark	5	1,002
Limestone "sand," hard, dark	25	1,027

Silurian System.

Limestone 'sand,'' light, soft	29½	1,056½
Shale, hard, blue, hard	12	1,068½
Total depth		1,068½

Log No. 629

Flahaven, No. 100.   Commenced: Oct. 21, 1919.   Completed: Nov. 24, 1919.

Strata.

Pennsylvanian System.	Thickness	Depth
Soil, brown, soft	6	6
Sand, red, soft	70	76
Shale, hard, blue, soft	131	207
Mississippian System.		
Limestone (Big Lime) hard, white	117	324
Shale, hard, light, soft	150	474
Shale, hard, blue, soft	330	804
Shale (red rock), soft	20	824
Shale, hard, white, soft	20	844
Devonian System.		
Shale, black, soft (Chattanooga)	132	976
Shale (fire clay), white, soft	17	993
Limestone (cap rock), hard, black	2	995
Limestone "sand," gray, soft	8	1,003
Limestone "sand," dark, hard	17	1,020
Limestone "sand," gray, soft	7	1,027
Limestone "sand," dark, hard	19	1,046
Silurian System.		
Limestone "sand," brown, soft	16	1,062
Shale, hard, blue soft	11½	1,073½
Total depth		1,073½

Log No. 630

Flahaven, No. 103.   Commenced: Dec. 8, 1919.   Completed: Dec. 31, 1919.

Strata.

Pennsylvanian System.	Thickness	Depth
Soil and sand, white, gray, soft	180	180
Mississippian System.		
Limestone (Big Lime), white, hard	168	348
Shale, blue, soft	425	773

Devonian System.   Thickness Depth

 Shale, black, soft (Chattanooga) 165 938
 Shale (fire clay), white, soft 18 956
 Limestone ''sand,'' gray, soft 8 964
 Limestone ''sand,'' hard, dark 18 982
 Limestone ''sand,'' light, soft 7 989
 Limestone ''sand,'' hard, dark 12 1,001

Silurian System.

 Limestone ''sand,'' brown, soft 17 1,018
 Limestone ''sand,'' dark, soft 3 1,021
 Shale, blue, soft 13 1,034
  Total depth 1,034

Log No. 631

Flahaven, No. 104.   Commenced: Oct. 20, 1919.   Completed: Oct. 31, 1919.

 Strata.
Pennsylvanian System.   Thickness Depth

 Soil, brown, soft 14 14
 Shale' hard, blue, soft 114 128
 Sand, white, soft 12 140
 Shale, hard, blue, soft 44 184
 Sand, white, soft 18 202

Mississippian System.

 Shale, hard, blue, soft 17 219
 Limestone (Big Lime), hard, white 100 319
 Shale, hard, green, soft 10 329
 Shale, hard, blue, soft 475 804

Devonian System.

 Shale, black, soft (Chattanooga) 165 969
 Shale (fire clay), white, soft 20 989
 Limestone ''sand,'' gray, soft 5 994
 Limestone ''sand,'' hard, dark 5 999
 Limestone ''sand,'' gray, soft 5 1,004
 Limestone ''sand,'' hard, dark 43 1,047

Silurian System.

 Limestone "sand," light, soft 12½ 1,059½
 Shale, hard, blue, soft 8 1,067½
  Total depth 1,067½

Log No. 632

Flahaven, No. 107.   Commenced: Oct. 4, 1919.   Completed: Oct. 21, 1919.

Strata.

Pennsylvanian System.	Thickness	Depth
Soil, brown, soft	15	15
Shale, hard, blue, soft	50	65
Sandstone, white, soft	25	90

Mississippian System.		
Limestone (Big Lime), white, hard	95	185
Shale, hard, blue	447	632
Shale (red rock), hard	15	647
Shale, hard, blue	20	667

Devonian System.		
Shale, black, soft (Chattanooga)	147	814
Shale (fire clay), white, soft	15	829
Limestone (cap rock), hard, black	5	834
Limestone "sand," gray, dark, hard, soft	40	874
Limestone, gray, soft	15	889

Silurian System.		
Limestone "sand," gray, soft, (pay)	12	901
Limestone light, soft	6	907
Limestone, dark, soft	5	912
Shale, hard, blue, soft	6	918
Shale, hard, blue, soft	7½	925½
Total depth		925½

Log No. 633

Flahaven, No. 108.   Commenced: Oct. 7, 1919.   Completed: Nov. 22, 1919.

Strata.

Pennsylvanian System.	Thickness	Depth
Soil	20	20
Sandstone, light, soft	25	45
Shale, hard, blue, soft	180	225

Mississippian System.		
Limestone (Big Lime), white, hard	95	320
Shale, hard, blue, soft	465	785
Shale (red rock), soft	10	795
Shale, hard, blue, soft	20	815

Devonian System

	Thickness	Depth
Shale black, soft (Chattanooga)	138	953
Shale (fire clay), white, soft	20	973
Limestone (cap rock), hard, black	4	977
Limestone ''sand,'' gray, soft	8	985
Limestone ''sand,'' hard, dark	15	1,000
Limestone ''sand,'' light, hard	35	1,035

Silurian System.

Limestone ''sand,'' brown sugar, medium ..	18	1,053
Shale, blue, soft	6	1,059
Shale, hard, red, soft	6⅔	1,065⅔
Total depth		1,065⅔

Log No. 634

Flahaven, No. 109. Commenced: Oct. 25, 1919: Completed: Nov. 13, 1919.

Strata.

Pennsylvanian System.

	Thickness	Depth
Soil, brown, soft	14	14
Sand, light, soft	30	44
Shale, hard, blue, soft	156	200

Mississippian System.

Limestone (Big Lime), hard, white	100	300
Shale hard, green, soft	75	375
Shale, hard, blue, soft	390	765
Shale (red rock), soft	20	785
Shale, hard, blue, soft	15	800

Devonian System.

Shale, black, soft (Chattanooga)	144	944
Shale (fire clay), white, soft	18	962
Limestone (cap rock), hard, black	2	964
Limestone ''sand,'' gray, hard	8	972
Limestone ''sand,'' dark, hard	7	979
Limestone ''sand,'' light, soft	19	998
Limestone ''sand,'' hard, dark	15	1,013

Silurian System.

Limestone ''sand,'' light, soft, coarse	29	1,042
Shale, hard, blue, soft	13½	1,055½
Total depth		1,055½

Log No. 635

Flahaven Land Co., No. 1, lessor. Ohio Oil Co,. lessee (logs 1-80 following). Location: The following records (1-80) are of wells drilled by the Ohio Oil Co. on its 1000 acre lease from the Flahaven Land Co. This tract is a sub-division of the original Flahaven farm of 2,490 acres, and is located about one mile south of Greeley P. O., south of the juncture of Little Sinking and Big Sinking Creeks, Lee Co., Ky. . The general location is about 8 miles east of Old Landing. Commenced: Feb. 20, 1918. Completed: Mar. 23, 1918. Production: natural production first 24 hours estimated at 10 bbls. oil. Authority: Ohio Oil Co. for this and immediately following logs (1-80) of the Flahaven Land Co.

Strata.

	Thickness	Depth
Pennsylvanian System.		
Soil, gray, soft	14	14
Shale, hard, brown	136	150
Mississippian System.		
Limestone (Big Lime), hard, gray	165	315
Shale, hard, brown	485	800
Devonian System.		
Shale, brown, hard (Chattanooga)	135	935
Shale (fire clay), gray, soft	12	947
Limestone (cap rock), hard, black	10	957
Limestone "sand," hard, light	4	961
Limestone, black, hard	16	977
Limestone "sand," hard, light	10	987
Limestone "sand," dark gray, hard	7	994
Silurian System.		
Limestone "sand," brown, hard	22	1,016
Limestone "sand," hard, light, (pay)	3	1,019
Shale, hard, green	2	1,021
Total depth		1,021

Log No. 636

Flahaven, No. 3. Commenced: June 14, 1918. Completed: July 18; 1918. Production: commenced producing July 15, 1918; natural production for first and second 24 hours, 30 bbls.

Strata.

Pennsylvanian System.   Thickness Depth

	Thickness	Depth
Soil, gray, soft	10	10
Sand, hard, gray	8	18
Shale, hard, brown	92	110
Coal, soft, black	3	113

Mississippian System.

	Thickness	Depth
Shale, hard, green	12	125
Limestone (Big Lime), hard, gray	145	270
Shale, hard, brown	480	750

Devonian System.

	Thickness	Depth
Shale, hard, brown (Chattanooga)	138	888
Shale (fire clay), gray, soft	16	904
Limestone (cap rock), hard, black	6	910
Limestone, hard, black	11	921
Limestone, hard, gray	19	940
Limestone "sand," hard, gray, (oil show)	3	943
Limestone, hard, gray	10	953

Silurian System.

	Thickness	Depth
Limestone "sand," hard, dark gray, (oil)	6	959
Limestone "sand," hard, dark gray	6	965
Limestone "sand," light gray, hard	4	969
Shale, hard, blue	½	969½
Total depth		969½

Casing record:

Size	Length
10"	18'
8¼"	45'
6¼"	275'
2"	960
⅝"	950'

Log No. 637

Flahaven, No. 4. Commenced: June 8, 1918. Completed: June 25, 1918. Production: commenced producing July 24, 1918; natural production first 24 hours, 30 bbls; natural production after second 24 hours, 15 bbls.; production after first 48 hours, after shot, 100 bbls.

Strata.

Mississippian System.   Thickness Depth

	Thickness	Depth
Soil, gray, soft	5	5
Limestone (Big Lime), hard, gray	143	148
Shale, hard, green	50	198

Mississippian System.	Thickness	Depth
Shale, hard, brown	392	590
Shale (red rock), hard	12	602

Devonian System.		
Shale, brown, hard (Chattanooga)	140	742
Shale (fire clay.), soft, gray	15	757
Limestone (cap rock), hard, black	3	760
Limestone ''sand,'' hard, brown, (oil)·	10	770
Limestone, hard, black	12	782
Limestone ''sand,'' hard, dark gray	8	790
Limestone ''sand,'' hard, brown	8	798
Limestone ''sand,'' hard, dark gray	7	805

Silurian System.		
Shale, hard, brown	4	809
Limestone ''sand,'' hard, dark gray, (some pay)	12	821
Shale, hard, blue	3	824
Total depth		824

Log No. 638

Flahaven, No. 5. Commenced: Aug. 19, 1918. Completed: Sept. 3, 1918. Production: commenced producing Sept. 4, 1918; natural production for the first 24 hours, 100 bbls.; natural production for the second 24 hours, 60 bbls. Production after shot was 150 bbls. for the first 24 hours, and 100 bbls. for the second 24 hours. The color of the oil was green.

Strata:

Pennsylvanian System.	Thickness	Depth
Soil, gray, soft	8	8
Limestone (Big Lime), hard, white	122	130
Shale, hard, blue	460	590
Shale (red rock), soft	12	602

Devonian System.		
Shale, brown, soft (Chattanooga)	156	758
Shale (fire clay), white, soft	16	774
Limestone (cap rock), hard, black	2	776
Limestone ''sand,'' hard, brown	10	786
Limestone ''sand,'' hard, dark	8	794
Limestone ''sand,'' hard, dark, (oil show)	7	801
Limestone ''sand,'' hard, light	11	812
Limestone, hard, gray	14	826

Silurian System.

	Thickness	Depth
Limestone "sand," hard, light	14	840
Shale, hard, blue	1	841
Total depth		841

Log No. 639

Flahaven, No. 6. Commenced: July 24, 1918. Completed: Aug.. 13, 1918. Production: commenced producing Aug. 20, 1918; natural production at end of 48 hours, 40 bbls.; natural production at end of 48 hours after shot, 120 bbls.

Strata.

Mississippian System.

	Thickness	Depth
Soil, soft, gray	8	8
Limestone (Big Lime), hard, white	140	148
Shale, hard, blue	457	605
Shale (red rock), soft	12	617
Shale, hard, blue	18	635

Devonian System.

	Thickness	Depth
Shale, brown, soft (Chattanooga)	147	782
Shale (fire clay), white, soft	9	791
Limestone (cap rock), hard, dark	7	798
Limestone "sand," hard, brown, (oil)	10	808
Limestone, hard, dark	10	818
Limestone "sand," hard, light, (oil show)	8	826
Limestone "sand," hard, light	7	833
Limestone "sand," hard, dark, (no pay)	8	841
Total depth		841

Log No. 640

Flahaven, No. 7. Commenced Sept. 6, 1918. Completed: Sept. 23, 1918.

Strata.

Mississippian System.

	Thickness	Depth
Soil, gray, soft	8	8
Shale, hard, blue	27	35
Limestone (Big Lime), hard, white	115	150
Shale, hard, blue	470	620
Shale (red rock), soft	12	632
Shale, hard, blue	28	660

Devonian System.   Thickness Depth

    Shale, brown, soft (Chattanooga)   140   800
    Shale (fire clay), white, soft.............   12   812
    Limestone (cap rock), hard, dark   5   817
    Limestone "sand," hard, brown, (oil)   9   826
    Limestone "sand," hard, dark   10   836
    Limestone "sand," hard, brown, (oil) ....   · 10   846
    Limestone, hard, gray   14   860

Silurian System.

    Limestone "sand," hard, light   15   875
    Shale, hard, blue   6   881
        Total depth   881

Log No. 641

Flahaven, No. 8. Commenced: June 30, 1918. Completed: July 28, 1918. Production: commenced producing Aug. 14' 1918; natural production at the end of 48 hours, 37 bbls.; natural production after shot at end of 48 hours, 125 bbls.

    Strata.
Mississippian System.   Thickness Depth

    Soil, brown, soft   6   6
    Shale, hard, brown   39   45
    Limestone (Big Lime), hard, gray   155   200
    Shale, hard, blue   450   650
    Shale (red rock), hard   15   665
    Shale, hard, brown (Sunbury)   15   680

Devonian System.

    Shale, brown, hard (Chattanooga)   138   818
    Shale (fire clay), gray, soft   12   830
    Limestone (cap rock), hard, black   10   840
    Limestone "sand," hard, brown, (1st pay) ..   10   850
    Limestone, hard, black   16   866
    Limestone "sand," hard, dark gray, (2nd pay)   10   876
    Limestone, hard, gray, sandy, (no pay)   6   882

Silurian System.

    Limestone "sand," hard, gray, (pay oil same
      in hole)   6   888
    Limestone, hard, brown, sandy   3   891
    Limestone "sand," hard, gray, (little pay) ..   13   904
    Shale, hard, blue   2   906
        Total depth   906

Log No. 642

Flahaven, No. 9. Commenced: August 26, 1918. Completed: September 14, 1918. Production: Commenced producing September 15, 1918; natural production after first 48 hours, 40 bbls.; natural production 48 hours after shot, 150 bbls.

Strata.

	Thickness	Depth
Mississippian System.		
Soil, gray, soft	7	7
Shale, hard, blue	33	40
Gravel, soft, white	5	45
Limestone (Big Lime), hard, white	155	200
Shale, hard, blue	468	668
Shale (red rock), soft	12	680
Shale, hard, blue	20	700
Devonian System.		
Shale, brown, soft (Chattanooga)	143	843
Shale (fire clay), white, soft	12	855
Limestone (cap rock), hard, black	10	865
Limestone "sand," hard, gray, (oil)	10	875
Limestone "sand," hard, dark	14	889
Limestone "sand," hard, light	6	895
Shale, hard, blue	5	900
Silurian System.		
Limestone, hard, dark	13	913
Limestone "sand," hard, light	12	925
Shale, hard, blue	3	928
Total depth		928

Log No. 643

Flahaven, No. 10. Commenced: October 2, 1918. Completed: November 27, 1918. Production: Commenced producing December 7, 1918; production first 24 hours after shot, 30 bbls.

Strata.

	Thickness	Depth
Pennsylvanian System.		
Soil, soft, gray	8	8
Shale, hard, blue	82	90
Mississippian System.		
Limestone (Big Lime), hard, white	144	234
Shale, hard, blue	466	700
Shale (red rock), soft	16	716
Shale, hard, blue	20	736

Devonian System.

	Thickness	Depth
Shale, brown, soft (Chattanooga)	134	870
Shale (fire clay), white, soft	15	885
Limestone (cap rock), hard, black	4	889
Limestone "sand," hard, brown, (oil)	10	899
Limestone "sand," hard, dark	17	916
Limestone "sand," hard, light	8	924
Limestone "sand," hard, light, (oil)	6	930

Silurian System.

Limestone, hard, gray	12	942
Limestone "sand," hard, white	8	950
Limestone "sand," fine, hard, white	4	954
Total depth		954

Log No. 644

Flahaven, No. 11. Commenced: December 9, 1918. Completed: January 4, 1919. Production: Commenced producing January 18, 1919; production after first 48 hours after shot, 10 bbls.

Strata.

Pennsylvanian System.

	Thickness	Depth
Soil, gray, soft	10	10
Shale, hard, blue	90	100

Mississippian System.

Limestone (Big Lime), hard, white	148	248
Shale, hard, blue	472	720
Shale (red rock), soft	15	735
Shale, hard, blue	20	755

Devonian System.

Shale, brown, soft (Chattanooga)	140	895
Shale (fire clay), white, soft	10	905
Limestone (cap rock), hard, black	4	909
Limestone "sand," hard, dark, (oil 914)	11	920
Limestone "sand," hard, light	6	926
Limestone "sand," hard, white	14	940
Limestone "sand," fine, hard, dark	14	954

Silurian System.

Limestone and sand, hard, dark	11	965
Limestone "sand," hard, white	11	976
Shale, hard, blue	5	981
Total depth		981

Log No. 645

Flahaven, No. 12. Commenced: March 11, 1920. Completed: April 20, 1920. Production: Commenced producing April 23, 1920; production first 48 hours after shot, 2 bbls.

Strata.

Pennsylvanian System.	Thickness	Depth
Soil, yellow, soft	10	10
Sand, hard, white	65	75
Mississippian System.		
Shale, hard, blue	25	100
Limestone (Big Lime), hard, white	110	210
Shale, green, soft	15	225
Shale, hard, blue	500	725
Devonian System.		
Shale, brown, soft (Chattanooga)	120	845
Shale (fire clay), white, soft	20	865
Limestone (cap rock), hard, black	5	870
Limestone "sand," hard, brown	3	873
Limestone, hard, dark	13	886
Limestone "sand," light, hard	3	889
Limestone "sand," hard, gray	7	896
Limestone, hard, dark	6	902
Limestone "sand," hard, brown	9	911
Total depth		911

Log No. 646

Flahaven, No. 13. Commenced: January 1, 1920. Completed: January 27, 1920. Production: Commenced producing February 2, 1920; production 48 hours after shot, 29 bbls.

Strata.

Pennsylvanian System.	Thickness	Depth
Soil, black, soft	10	10
Sand, white	30	40
Shale, hard	70	110
Mississippian System.		
Limestone (Big Lime), hard	140	250
Shale, hard, soft	496	746

Devonian System.	Thickness	Depth
Shale, brown, soft (Chattanooga)	139	885
Shale (fire clay), white, soft	18	903
Limestone (cap rock), hard, black	2	905
Limestone "sand," brown, soft, (oil)	14	919
Total depth		919

Log No. 647

Flahaven, No. 14. Commenced: August 25, 1918. Completed: September 15, 1918. Production: Commenced producing September 16, 1918; production after the first 24 hours after shot, 150 bbls.

Strata.		
Pennsylvanian System.	Thickness	Depth
Sand, hard, brown	15	15
Shale, black, soft	15	30
Mississippian System.		
Limestone, white, soft, (Big Lime)	35	65
Limestone, white, hard (Big Lime)	70	135
Shale, green, soft	30	165
Limestone, hard, white	25	190
Unnamed substance	18	208
Shale, white, soft	392	600
Shale (red rock), soft	25	625
Shale, gray	5	630
Devonian System.		
Shale, brown, soft (Chattanooga)	145	775
Shale (fire clay), soft	10	785
Limestone (cap rock), black	5	790
Limestone "sand," gray, (filled up with oil 125 feet)	5	795
Limestone "sand," (oil)	7	802
Limestone "sand," gray	5	807
Limestone "sand," hard, dark	3	810
Limestone, dark	4	814
Limestone, light gray	5	819
Limestone, dark	2	821
Limestone, light gray	6	827
Limestone	10	837
Silurian System.		
Limestone, (oil)	8	845
Limestone "sand," light	5	850
Shale, hard	2	852
Total depth		852

Log No. 648

Flahaven, No. 15. Commenced: July 25, 1918. Completed: August 9, 1918. Production: Commenced producing August 13, 1918; production after first 48 hours after shot, 5 bbls.

Strata.

	Thickness	Depth
Pennsylvanian System.		
Soil, pink, soft	10	10
Sand, hard, gray	6	16
Shale, hard, black	10	26
Sand, hard, gray	74	100
Shale, hard, gray, soft	35	135
Mississippian System.		
Limestone (Big Lime), hard, gray	120	255
Shale, hard, green, soft	10	265
Shale, hard, gray	480	745
Devonian System.		
Shale, brown, hard (Chattanooga)	141	886
Shale (fire clay), gray, soft	15	901
Limestone (cap rock), hard, black	5	906
Limestone "sand," brown, hard, (oil)	5	911
Limestone, hard, black	4	915
Limestone "sand," hard, black, (no pay)	2	917
Limestone, hard, gray	5	922
Limestone "sand," hard, gray, (no pay)	9	931
Limestone "sand," hard, gray, (best oil)	11	942
Limestone, hard, gray	6	948
Limestone, hard, brown	6	954
Silurian System.		
Limestone "sand," hard, light, (no water)	6	. 960
Limestone "sand," hard, brown, (no pay)	3	963
Shale, hard, blue	2	965
Total depth		965

Log No. 649

Flahaven, No. 16. Commenced: September 26, 1918. Completed: October 17, 1918. Production: Commenced producing October 19, 1918; natural production after first 48 hours, 75 bbls. oil; production after 48 hours after shot, 100 bbls.

Strata.

Pennsylvanian System.	Thickness	Depth
Shale, hard	12	12
Sand	18	30

Mississippian System.		
Shale	15	45
Limestone (Big Lime)	125	170
Shale, hard	25	195
Limestone	25	220
Shale	410	630
Shale (red rock)	15	645
Shale, hard	20	665

Devonian System.		
Shale, brown (Chattanooga)	135	800
Fire clay	17	817
Limestone (cap rock)	5	822
Limestone "sand," (first pay)	10	832
Limestone "sand," dark	4	836
Limestone, sandy	9	845
Limestone	6	851
Shale, hard	5	856
Limestone	8	864
Limestone "sand"	16	880
Total depth		880

Log No. 650

Flahaven, No. 17. Commenced: November 6, 1918. Completed: November 30, 1918. Production: 24 hours after shot, 80 bbls.

Strata.

Mississippian System.	Thickness	Depth
Soil, gray, soft	10	10
Limestone (Big Lime), hard, white	90	100
Shale, hard, blue, soft	485	585
Shale (red rock), soft	5	590
Shale, hard, blue, soft	10	600

Devonian System.		
Shale, brown, soft (Chattanooga)	140	740
Shale (fire clay), white, soft	14	754
Limestone (cap rock), hard, black	4	758
Limestone "sand," hard, dark	2	760

THE CORNIFEROUS "SAND."

An exposure of the Onondage Limestone (Corniferous "sand") on the L. & N. R. R., northwest of Irvine, Estill County, Kentucky. Note the hammer on the second ledge for size.

Devonian System.	Thickness	Depth
Limestone "sand," hard, brown (oil)	10	770
Limestone "sand," hard, dark	20	790
Limestone "sand," hard, brown, (oil)	6	796
Limestone, hard, gray	4	800

Silurian System.		
Limestone "sand," hard, brown, (oil)	8	808
Limestone "sand," hard, white	4	812
Limestone "sand," hard, dark	12	824
Shale, hard, blue, soft	5	829
Total depth		829

Log No. 651

Flahaven, No. 18. Commenced: November 1, 1918. Completed: November 14, 1918. Production: Commenced producing November 15, 1918; natural production after first 48 hours, 75 bbls.; production after first 24 hours after shot, 180 bbls. oil.

Strata.

Pennsylvanian System.	Thickness	Depth
Soil, gray, soft	12	12
Sand, brown, hard	23	35
Shale, hard, brown	15	50

Mississippian System.		
Limestone (Big Lime), hard, gray	140	190
Shale, hard	20	210
Shale, soft, (soapstone)	430	640
Shale (red rock)	12	652
Shale, hard	10	662

Devonian System.		
Shale, black (Chattanooga)	148	810
Shale (fire clay)	20	830
Limestone (cap rock)	5	835
Limestone "sand," gray	10	845
Limestone, black	5	850
Limestone "sand"	40	890
Shale, hard	4	894
Total depth		894

NOTE—The Devonian-Silurian contact is toward the bottom of the 40 feet of limestone above 890.

Log No. 652

Flahaven, No. 19. Commenced: December 5, 1918. Completed: January 4, 1919. Production: commenced producing January 16, 1919; production first 48 hours after shot, 145 bbls. Shot January 6, 1919.

Strata.

	Thickness	Depth
Pennsylvanian System.		
Soil, soft, dark	28	28
Sandstone, yellow, soft	2	30
Shale, soft, blue	35	65
Mississippian System.		
Limestone (Big Lime), hard, white	135	200
Shale, hard, blue, soft	490	690
Shale (red rock), soft	12	702
Shale, hard, blue, soft	18	720
Devonian System.		
Shale, brown, soft (Chattanooga)	119	839
Shale (fire clay), white, soft	20	859
Limestone (cap rock), hard, black	5	864
Limestone "sand," dark, hard, (oil)	10	874
Limestone "sand" and limestone, hard, light	20	894
Limestone "sand," hard, white	8	902
Limestone, hard, dark	4	906
Limestone "sand," hard, white	8	914
Silurian System.		
Limestone "sand," hard, brown, (no oil)	6	920
Shale, hard, blue, soft	3	923
Total depth		923

Log No. 653

Flahaven, No. 20. Commenced: Feb. 20, 1919. Completed: Mar. 12, 1919. Production: commenced producing Mar. 20, 1919; production first 48 hours after shot, 100 bbls. Shot March 23, 1919.

Strata.

	Thickness	Depth
Pennsylvanian System.		
Soil, gray, soft	14	14
Sandstone, yellow, soft	16	30
Shale, blue, soft	60	90

Mississippian System.	Thickness	Depth
Limestone (Big Lime), hard, white	120	210
Shale, blue, hard, soft	470	680
Shale (red rock), soft	10	690

Devonian System.		
Shale, brown, soft (Chattanooga)	150	840
Shale (fire clay), white, soft	20	860
Limestone (cap rock), hard, black	4	864
Limestone "sand," hard, white :............	20	884
Limestone "sand," brown, hard	20	904
Total depth		904

Log No. 654

. Flahaven, No. 21. Commenced: Dec. 23, 1918. Completed: Jan. 21, 1919. Production: commenced producing Jan. 25, 1919; production after first 48 hours after shot, 20 bbls. Shot Jan. 24, 1919.

Strata.

Pennsylvanian System.	Thickness	Depth
Sand, white	20	20
Shale, gray	40	60
Shale, gray	40	100
Shale	100	200

Mississippian System.		
Limestone (Big Lime)	150	350
Shale, gray ....'	25	375
Limestone	25	400
Shale, soft	400	800
Shale (red rock)	15	815

Devonian System.		
Shale, brown (Chattanooga)	171	986
Shale (fire clay)	15	1,001
Limestone "sand," (pay)	13	1,014 -
Limestone, black	6	1,020
Limestone, dark	38	1,058
Total depth		1,058

Log No. 655

Flahaven, No. 22. Commenced: Feb. 5, 1919. Completed: Feb.
20, 1919. Production: second 24 hours after shot, 75 bbls. Shot Feb.
4, 1919.

Strata.

Pennsylvanian System.	Thickness	Depth
Sand, brown, soft	50	50
Shale, hard, black	200	250

Mississippian System.		
Limestone (Big Lime), hard, white	150	400
Shale, soft, white (soapstone)	440	840
Shale (red rock), soft	15	855
Shale, hard, white	15	870

Devonian System.		
Shale, brown, soft (Chattanooga)	140	1,010
Shale (fire clay), white	20	1,030
Shale, hard, black	4	1,034
Limestone (cap rock)	5	1,039
Limestone "sand," (oil)	5	1,044
Shell	2	1,046
Limestone "sand," (oil)	6	1,052
Limestone "sand," dark gray	2	1,054
Limestone "sand," gray	4	1,058
Limestone "sand," limy	22	1,080
Total depth		1,080

Log No. 656

Flahaven, No. 23. Commenced: Feb. 7, 1919. Completed: Mar.
17, 1919. Production: commenced producing Mar. 22, 1919; production after 48 hours after shot, 75 bbls. Shot Mar. 18, 1919.

Strata.

Pennsylvanian System.	Thickness	Depth
Soil, soft, dry	20	20
Sandstone, soft, yellow	25	45
Shale, soft, blue	40	85
Shale, hard, blue, soft	120	205

Mississippian System.		
Limestone (Big Lime), hard, white	140	345
Shale, hard, blue, soft	500	845

Devonian System.

	Thickness	Depth
Shale, brown, soft (Chattanooga)	130	975
Shale (fire clay), white, soft	14	989
Limestone (cap rock), hard, black	3	992
Limestone ''sand,'' hard, brown	9	1,001
Limestone ''sand,'' hard, dark	20	1,021
Limestone ''sand,'' white, hard	7	1,028
Limestone ''sand'' and lime, hard, dark	3	1,031
Total depth		1,031

Log No. 657

Flahaven, No. 24. Commenced: Mar. 26, 1919. Completed: Apr. 9, 1919. Production: commenced producing April 14, 1919; production after 48 hours after shot, 100 bbls. Shot April 11, 1919.

Strata.

Pennsylvanian System.

	Thickness	Depth
Sand, white, soft	30	30
Shale	175	205

Mississippian System.

Limestone (Big Lime)	135	340
Shale, green	17	357
Shale, soft, sandy	501	858

Devonian System.

Shale, brown (Chattanooga)	137	995
Limestone (cap rock)	8	1,003
Limestone ''sand,'' (oil)	15	1,018
Limestone, black	2	1,020
Limestone ''sand,'' white	17	1,037
Total depth		1,037

Log No. 658

Flahaven, No. 25. Commenced: May 30, 1919. Completed: June 14, 1919. Production: commenced producing June 17, 1919: production 48 hours after shot, 50 bbls. Shot June 16, 1919.

Strata.

Pennsylvanian System.

	Thickness	Depth
Soil, red, soft	15	15
Shale, black, soft	170	185

Mississippian System.	Thickness	Depth
Limestone (Big Lime), hard, white	160	345
Shale, soft, white	480	825
Shale (red rock), soft	5	830
Devonian System.		
Shale, brown, soft (Chattanooga)	140 ·	970
Shale (fire clay), gray, soft	22	992
Limestone (cap rock), hard, black	3	995
Limestone "sand," light brown, hard	12	1,007
Limestone, hard, black	4	1,011
Limestone "sand," hard, gray	14	1,025
Total depth		1,025

Log No. 659

Flahaven, No. 26. Commenced: April 29, 1919. Completed: May 12, 1919. Production: commenced producing May 15, 1919; production 48 hours after shot, 150 bbls. Shot May 13, 1919.

Strata.

Pennsylvanian System.	Thickness	Depth
Sand	12	12
Shale, soft	193	205
Mississippian System.		
Limestone (Big Lime), hard	135	340
Shale, soft	502	842
Devonian System.		
Shale, brown, soft (Chattanooga)	135	977
Shale (fire clay), soft	15	992
Limestone (cap rock), hard	2	994
Limestone "sand," (oil)	14	1,008
Limestone, dark	4	1,012
Limestone, gray	15	1,027
Limestone "sand," light	2	1,029
Total depth		1,029

Log No. 660

Flahaven, No. 27. Commenced: Dec. 20, 1918. Completed: Mar. 3, 1919. Production: commenced producing Mar. 17, 1919; production 48 hours after shot, 5 bbls. Shot March 11, 1919.

Strata.

Mississippian System.	Thickness	Depth
Soil, gray, soft	1	1
Limestone (Big Lime), hard, white	85	86
Shale hard, blue, soft	514	600

Devonian System.	Thickness	Depth
Shale, brown, soft (Chattanooga)	136	736
Shale (fire clay), white, soft	16	752
Limestone (cap rock), hard, dark	2	754
Limestone ''sand,'' hard, dark	20	774
Limestone ''sand,'' hard, white	41	815
Total depth		815

NOTE—The Devonian-Silurian contact is toward the base of the last 41 feet.

Log No. 661

Flahaven, No. 28. Commenced: Jan. 10, 1919. Completed: Jan. 28, 1919. Production: commenced producing Jan. 31, 1919; production 48 hours after shot, 125 bbls. Shot Jan. 29, 1919.

Strata.

Mississippian System.	Thickness	Depth
Soil, gray, soft	15	15
Limestone (Big Lime), hard, white	105	120
Shale, hard, blue, soft	500	620
Devonian System.		
Shale, brown, soft (Chattanooga)	130	750
Shale (fire clay), blue, soft	12	762
Limestone (cap rock), hard, black	2	764
Limestone ''sand,'' hard, brown, (oil)	10	774
Limestone ''sand,'' hard, dark, fine	31	805
Limestone ''sand,'' hard, light, (oil show) ..	5	810
Limestone ''sand,'' hard, dark	5	815
Silurian System.		
Limestone ''sand,'' hard, brown, (oil show) ..	5	820
Shale, hard, blue, soft	4	824
Total depth		824

Log No. 662

Flahaven, No. 29. Commenced: Feb. 8, 1919. Completed: Mar. 5, 1919. Production: commenced producing Mar. 18, 1919; production 48 hours after shot, 115 bbls. Shot Mar. 11, 1919.

Strata.

Mississippian System.	Thickness	Depth
Soil, gray, soft	20	20
Limestone (Big Lime), hard, white	100	120
Shale, hard, blue, soft	490	610

Devonian System.

	Thickness	Depth
Shale, brown, soft (Chattanooga)	144	754
Shale (fire clay), white, soft	16	770
Limestone (cap rock), hard, dark	4	774
Limestone "sand," hard, dark, (oil)	12	786
Limestone "sand," hard, dark	47	833
Shale, hard, blue, soft	2	835
Total depth ....·...........		835

NOTE—The Devonian-Silurian contact is toward the base of the 47 feet of limestone above 833 feet in depth.

Log No. 663

Flahaven, No. 31. Commenced: Mar. 25, 1919. Completed: Apr. 7, 1919. Production: commenced producing Apr. 7, 1919; production 48 hours after shot, 144 bbls. Shot Apr. 8, 1919.

Strata.

Pennsylvanian System.

	Thickness	Depth
Soil, gray, soft	15	15
Shale, hard, blue, soft	128	143
Mississippian System.		
Limestone (Big Lime), hard, white	126	269
Shale, hard, blue, soft	486	755
Shale (red rock), soft	10	765
Shale, hard, blue, soft	10	775
Devonian System.		
Shale, brown, soft (Chattanooga)	130	905
Shale (fire clay), white, soft	15	920
Limestone (cap rock), hard, dark	2	922
Limestone "sand," brown, hard, (oil)	15	937
Limestone "sand," hard, dark, (dry)	3	940
Limestone "sand," brown, hard, (oil)	5	945
Limestone "sand," and lime, hard, dark	12	957
Total depth		957

Log No. 664

Flahaven, No. 32. Commenced: April 7, 1919. Completed: April 29, 1919. Production: commenced producing May 4, 1919; production 48 hours after shot, 80 bbls. Shot April 28, 1919.

Strata.

Pennsylvanian System.

	Thickness	Depth
Soil, gray, soft	20	20
Shale, hard, blue, soft	110	130

Mississippian System.	Thickness	Depth
Limestone (Big Lime), hard, white	120	250
Shale, hard, blue, soft	470	720
Shale (red rock), soft	10	730
Shale, hard, blue, soft	25 -	755
Devonian System.		
Shale, brown, soft (Chattanooga)	135	890
Shale (fire clay), white, soft	16	906
Limestone (cap rock), hard, black	4	910
Limestone "sand," hard, brown, (oil)	14	924
Limestone "sand," hard, dark, (no oil)	18	942
Total depth		942

Log No. 665

Flahaven, No. 33. Commenced: April 23, 1919. Completed May 12, 1919. Production: commenced producing May 16, 1919; production after 48 hours after shot, 100 bbls. Shot May 13, 1919.

Strata.		
Pennsylvanian System.	Thickness	Depth
Soil, gray, soft	25	25
Shale, hard, blue, soft	75	100
Sandstone, red, soft	20	120
Mississippian System.		
Shale, hard, blue, soft	20	140
Limestone (Big Lime), hard, white	120	260
Shale, hard, blue, soft	470	730
Shale (red rock), soft	12	742
Shale, hard, blue, soft	27	769
Devonian System.		
Shale, brown, soft (Chattanooga)	131	900
Shale (fire clay), white, soft	16	916
Limestone (cap rock), hard, dark	3	919
Limestone "sand," brown, hard, (oil)	13	932
Limestone "sand," hard, dark, (no oil)	25½	957½
Total depth		957½

Log No. 686

Flahaven, No. 34. Commenced: April 23, 1919. Completed: May 7, 1919. Production: commenced producing May 10, 1919; production 48 hours after shot, 140 bbls. Shot May 8, 1919.

Strata.		
Pennsylvanian System.	Thickness	Depth
Soil, gray, soft	1	1
Sandstone, yellow, soft	49	50
Shale, hard, blue, soft	135	185

Mississippian System. — Thickness / Depth

	Thickness	Depth
Limestone (Big Lime), hard, gray	145	330
Shale, hard, blue, soft	470	800
Shale (red rock) soft, ...	10	810
Shale, hard, blue, soft	10	820
Devonian System.		
Shale, brown, soft (Chattanooga)	140	960
Shale (fire clay), white, soft	13	973
Limestone (cap rock), hard, dark	4	977
Limestone "sand," brown, hard, (oil)	11	988
Limestone "sand," hard, dark, (no oil)	12	1,000
Limestone "sand," hard, gray, (oil)	7	1,007
Limestone "sand," hard, white, (no oil)	4	1,011
Total depth		1,011

Log No. 667

Flahaven, No. 35. Commenced: May 10, 1919. Completed: May 27, 1919. Production: commenced producing June 4, 1919; production 48 hours after shot, 75 bbls. Shot May 28, 1919

Strata.

Pennsylvanian System.	Thickness	Depth
Soil, gray, soft	12	12
Shale, hard, blue, soft	173	185
Mississippian System.		
Limestone (Big Lime), hard, white	135	320
Shale, hard, blue, soft	465	785
Shale (red rock), soft	15	800
Shale, hard, blue, soft	15	815
Devonian System.		
Shale, brown, soft (Chattanooga)	150	965
Shale (fire clay), white, soft	15	980
Limestone (cap rock), hard, dark	3	983
Limestone "sand," brown, hard, (oil)	17	1,000
Limestone "sand," hard, dark, (no oil)	18	1,018
Total depth		1,018

Log No. 668

Flahaven, No. 36. Commenced: April 18, 1919. Completed: May 6, 1919. Production: commenced producing May 9, 1919; production 48 hours after shot, 155 bbls. Shot May 7, 1919.

Strata.

Pennsylvanian System.	Thickness	Depth
Soil, gray, soft	20	20
Shale, hard, blue, soft	140	160

Mississippian System.	Thickness	Depth
Limestone (Big Lime), gray, hard	120	280·
Shale, hard, blue, soft	476	756
Shale (red rock), soft	14	770
Shale, hard, blue, soft	10	780
Devonian System.		
Shale, brown, soft (Chattanooga)	140	920
Shale (fire clay), white, soft	15	935
Limestone (cap rock), hard dark	4	939
Limestone "sand," brown, hard, (oil)	15	954
Limestone "sand," gray, hard, (dry)	12	966
Total depth		966

Log No. 669

Flahaven, No. 38. Commenced: May 15, 1919. Completed: June 4, 1919. Production: commenced producing June 11, 1919; production 48 hours after shot, 50 bbls. Shot June 4, 1919.

Strata.

Pennsylvanian System.	Thickness	Depth
Soil, gray, soft	16	16
Shale, hard, blue, soft	64	80
Mississippian System.		
Limestone (Big Lime), white, hard	140	220
Shale, hard, blue, soft	450	670
Shale (red rock), soft	10	680
Shale, hard, blue, soft	15	695
Devonian System.		
Shale, brown, soft (Chattanooga)	137	832
Shale (fire clay), white, soft	12·	844
Limestone (cap rock), hard, dark	3	847
Limestone "sand," brown, hard, (oil)	14	861
Limestone "sand," light, hard, (no oil)	8	869
Limestone and sand, hard, dark, (no oil)	12½	881½
Total depth		881½

Log No. 670

Flahaven, No. 39. Commenced: June 9, 1919. Completed: June 27, 1919. Production: commenced producing July 1, 1919; production 48 hours after shot, 10 bbls. Shot June 27, 1919.

Strata.

Pennsylvanian System.	Thickness	Depth
Shale, hard, white, soft	20	20
Sand, hard	40	60
Shale, hard, blue, soft	120	180

Mississippian System.	Thickness	Depth
Limestone (Big·Lime), hard, white	150	330
Shale, hard, green, soft	20	350
Shells, gritty, white	20	370
Shale, hard, dark, soft	448	818

Devonian System.	Thickness	Depth
Shale, brown (Chattanooga)	140	958
Shale (fire ·clay), white	20	978
Shale (red rock)	2	980
Limestone (cap rock), hard, dark	3	983
Limestone ''sand,'' soft	31	1,014
Total depth		1,014

Log No. 671

Flahaven, No. 40. Commenced: May 30, 1919. Completed: June 6, 1919. Production: commenced producing July 12, 1919; production after 48 hours after shot, 1 bbl. Shot June 17, 1919.

Strata.

Pennsylvanian System.	Thickness	Depth
Soil, gray, soft	20	20
Shale, hard, blue, soft	80	100
Sandstone, gray, soft	32	132

Mississippian System.	Thickness	Depth
Limestone (Big Lime), hard, white	138	270
Shale, hard, blue, soft	505	775
Shale (red rock), soft	5	780
Shale, hard, blue, soft	5	785

Devonian System.	Thickness	Depth
Shale, brown, soft (Chattanooga)	125	910
Shale (fire clay), white, soft	12	922
Limestone (cap rock), hard, dark	3	925
Limestone ''sand,'' brown, hard	12	937
Limestone ''sand,'' white, hard, (no oil)	16	953
Total depth		953

Log No. 672

Flahaven, No. 41. Commenced: May 26, 1919. Completed: June 7, 1919. Production: commenced producing June 14, 1919; production after 48 hours after shot, 50 bbls. Shot June 9, 1919.

Strata.

Pennsylvanian System.	Thickness	Depth
Soil, gray, soft	12	12
Shale, hard, blue, soft	13	25
Sandstone, gray, soft	75	100
Shale, hard, blue, soft	115	215

Mississippian System.	Thickness	Depth
Limestone (Big Lime), hard, white	135	350
Shale, hard, blue, soft	500	850
Devonian System.		
Shale, brown, soft (Chattanooga)	130	980
Shale (fire clay), white, soft	13	993
Limestone (cap rock), hard, dark	2	995
Limestone "sand," brown, hard, (oil)	12	1,007
Limestone, hard, dark	4	1,011
Limestone "sand," hard and dark, (oil)	7	1,018
Limestone "sand," light, hard, (no oil)	14	1,032
Total depth		1,032

Log No. 673

Flahaven, No. 42. Commenced: June 27, 1919. Completed: July 21, 1919. Production: commenced producing July 28, 1919; production 48 hours after shot, 135 bbls. Shot July 22, 1919.

Strata.		
Pennsylvanian System.	Thickness	Depth
Sand, red, medium	40	40
Shale, dark, soft	20	60
Mississippian System.		
Limestone, hard, white ....,.....	20	80
Shale, white, soft	20	100
Limestone (Big Lime), hard, white	105	205
Shale, soft, green	20	225
Shale, white, medium	476	701
Devonian System.		
Shale, brown, soft (Chattanooga)	130	831
Shale, white, soft	20	851
Limestone "sand," dark, soft	10	861
Total depth		861

Log No. 674

Flahaven, No. 43. Commenced: May 21, 1919. Completed: June 12, 1919.

Strata.		
Pennsylvanian System.	Thickness	Depth
Soil, gray, soft	20	20
Shale, hard, blue, soft	110	130
Sandstone, gray, soft	30	160

Mississippian System.	Thickness	Depth
Shale, hard, blue	20	180
Limestone (Big Lime), white, hard	120	300
Shale, hard, blue, soft	480	780
Shale (red rock), soft	10	790
Devonian System.		
Shale, brown, soft (Chattanooga)	150	940
Shale (fire clay), white, soft	25	965
Limestone (cap rock), hard, dark	2	967
Limestone "sand," hard, dark, (oil)	15	982
Limestone "sand," hard, dark, (dry)	4	986
Limestone "sand," hard, white, (dry)	8	994
Total depth		994

Log No. 675

Flahaven, No. 44. Commenced: June 16, 1919. Completed: July 8, 1919. Production: commenced producing July 12, 1919; production 48 hours after shot, 15 bbls. Shot July 9, 1919.

Strata.

Pennsylvanian System.	Thickness	Depth
Soil, gray, soft	20	20
Shale, hard, blue, soft	145	165
Sandstone, gray, soft	15	180
Mississippian System.		
Limestone (Big Lime), hard, white	140	320
Shale, hard, blue, soft	430	750
Shale (red rock), soft	15	765
Shale, hard, blue, soft	55	820
Devonian System.		
Shale, brown, soft (Chattanooga)	130	950
Shale (fire clay), white, soft	18	968
Limestone (cap rock), hard, dark	4	972
Limestone "sand," brown, hard	12	984
Limestone "sand," hard, dark, (no oil)	3	987
Total depth		987

Log No. 676

Flahaven, No. 47. Commenced: July 5, 1919. Completed: July 16, 1919. Production: commenced producing July 19, 1919; production 48 hours after shot, 10 bbls. Shot July 17, 1919, between 899 and 909 feet.

Strata.

Pennsylvanian System.	Thickness	Depth
Shale, black, soft	100	100

Mississippian System.	Thickness	Depth
Limestone (Big Lime), hard, white	140	240
Shale, soft, white	490	730
Devonian System.		
Shale, brown, soft (Chattanooga)	160	890
Shale (fire clay), gray, soft	6	896
Limestone (cap rock), hard, black	2	898
Limestone "sand," brown, hard	11	909
Total depth		909

Log No. 677

Flahaven, No. 48. Commenced: July 14, 1919. Completed: Sept. 19, 1919. Production: commenced producing Sept. 20, 1919; production 48 hours after shot, 6 bbls. Shot Sept. 18, 1919, between 958 and 968 feet.

Strata.

Pennsylvanian System.	Thickness	Depth
Soil, black, soft	30	30
Sand, white, soft	100	130
Mississippian System.		
Shale, hard, white	50	180
Limestone (Big Lime), hard, white	110	290
Shale, hard, white, soft	510	800
Devonian System. .		
Shale, brown, soft (Chattanooga)	135	935
Shale (fire clay), white, soft	21	956
Limestone (cap rock), hard, black	2	958
Limestone "sand," gray, soft, (oil)	10	968
Total depth		968

Log No. 678

Flahaven, No. 49. Commenced: Aug. 11, 1919. Completed: Aug. 19, 1919. Production: commenced producing Aug. 27, 1919; production 48 hours after shot, 4 bbls. Shot Aug. 20, 1919, between 1012 and 1026 feet.

Strata.

Pennsylvanian System.	Thickness	Depth
Soil, black, soft	12	12
Sand, brown, soft	78	90
Shale, brown, soft	109	199

Mississippian System.   Thickness Depth

    Limestone (Big Lime), hard, white 161   360
    Shale, white, soft 480   840
    Shale (red rock), soft 5   845
    Shale, white, soft 15   860
    Shale, brown, soft (Chattanooga) 132   992
    Shale (fire clay), white, soft 15   1,007
    Limestone (cap rock), hard, dark 5   1,012
    Limestone "sand," dark, soft, (pay) 14   1,026
        Total depth 1,026

Log No. 679

Flahaven, No. 51. Commenced: July 29, 1919. Completed: Aug. 20, 1919. Production: commenced producing Aug. 26, 1919; production 48 hours after shot, 4 bbls. Shot Aug. 21, 1919, between 944 and 959 feet.

    Strata.

Pennsylvanian System.   Thickness Depth

    Sandstone (mountain), white, soft 145   145

Mississippian System.

    Limestone (Big Lime), hard, white 125   270
    Shale, hard, white, soft 511   781

Devonian System.

    Shale, brown, soft (Chattanooga) 140   921
    Shale (fire clay), white, soft 20   941
    Limestone (cap rock), hard, black 3   944
    Limestone "sand," brown, soft, (oil) 15   959
        Total depth 959

Log No. 680

Flahaven, No. 52. Commenced: July 5, 1919. Completed: Aug. 12, 1919. Production: commenced producing Aug. 18, 1919; production 48 hours after shot, 10 bbls. Shot Aug. 15, 1919, between 1003 and 1016 feet.

    Strata.

Pennsylvanian System.   Thickness Depth

    Sandstone (mountain), white, soft 80   80
    Shale, hard, white, soft 120   200

IRREGULAR SEDIMENTATION IN THE POTTSVILLE.
The whimsical play of off shore currents in Pottsville seas or lagoons developed the uneven characteristic of the sandstone ledge shown in detail above. But it did not injure at all its possibilities

Mississippian System.	Thickness	Depth
Limestone (Big Lime), hard, white	142	342
Shale, hard, white, soft	502	844
Devonian System.		
Shale, brown, soft (Chattanooga)	137	981
Shale (fire clay), white, soft	20	1,001
Limestone (cap rock), hard, black	2	1,003
Limestone "sand," brown, soft, (oil)	13	1,016
Total depth		1,016

Log No. 681

Flahaven, No. 53. Commenced: Aug. 8, 1919. Completed: Aug. 29, 1919. Production: commenced producing Sept. 5, 1919; production 48 hours after shot, 45 bbls. Shot Aug. 30, 1919, between 748 and 763 feet.

Strata.		
Pennsylvanian System.	Thickness	Depth
Limestone, hard, white	20	20
Cavity	5	25
Limestone, hard, white	5	30
Shale, yellow, soft, muddy, caving	5	35
Limestone, hard, white	5	40
Quicksand, brown, soft	3	43
Limestone (Big Lime), hard, white	62	105
Shale, hard, white, soft	465	570
Devonian System.		
Shale, brown, soft (Chattanooga)	155	725
Shale (fire clay), white, soft	21	746
Limestone (cap rock), hard, black	2	748
Limestone "sand," brown, soft, (oil)	15	763
Total depth		763

Log No. 682

Flahaven, No. 55. Commenced: July 28, 1919. Completed: Aug. 9, 1919. Production: commenced producing Aug. 17, 1919; production 48 hours after shot, 80 bbls. Shot Aug. 11, 1919, between 926 and 938 feet.

Strata.		
Pennsylvanian System.	Thickness	Depth
Sandstone (mountain), white, soft	125	125
Shale, hard, white, soft	25	150

Mississippian System.	Thickness	Depth
Limestone (Big Lime), hard, white	145	295
Shale, hard, white, soft	460	755
Shale (red rock), soft	13	768
Devonian System.		
Shale, brown, soft (Chattanooga)	135	903
Shale (fire clay), white, soft	20	923
Limestone (cap rock), hard, black	3	926
Limestone ''sand,'' brown, hard, (oil)	12	938
Total depth		938

Log No. 683

Flahaven, No. 57. Commenced: Aug. 20, 1919. Completed: Aug. 28, 1919. Production: commenced producing Sept. 2, 1919; production 48 hours after shot, 60 bbls. Shot Aug. 30, 1919, between 1018 and 1028 feet.

Strata. Pennsylvanian System.	Thickness	Depth
Sandstone, yellow, soft	40	40
Shale, hard, dark, soft	110	150
Shale and shells, light and hard	85	235
Shale (fire clay), white, soft	10	245
Mississippian System.		
Limestone (Big Lime), hard, white	135	380
Shale, hard, green, soft	20	400
Shale, hard, white	200	600
Sand, hard	40	640
Shale, hard, soft	220	860
Shale (red rock)	10	870
Devonian System.		
Shale, brown (Chattanooga)	138	1,008
Shale (fire clay)	8	1,016
Limestone (cap rock), hard, dark	2	1,018
Limestone ''sand,'' light, soft, (oil)	10	1,028
Total depth		1,028

Log No. 684

Flahaven, No. 58. Commenced: Aug. 22, 1919. Completed: Sept. 3, 1919. Production: commenced producing Sept. 8, 1919, production 48 hours after shot, 30 bbls. Shot Sept. 4, 1919, between 816 and 828 feet.

Strata.

Mississippian System.	Thickness	Depth
Soil, dark, soft	40	40
Limestone (Big Lime), hard, white	130	170
Shale, hard, white, soft	485	655
Devonian System.		
Shale, brown, soft (Chattanooga)	139	794
Shale (fire clay), white, soft	20	814
Limestone (cap rock), hard, black	2	816
Limestone ''sand,'' brown, soft, (oil) ......:	12	828
Total depth		828

Log No. 685

Flahaven, No. 59. Commenced: Sept. 6, 1919. Completed: Sept. 20, 1919. Production: commenced producing Sept. 24, 1919; production 48 hours after shot, 35 bbls. Shot Sept. 22, 1919, between 997 and 1006 feet.

Strata.

Pennsylvanian System.	Thickness	Depth
Soil, black, soft	14	14
Sandstone, red, soft	181	195
Mississippian System.		
Limestone (Big Lime), hard, white	155	350
Shale, soft, white	499	849
Shale (red rock), soft	5	854
Devonian System.		
Shale, brown, soft (Chattanooga)	125	979
Shale (fire clay), white, soft	15	994
Limestone (cap rock), hard, black	2	996
Limestone ''sand,'' brown, soft	10	1,006
Total depth		1,006

Log No. 686

Flahaven, No. 61. Commenced: Sept. 1, 1919. Completed: Oct. 1, 1919. Production: commenced producing Oct. 12, 1919. Shot Oct. 6, 1919, between 996 and 1004 feet.

Strata.

Pennsylvanian System.	Thickness	Depth
Soil, black, soft	40	40
Sand, white, soft	80	120
Shale, hard, white, soft	60	180

Mississippian System.	Thickness	Depth
Limestone (Big Lime), hard, white	130	310
Shale, hard, white, soft	540	850
Devonian System.		
Shale, brown, soft (Chattanooga)	120	970
Shale (fire clay), white soft	22	992
Limestone (cap rock), hard, black	2	994
Limestone ''sand,'' hard, dark, (oil)	8	1,002
Limestone, hard, white	30	1,032
Limestone ''sand,'' hard, light	5	1,037
Limestone ''sand,'' hard, dark	16	1,053
Shale, hard, black, soft	6	1,059
Total depth		1,059

Log No. 687

Flahaven, No. 62. Commenced: Sept. 8, 1919. Completed: Sept. 25, 1919. Production: commenced producing Sept. 30, 1919; production 48 hours after shot, 8 bbls. Shot Sept. 26, 1919, between 975 and 985 feet.

Strata.

Pennsylvanian System.	Thickness	Depth
Soil, black, soft	94	94
Sand, white, soft	40	134
Shale, hard, white, soft	66	200
Mississippian System.		
Limestone (Big Lime), hard, white	125	325
Shale, hard, white, soft	490	815
Devonian System.		
Shale, brown, soft (Chattanooga)	138	953
Shale (fire clay), white, soft	20	973
Limestone (cap rock), hard, black	2	975
Limestone ''sand,'' white, soft, (oil)	10	985
Total depth		985

Log No. 688

Flahaven, No. 63. Commenced: Sept. 29, 1919. Completed: Dec. 25, 1919. Production: commenced producing Dec. 25, 1919; production 48 hours after shot, 15 bbls. Shot Dec. 23, 1919, between 992 and 1002 feet.

Strata.

Pennsylvanian System.	Thickness	Depth
Soil, black, soft	140	140
Sand, white, soft	60	200

Mississippian System.	Thickness	Depth
Limestone (Big Lime), hard, white	140	340
Shale, hard, white, soft	500	840
Devonian System.		
Shale, brown, soft (Chattanooga)	128	968
Shale (fire clay), white, soft	20	988
Limestone (cap rock), hard, black	2	990
Limestone ''sand,'' brown, soft, (oil)	6	996
Limestone, hard, white	6	1,002
Total depth		1,002

Log No. 689

Flahaven, No. 64. Commenced: Sept. 19, 1919. Completed: Sept. 29, 1919. Production: commenced producing Oct. 7, 1919; production 48 hours after shot, 40 bbls. Shot Sept. 30, 1919, between 1012 and 1022 feet.

Strata.

Pennsylvanian System.	Thickness	Depth
Sandstone, yellow, soft	30	30
Shale, hard, dark	190	220
Mississippian System.		
Limestone (Big Lime), hard, white	130	350
Shells and shale, hard	224	574
Shale, hard, dark, soft	276	850
Shale (red rock)	10	860
Devonian System.		
Shale, brown (Chattanooga)	.120	980
Shale (fire clay), light	28	1,008
Limestone (cap rock), hard, dark	2	1,010
Limestone ''sand,'' brown, soft, (oil)	12	1,022
Total depth		1,022

Log No. 690

Flahaven, No. 65. Commenced: Oct. 24, 1919. Completed: Nov. 25, 1919. Production: commenced producing Nov. 30, 1919; production 48 hours after shot, 10 bbls. Shot Nov. 26, 1919, between 902 and 912 feet.

Strata.

Pennsylvanian System.	Thickness	Depth
Soil, black, soft	100	100
Sand, white, soft	15	115

Mississippian System.	Thickness	Depth
Limestone (Big Lime), hard, white	161	276
Shale, hard, white, soft	464	740
Devonian System.		
Shale, brown, soft (Chattanooga)	140	880
Shale (fire clay), white, soft	20	900
Limestone (cap rock), black, hard	2	902
Limestone ''sand,'' brown, soft, (oil)	10	912
Total depth		912

Log No. 691

Flahaven, No. 67. Commenced: Sept. 14, 1919. Completed: Oct. 11, 1919; Production: Commence producing Oct. 17, 1919, production 48 hours after shot, 12 bbls. Shot Oct. 13, 1919,, between 1145 and 1157 feet.

Strata.

Pennsylvanian System.	Thickness	Depth
Soil, red, soft	12	12
Sand, hard, red	160	172
Shale, hard, white, soft	40	212
Sand, red, soft	28	240
Shale, hard, gray, soft	130	370
Sand, hard, black	30	400
Mississippian System.		
Limestone (Big Lime), hard, white	120	520
Shale, hard, green, soft	50	570
Limestone, hard, white	20	590
Shale, hard, white, soft	405	995
Devonian System.		
Shale, brown, soft (Chattanooga)	130	1,125
Shale (fire clay), white, soft	17	1,142
Limestone (cap rock), hard, black	3	1,145
Limestone ''sand,'' brown, hard	12	1,157
Total depth		1,157

Log No. 692

Flahaven, No. 68. Commenced: Sept. 25, 1919. Completed: Oct. 21, 1919.

Strata.

Pennsylvanian System.	Thickness	Depth
Sand, red, medium	205	205
Limestone, hard, white	10	215
Shale, dark, medium	85	300
Sand, hard, white	30	330
Shale, dark, soft	30	360

Mississippian System.	Thickness	Depth
Limestone (Big Lime), hard, white	125	485
Shale, white, medium	504	989
Devonian System.		
Shale, brown, medium (Chattanooga)	130	1,119
Shale, white, soft	18	1,137
Limestone (cap rock), hard ,dark	2	1,139
Limestone ''sand,'' hard, gray	65	1,204
Shale, dark, soft	4	1,208
Total depth		1,208

Log No. 693

Flahaven, No. 69. Commenced: Oct. 6, 1919. Completed: Nov. 7, 1919. Production: commenced producing Nov. 12, 1919; production 48 hours after shot, 6 bbls. Shot Nov. 8, 1919, between 994 and 1005 feet.

Strata.

Pennsylvanian System.	Thickness	Depth
Gravel and shale, brown, soft	55	55
Shale, hard, dark, soft	140	195
Mississippian System.		
Limestone (Big Lime), gray, hard	135	330
Shale, gray, soft	506	836
Devonian System.		
Shale, brown, soft (Chattanooga)	140	976
Shale (fire clay), white, soft	15	991
Limestone (cap rock), hard, dark	3	994
Limestone ''sand,'' brown, soft, (oil)	11	1,005
Total depth		1,005

Log No. 694

Flahaven, No. 70. Commenced: Oct. 24, 1919. Completed: Nov. 13, 1919. Production: commenced producing Nov. 11, 1919; production 48 hours after shot, 10 bbls. Shot Nov. 15, 1919, between 1003 and 1014 feet.

Strata.

Pennsylvanian System.	Thickness	Depth
Sand, white, soft	14	14
Shale, hard, dark	40	54
Sand, hard, dark	141	195

Mississippian System.	Thickness	Depth
Limestone (Big Lime), hard, white	130	325
Shale, hard, and soapstone, white ,.........	530	855
Shale (red rock), soft	10	865
Devonian System.		
Shale, brown, soft (Chattanooga)	120	985
Shale (fire clay), white, soft	15	1,000
Limestone (cap rock), hard, dark	2	1,002
Limestone "sand," brown, soft	12	1,014
Total depth		1,014

Log No. 695

Flahaven, No. 72. Commenced: Oct. 2, 1919. Completed: Oct. 31, 1919. Production: commenced producing Nov. 6, 1919; production 48 hours after shot, 8 bbls. Shot Nov. 1, 1919.

Strata.		
Pennsylvanian System.	Thickness	Depth
Soil, black, soft	80	80
Sand, white, soft	35	115
Mississippian System.		
Limestone (Big Lime), hard, white	130	245
Shale, hard, white, soft	472	717
Devonian System.		
Shale, brown, soft (Chattanooga)	140	857
Shale (fire clay), white, soft	20	877
Limestone (cap rock), hard, black	1	878
Limestone "sand," brown, soft, (oil)	10	888
Total depth		888

Log No. 696

Flahaven, No. 75. Commenced: Nov. 21, 1919. Completed: Jan. 1, 1920. Shot Jan. 2, 1920, between 1101 and 1117 feet.

Strata.		
Pennsylvanian System.	Thickness	Depth
Soil, black, soft	6	6
Sand, white, soft	119	125
Shale, hard, white, soft	165	290
Mississippian System.		
Limestone (Big Lime), hard, white	130	420
Shale, hard, white, soft	490	910

Devonian System.

	Thickness	Depth
Shale, brown, soft (Chattanooga)	140	1,050
Shale (fire clay), white, soft	21	1,071
Limestone (cap rock), hard, black	2	1,073
Limestone, hard, black	11	1,084
Limestone "sand," hard, white	32	1,116
Limestone "sand," white, soft, (oil)	11	1,127
Total depth		1,127

Log No. 697

Flahaven, No. 76. Commenced: Feb. 21, 1920. Completed: April 5, 1920. Production: 48 hours after shot, 3 bbls. Shot April 5, 1920, between 961 and 971 feet.

Strata.

Pennsylvanian System.

	Thickness	Depth
Soil, soft	90	90
Sand, hard, white	105	195
Mississippian System.		
Limestone (Big Lime), hard, white	105	300
Shale, hard, green, soft	20	320
Shale, hard, blue, soft	503	823
Devonian System.		
Shale, black, soft (Chattanooga)	110	933
Limestone, hard, dark	10	943
Shale (fire clay), light, hard	14	957
Limestone (cap rock), hard, black	2	959
Limestone "sand," hard, light and dark, (oil)	12	971
Total depth		971

NOTE—The single occurrence of 10 feet of limestone between the "fire clay" shale and the black shale of the Devonian is unusual.

Log No. 698

Flahaven, No. 77. Commenced: Nov. 13, 1919. Completed: Dec. 18, 1919. Production: commenced producing Dec. 23, 1919; production 48 hours after shot, 15 bbls. Shot Dec. 20, 1919, between 1119 and 1130 feet.

Strata.

Pennsylvanian System.

	Thickness	Depth
Sand, red, medium	195	195
Shale, dark, medium	145	340
Mississippian System.		
Limestone (Big Lime), hard, white	120	460
Shale, green, soft	20	480
Shale, white, medium	487	967

Devonian System.	Thickness	Depth
Shale, brown, medium (Chattanooga)	130	1,097
Shale, white, soft	20	1,117
Limestone (cap rock), hard, dark	2	1,119
Limestone "sand," brown, hard, (oil)	11	1,130
Total depth		1,130

Log No. 699

James M. Olinger, No. 3, lessor. Commenced: July 16, 1918. Completed: Aug. 22, 1918. Shot: Aug. 19, 1918. Authority: Ohio Oil Co.

Strata.		
Pennsylvanian System.	Thickness	Depth
Soil, brown, soft	30	30
Sand, hard, brown	20	50
Shale, hard, and shells, gray	269	319
Mississippian System.		
Limestone, hard, white	90	409
Shale, hard, and shells, blue	425	834
Devonian System.		
Shale, brown (Chattanooga)	160	994
Shale, blue, soft	15	1,009
Limestone (cap rock)	25	1,034
Limestone "sand," brown	21	1,055
Total depth		1,055

CHAPTER VII.

LOGAN COUNTY.

Production: **Oil and Gas.** Producing Sands: "Shallow" (Mississippian), Corniferous (Devonian), Niagaran (Silurian).

Log No. 700

M. E. Hall, No. 1, lessor. Authority: The Bertram Developing Co.
Strata.

Mississippian System.	Thickness	Depth
Soil	4	4
Limestone, white	396	400
Limestone, chocolate	400	800
Limestone, dark	241	1,041
Devonian System.		
Shale (Chattanooga)	77	1,118
Limestone "sand," (water)	20	1,138
Limestone, variable in color	123	1,261
Limestone "sand" and shale	4	1,265
Limestone, variable in color	127	1,392
Total depth		1,392

Log No. 701

Flowers, No. 1, lessor. Location: 2 miles south of Russellville. Completed: Feb. 25, 1921. 1st shot, 40 qts. 1st showing, 565-585. 6¼ in. casing, 395. Authority: C. A. Phelps, Bowling Green, Ky.

Strata.

Mississippian System.	Thickness	Depth
Limestone, shale, etc.	1,031	1,031
Devonian System.		
Shale, black	94	1,125
Limestone, white	20	1,145
Limestone (cap rock)	30	1,175
Total depth		1,175

Log No. 702

Shaker, No. 1, lessor. Authority: C. A. Phelps, Bowling Green Ky.

Strata.

Mississippian System.	Thickness	Depth
Limestones, shales, etc.	1,038	1,038
Devonian System.		
Shale, black (Chattanooga)	79	1,117

Limestone (cap rock)	26	1,143
Limestone "sand," (oil) good showing)	15	1,158
Total depth		1,158

Log No. 703

Nourse, No. 1, lessor. Location: 3 miles east of Russellville. Authority: C. A. Phelps, Bowling Green, Ky.

Strata.

Mississippian System.	Thickness	Depth
Limestones, shales, etc.	1,090	1,090

Devonian System.		
Shale, black (Chattanooga)	85	1,175
Limestone, white	50	1,225
Limestone, white and blue	10 ·	1,235
Limestone, gray	34	1,269
Limestone "sand," (pay) ....\.......	5	1,274
Total depth		1,274

Log No. 704

Johnson, No. 1, lessor. Location: 2½ miles northeast of South Bend. Shot 60 qts., 1,102-1,115. 6½" casing, 429. Shot 60 qts., 1,181-1,200. Authority: C. A. Phelps, Bowling Green, Ky.

Strata.

Mississippian System.	Thickness	Depth
Limestones, shales, etc.	930	930
Devonian System.		
Shale, black ...., ...	72	1,002
Limestone, white	30	1,032
Limestone, hard	43	1,075
Silurian System.		
Limestone "sand," (white water)	10	1,085
Limestone "sand," (gas)	3	1,088
Limestone, gray	19	1,107
Limestone, gray and brown, (oil show)	5	1,112
Limestone, gray	33	1,145
Shale (red rock), limy	8	1,153
Shale (red rock), limy	12	1,165
Limestone, gray, (oil show)	17	1,182
Limestone, brownish gray	2	1,184
Limestone, brownish gray	3	1,187
Limestone, brownish gray	20	1,207
Limestone, gray and blue	83	1,290
Total depth		1,290

Log No. 705

Otis Matlock, No. 1, lessor. Location: 3½ miles southwest of Auburn P. O. Commenced: Feb. 5, 1921. Completed: Mar. 24, 1921.

Contractors: Overton & Ward. Drillars: Ward & Jarrett. Shot, 60 qts. Authority: N. Garland, driller.

Strata.

Mississippian System.

	Thickness	Depth
Clay	10	10
Limestone, gray	16	26
Cavity, mud	2	28
Limestone, gray	112	140
Limestone, brown	45	185
Limestone, gray	145	330
Limestone, brown	270	600
Limestone, black	40	640
Limestone, brown, and flint, white	65	705
Limestone, black	40	745

Mississippian System.

	Thickness	Depth
Limestone, brown, and flint, white	45	790
Limestone, black	30	820
Limestone, brown	45	865
Limestone, blue	30	895
Limestone, brown	15	910
Shale, green (New Providence)	28	938

Devonian System.

	Thickness	Depth
Shale, black (Chattanooga)	73	1,011
Limestone (cap rock), white	19	1,030
Limestone, brown, and flint, brown	25	1,055

Silurian System.

	Thickness	Depth
Limestone, blue	5	1,060
Limestone, gray	8	1,068
Limestone, grayish brown, and sand	6	1,074
Limestone, gray, and sand	12	1,086
Limestone, gray	4	1,090
Total depth		1,090

Fresh water, 27 and 70 feet. Sulphur water, 185 feet. Sulphur gas, 330 feet. Show of oil, 1068-1086, with little gas. 28 feet, 8¼ in. casing; 227 feet, 6¼ in. casing.

One mile south of this well is the Fisher well, on Curtis Lease.

LINCOLN COUNTY.

Production: **Oil and Gas.** Producing Sands "Shallow Gas Sand" (Mississippian), Corniferous (Devonian), "Second Sand" (Silurian).

Log No. 706

David G. Elliott, No. 1, lessor. Roeser & Shoenfelt, lessee. Location: near Casey County line, ¼ mile south of Green River. Commenced: Spring, 1920. Contractor: W. H. Mahon.

Strata.

Mississippian System.	Thickness	Depth
Clay	54	54
Devonian System.		
Shale, black ...:..........................	44	98
Limestone	18	116
Silurian System.		
Shale	49½	166½
Limestone	14½	181
Shale	159	340
Ordovician System.		
Limestone	276	616
Limestone, sandy, brown, soft, Correlatives of Sunnybrook Sand	22	638
Limestone, Correlatives of Sunnybrook Sand...	49	687
Limestone, sandy, brown, soft, Correlatives of Sunnybrook Sand	18	705
Limestone, blue, Correlatives of Sunnybrook Sand	105	810
Limestone, sandy, brown, Correlatives of Sunnybrook sand	45	855
Incomplete depth		855

Incomplete record, dry to 855; did not need to case.

Log No. 707

J. Hollar, No. 1, lessor. Daniel Boone Oil Co., lessee. Location: Green River District. Commenced: May 7, 1919. Authority: Daniel Boone Oil Co.

Strata.

Mississippian System.	Thickness	Depth
Gravel	6	6
Shale, sandy, soft	46	52
Devonian System.		
Shale, black(Chattanooga)	47	99
Limestone	2	101
Limestone ''sand,'' (oil show)	20	121
Total depth		121

Log No. 708

Sarah Hubble, No. 1, lessor. Daniel Boone Oil Co., lessee. Location: Green River District. Commenced: May 4, 1919. Production: Dry. Authority: Daniel Boone Oil Co.

Strata.

Mississippian System.	Thickness	Depth
Gravel	10	10
Shale, sandy, soft	35	45
Devonian System.		
Shale, black, (Chattanooga)	30	75
Limestone	25	100
Limestone ''sand,'' (dry)	18	118
Total depth		118

Log No. 709

Sanders, No. 1, lessor. Daniel Boone Oil Co., lessee. Location: Hurricane Creek. Drilled in the spring of 1919. Authority: Daniel Boone Oil Co.

Strata.

Mississippian System.	Thickness	Depth
Soil	12	12
Sand	38	50
Shale, hard	.25	75
Limestone, sandy	15	90
Limestone, sandy	10	100
Shale, hard	160	260
Devonian System.		
Shale	34	294
Shale (fire clay)	3	297
Limestone, shelly	7	304
Shale (fire clay)	2	306
Limestone ''sand,'' (show of oil)	46	352
Total depth		352

Log No. 710

Albert Schuler, No. 1, lessor. Daniel Boone Oil Co., lessee. Location: on Buck Creek. Drilled during 1918. Production: gas from 192 to 200 feet; oil at 185 feet. Authority: Daniel Boone Oil Co.

Strata.

Mississippian System.	Thickness	Depth
Gravel	14	14
Gravel	28	42
Shale, sandy	18	60
Limestone, sandy	15	75
Devonian System.		
Shale, hard	30	105

Shale, ...	45	150
Limestone, sandy	54	204
Total depth		204

NOTE—The Devonian-Silurian contact in this well occurs in the last 54 feet of limestone. The well finished in the Silurian.

Log No. 711

Albert Schuler, No. 2, lessor. Daniel Boone Oil Co., lessee. Location: on Buck Creek. Drilled in the spring of 1918. Production: Gas at 190 feet. Authority: Daniel Boone Oil Co.
Strata.

Mississippian System.	Thickness	Depth
Shale, hard, white (water at 35)	50	50
Shale, hard, blue	55	105
Devonian System.		
Shale, black (Chattanooga)	52	157
Limestone and "sand"	60	217
Shale, hard, white	6	223
Limestone	27	250
Total depth		250

3" casing 20 feet.

NOTE—The Devonian-Silurian contact occurs in the 60 feet of limestone above 217 feet. The well finished in the Silurian.

Log No. 712

Albert Schuler, No. 3, lessor. Daniel Boone Oil Co., lessee. Location: on Buck Creek. Drilled in the spring of 1918. Production: Oil from 173 to 177 feet, water at 150 feet, gas at 190 feet. Authority: Daniel Boone Oil Co.
Strata.

Mississippian System.	Thickness	Depth
Shale, hard	108	108
Devonian System.		
Shale (Chattanooga)	52	160
Limestone "sand"	3	163
Shale (fire clay)	4	167
Limestone "sand"	38	205
Total depth		205

NOTE—The Devonian-Silurian contact is toward the base of the last 38 feet of limestone.

Log No. 713

Albert Schuler, No. 4, lessor. Daniel Boone Oil Co., lessee. Location: on Buck Creek. Drilled in 1918. Production: oil at 180 feet. Authority: Daniel Boone Oil Co.

Strata.

Mississippian System.	Thickness	Depth
Gravel	6	6
Limestone	30	36
Shale, hard	79	115
Devonian System.		
Shale (Chattanooga)	44	159
Shale (fire clay)	8	167
Limestone	9	176
Limestone ''sand''	32	208
Total depth		208

NOTE—The Devonian-Silurian contact occurs in the lower part of the last 32 feet.

Log No. 714

Albert Schuler, No. 5 lessor. Daniel Boone Oil Co., lessee. Location: on Buck Creek. Drilled in 1918. Authority: Daniel Boone Oil Co.

Strata.

Mississippian System.	Thickness	Depth
Soil and shale, hard	110	110
Devonian System.		
Shale (Chattanooga)	48	158
Shale, hard, white	14	172
Limestone ''sand'' (gas at 192)	33	205
Silurian System.		
Limestone	19	224
Total depth		224

Log No. 715

Albert Schuler, No. 6, lessor. Daniel Boone Oil Co., lessee. Location: on Buck Creek. Drilled in 1918. Production: gas at 188 feet. Authority: Daniel Boone Oil Co.

Strata.

Mississippian System.	Thickness	Depth
Shale, hard	102	102
Devonian System.		
Shale (Chattanooga)	50	152
Shale (fire clay) and limestone	10	162
Limestone	48	210
Total depth		210

NOTE—The Devonian-Silurian contact is within the 48 feet of. limestone above 210 feet.

Log No. 716

Albert Schuler, No. 7, lessor. Daniel Boone Oil Co., lessee. Location: on Buck Creek. Drilled in 1918. Authority: Daniel Boone Oil Co.

Strata.

Mississippian System.	Thickness	Depth
Shale, hard	147	147

Devonian System.

	Thickness	Depth
Shale (Chattanooga) (water 180)	44	191
Clay and shale, hard	13	204
Limestone ''sand''	35	239

Silurian System.

	Thickness	Depth
Shale, hard	6	245
Total depth		245

Lög No. 717

Albert Schuler, No. 8, lessor. Daniel Boone Oil Co., lessee. Location: on Buck Creek. Drilled in 118. Authority: Daniel Boone Oil Co.

Strata.

Mississippian System.	Thickness	Depth
Shale, sandy	125	125
Shale, hard	140	265

Devonian System.

	Thickness	Depth
Shale (Chattanooga)	50	315
Clay	10	325
Limestone, (dry)	71	396
Total depth		396

MADISON COUNTY.

Production: **Small oil and gas.** Producing **Sand: Corniferous (Devonian)** exposed.

Log No. 718

Snyder, No. 1, lessor. Atlanta Oil & Gas Co., lessee. Location: 1½ miles from Berea. Production: gas and oil show; well abandoned. Authority: Atlanta Oil & Gas Co.

Strata.

Devonian System.	Thickness	Depth
Shale, black (Chattanooga)	36	36
Limestone "sand" brown	13	49
Shale, gray	5	54
Limestone, (cased)	4	58
Shale	2	60
Sand	5	65
Shale, white	3	68
Sand	2	70
Shale	4	74
Limestone "sand" (small oil show)	5	79
Silurian System.		
Limestone	11	90
Shale (fire clay)	36	126
Total depth		126

Log No. 719

Winn, No. 1, lessor. Atlanta Oil & Gas Co., lessee. Location: ½ mile from Berea. Authority: Atlanta Oil & Gas Co.

Strata.

Devonian System.	Thickness	Depth
Shale, black (Chattanooga)	26	26
"Sand," dark	13	39
Shale	5	44
Limestone "sand," (oil show)	2	46
Shale, white	24	70
Limestone "sand"	20	90
Shale	2	92
Silurian System.		
Limestone "sand"	3	95
Shale, (oil show)	5	100
Total depth		100

MAGOFFIN COUNTY.

**Production: Oil and Gas. Producing Sands: Pottsville (Pennsylvanian),
Maxton, Big Lime and Wier (Mississippian).**

Log. No. 720

Harris Howard, lessor, No. 1. Bedrock Oil Co., lessee. Location:
Meadows Branch of Upper Licking River. Elevation: ???

Strata.	Thickness	Depth
Pennsylvania System.		
Soil	26	26
Shale	34	60
Coal	3	63
Shale	104	167
Coal	3	170
Sand	15	185
Sand, white (show of oil)	10	195
Sand	80	275
Sand and shale	25	300
Shales	20	320
Shales	155	475
Sand, white (gas at 500')	25	500
Sand, white (show of oil)	50	550
Sand, salt water at 570'	20	570
Shale	70	640
Mississippian System.		
Shale	100	740
Limestone, white (Big Lime)	95	835
Shales	80	915
Sand (show of oil)	50	965
Shales, Sunbury	195	1,160
Sand, salt water near top	80	1,240
Shales, sandy shells	70	1;310
Soft black shale (Sunbury)	40	1,350
Hard, yellow, sandy, shale (Berea)	40	1,390

LEDGE OF POTTSVILLE CONGLOMOMERATE.
"rock houses." The Pottsville is an excellent oil reservoiring sand and is a possible producer at many points in Kentucky.
This basal sandstone of the coal measures series exhibits where ever it outcrops a marked tendency to form cliff
Photo on the south limb of Rough Creek Anticline Webster.

Devonian System.	Thickness	Depth
Shale, soft black (Chattanooga) .... :..	360	1,750
Shale, gray	116	1,866
Sandy lime, hard on top, sweet gas	8	1,874
Limestone, soft and hard streaks, gray, 1870-1874..	19	1,893
Sand, white, some limestone	11	1,904
Limestone, hard and soft streaks	12	1,916
Limestone, sandy, little H2SO4	9	1,925
Limestone, hard and soft alternately	37	1,962
Total depth		1,962

NOTE—At 1,240 casing was drawn, and salt water filled well to within 300 feet of the top.

Log No. 721

Clay Adams, No. 1. C. K. Dresser, Bradford, Pa., lessee. Location: Head of Raccoon Creek. Production: 5 bbls. prior to shot. Completed to Wier sand, October 2, 1920. Authority: W. G. Roeder, Lexington, Kentucky.

Strata.

Pennsylvanian System.	Thickness	Depth
Sandstone, shale and coal	420	420
Sand, brown	12	432
Sand (hole full of water)	38	470
Sand, settling	60	530
Mississippian System.		
Limestone, white	60	590
Limestone, white mud	60	650
Shale, blue	80	730
Shale, hard black	22	752
Shale, blue	135	887
Sand, coarse (gas)	8	895
Limestone, sandy ..:.......................	5	900
Limestone and shale, broken	4	904
Shale, sandy	4	908
Limestone, dark gray	3	911
Limestone, sandy ....:.	8	919
Sand, hard, dark	2	921
Shale, gray black	20˙	941
Shale, sandy	5	946
Sand, coarse, light gray	12	958
Sand, light gray	18	976
Sand, light gray, Wier	1	977
Sand, ½ bbl. oil, Wier	14	991
Sand, oil, Wier	15½	1,006½
Sand	60	1,066½
Total depth		1,066½

Log No. 722

Keaton No. 1 (?). Location: Mouth of Johnson Creek. Began: June 16, 1914. Finished: August 6, 1914. Production: Dry. Driller, E. Guignon. Authority: L. Beckner.

Strata.

	Thickness	Depth
Pennsylvanian System.		
Sand and gravel	46	46
Shells and shale, hard	54	100
Sand	10	110
Shells and shale, hard	210	320
Sand	10	330
Shale, hard	15	345
Sand	142	487
Shale, hard	76	563
Sand and shale	10	573
Sand	10	583
Mississippian System.		
Limestone (Big Lime), cased	132	715
Sandstone	60	775
Sandstone and shale	244	1,019
Shale, hard	129	1,148
Sand (Wier in part?)	107	1,255
Shale, black (Sunbury?)	30	1,285
Sand and shells (Berea?)	80	1,365
Devonian System.		
Shale	330	1,695
Shale, hard, white	70	1,765
Silurian System.		
Limestone (oil "sand"), brown	185	1,950
Limestone, gray	106	2,056
Limestone, sand, white, very hard	25	2,081
Limestone, sand, broken	22	2,103
Total depth		2,103

NOTE—The base of the Silurian and top of the Ordovician is included in 185 feet above 1,950. The record is a very poorly kept one.

Log No. 723

Willie Keaton, No. 1. Gypsey Oil & Gas Co., lessee. Location: Johnson Creek, near Nettie P. O., and the southern nipple of Morgan County. Production: Dry.

Strata.

Pennsylvanian System.	Thickness	Depth
Soil	18	18
Shale and shells	402	420
Sandstone	95	515
Shale	115	630
Sandstone	174	804

Mississippian System.		
Limestone (Little Lime)	6	810
Shale	2	812
Limestone, (Big Lime)	123	935
Shale (Waverly)	367	1,302
Shale, black	4	1,306
Sandstone (Wier sand?)	20	1,326
Shale, white	14	1,340
Sandstone (Berea Grit)	15	1,355
Shale, white	25	1,380

Devonian System.		
Shale, brown	298	1,678
Shale, white	40	1,718
Limestone (oil show at 1,838)	197	1,915
Limestone	74	1,989
Total depth		1,989

NOTE—The base of the Devonian System and top of the Silurian System is indefinite, being included within the 197 feet of limestone beneath the white shale. The well stopped in the top of a red shale which was not measured.

Log No. 724

James Love, No. 1. Browning Oil Co., lessee. T. H. Turner, trustee, lessor. Location: Mine Fork. Elevation: 1,160.

Strata.

Pennsylvanian System.	Thickness	Depth
Soil	21	21
Shale, dark 2' coal at 100'	334	355
Sand, white (water 465-495)	210	565
Shale	68	633
Sand, white	40	673

Mississippian System.		
Limestone (Little Lime)	35	708
Sand, white, soft	17	725
Shale, blue	3	728
Limestone (Big Lime)	40	768

Mississippian System.

	Thickness	Depth
Sand and shale	306	1,074
Sand, gray, gas	11	1,085
Shale, blue	38	1,123
Sand, gray, hard (show oil 1,127-1,132)......	24	1,147
Shale, Wier sand 42 feet	6	1,153
Sand, oil show, Wier sand 42 feet	6	1,159
Shale, Wier sand 42 feet	6	1,165
Sand, hard, Wier sand 42 feet	24	1,189
Shale, black (Sunbury)	12	1,201
Shale, black, sandy (Sunbury)	6	1,207
Sand, Berea	3	1,210
Sand, hard, gray, Berea	41	1,251
Sand, light break, Berea	6	1,257
Sand, break, Berea	3	1,260
Sand, hard, Berea	32	1,292

Devonian System.

	Thickness	Depth
Shale, black	408	1,700
Total depth		1,700

NOTE—White shale showed at 1,700, the bottom of the well. The drill stopped undoubtedly at or very close to the Devonian limestone.

Log No. 725

Browning Oil Co., No. 1, lessee.   John Mart Phipps, lessor.

Strata.

Pennsylvanian System.

	Thickness	Depth
Conductor 8"	24	24
Sand, gray (water at 150)	226	250
Shale, dark	30	280
Sand	50	330
Shale	10	340
Sand	20	360

Mississippian System.

	Thickness	Depth
Limestone	20	380
Shale, (Pencil Cave)	13	393
Limestone (Big Lime)	45	438
Shale, pea green	257	695
Shale, bluish black	53	748
Sandstone, gray, (oil and gas show)	41½	752½
Sand, gray	6½	759
Sand, gray, good show oil	4½	763½
Sand, gray, oil	9	772½
Shale, blue	20	792½
Sand, gray, oil	18	810½

Mississippian System.	Thickness	Depth
Sand, soft brown, good oil, Wier sand	1	811½
Sand, second pay, Wier sand	3	814½
Shale, blue, break, Wier sand	9½	824
Sand, some gas, Wier sand	2	826
Sand, gray-brown, gas, Wier sand	3½	829½
Sand, gray, no oil, Wier sand	4	833½
Sand, gray, little oil, Wier sand	2½	836
Shale, blue, Wier sand	6½	842½
Sand, gray-brown, (show oil?), Wier sand ...	11	853½
Shale, dull, Wier sand	6	859½
Shale, "Sunbury"	2½	862
Total depth		862

Log No. 726

T. M. Cooper, No. 1. Browning Oil Co., lessee. Location: Brushy Fork, Fork of Licking River. Salt water: one bailer per hour.
Strata.

Pennsylvanian System.	Thickness	Depth
Soil	5	5
Sand, coarse	45	50
Shale	100	150
Sand, coarse	50	200
Sand, fine, white	170	370
Shale, brown	90	460
Sand, gray	40	500
Mississippian System.		
Limestone and shale	32	532
Limestone, light brown	16	548
Limestone and shale	12	560
Limestone, black	20	580
Limestone, white	65	645
Shale, red, sandy	8	653
Shale, blue	232	885
Sand, broken, and shale	33	918
Sand and shale	44	962
Sand, white	12	974
Shale	6	980
Sand, white, Wier correllative	63	1,043
Shale, Wier correllative	4	1,047
Sand, white, Wier correllative	4	1,051
Sand, white, Wier correllative	13	1,064
Shale, gray and black (Sunbury)	36	1,100
Sandstone (Berea)	25	1,125
Shale, blue	10	1,135

Devonian System.   ·Thickness Depth

	Thickness	Depth
Shale, black	8	1,143
Shale, blue	10	1,153
Total depth		1,153

Log No. 727

L. C. Bailey, No. 1, lessor. Formerly owned by Browning Pet. Co., now by Cumberland Pet. Co. Production: Reported 40 bbls. Elevation: 1,045.

Strata.

Pennsylvanian System.   Thickness Depth

	Thickness	Depth
Sand and gravel	30	30
Sand	6	36
Shale	12	48
Coal	2	50
Sand	35	85
Sand	110	195
Shale	45	240
Sand	30	270
Shale	35	305
Sand, settling	160	465
Shale	85	530
Sand	40	570
Shale, blue	35	605
Sand, blue, hard	5	610
Shale, blue	8	618

Mississippian System.

	Thickness	Depth
Limestone, white (Little lime)	8	626
Shale, blue (Pencil cave)	18	644
Limestone (Big Lime)	61	705
Sandy shale, pea green	185	890
Shale, blue	85	975
Sand, gray-brown, oil, Wier sand	29	1,004
Shale, blue, Wier sand	14	1,018
Sand, oil, Wier sand	17	1,035
Shale, blue	10	1,045
Sand, gray-brown	8	1,053
Total depth		1,053

Log No. 728

Hostin Conley, lessor. Mine Fork Pet. Co., lessee. Location: Headwaters of Mine Fork Creek, on a branch of Litteral Fork. Elevation: 950.

Strata.

Pennsylvanian. System.	Thickness	Depth
Shale	68	68
Sand (show oil and gas)	14	82
Shale	390	472
Sand	84	556

Mississippian System.

	Thickness	Depth
Limestone (Little Lime)	22	578
Shale (Pencil Cave)	8	586
Limestone (Big Lime)	60	646
Shale sand	104	750
Shale	168	918
Sand, grayish brown (pay oil)	16	934
Shale	15	949
Sand, grayish brown, Wier sand	29	978
Shale, Wier sand	7	985
Sand (pay oil), Wier sand	18	1,003
Shale	10	1,013
Sand, gas in top	13	1,026
Total depth		1,026

Log No. 729

Crate Meade, No. 1, lessor. Browning Pet. Co., lessee. Location: Headwaters of Pigeon Creek, near Johnson County line. Production: 37 bbls. oil and 300,000 ft. gas. Elevation: 1,020.

Strata.

Pennsylvanian System.	Thickness	Depth
Sandstone, shales and coals	609	609

Mississippian System.

	Thickness	Depth
Limestone, (Big Lime)	53	662
Unrecorded sediments	279	941
Sand (1st), Wier sand	22	963
Shale (break), Wier sand..................	21	984
Sand, (2nd), Wier sand	12	996
Shale (break), Wier sand	17	1,013
Sand, Wier sand	3	1,016
Total depth		1,016

Log No. 730

R. B. Griffith, No. 3. Near Wheelersburg P. O. Production: Last two feet in gas sand. Estimated: 15 bbls. oil.

Strata.

Pennsylvanian System.	Thickness	Depth
Unrecorded sediments	624½	624½

Mississippian System.

	Thickness	Depth
Limestone (Big Lime), Top at		624½
Unrecorded sediments	258	882½
Sand (1st)	123	1,005½
Shale	21	1,026½
Sand (pay oil), Wier sand	29	1,055½
Shale, Wier sand	9	1,064½
Sand (pay oil), Wier sand	3	1,067½
Total depth		1,067½

Log No. 731

Milt Wheeler, No. 2, lessor. Bedrock Oil Company, lessee. Location: Litteral Fork near Wheelersburg. Production reported: 15 bbls. of oil.

Strata.

Pennsylvanian System.	Thickness	Depth
Unrecorded sediments	500	500

Mississippian System.

	Thickness	Depth
Limestone (Big Lime, top at 500)	...	500
Limestone (Big Lime) and sandy shale	340	840
Sand, 1st (Wier)	25	865
Shale	20	885
Sand	28	913
Shale	7	920
Sand, gas sand	13	933
Shale	5	938
Sand	3	941
Total depth		941

Log No. 732

Daniel Victoria, No. 1, lessor. Fred Courson, lessee. Location: on Brushy Fork.

Strata.

Pennsylvanian System.	Thickness	Depth
Unrecorded sediments	588	588

Mississippian System.

	Thickness	Depth
Limestone (Big Lime—top 588)		588
Cased at 594		
Limestone (Big Lime) and sandy shales	376	964
Sand (Wier)	76	1,040
Total depth		1,040

NOTE—Since 76 feet is somewhat too thick for the Wier sand normally, it is probable that the driller included by mistake at least one or two higher strata.

Log No. 733

John Blanton, No. 1, lessor. Structural Oil Co., lessee. Elevation: 960. Production reported: 15 bbls. oil. Shot, 80 qts.

Strata.

Pennsylvanian System.

	Thickness	Depth
Soil and blue shale	29	29
Sand	10	39
Shale, blue	61	100
Shale, black	60	160
Sand	30	190
Shale, blue	10	200
Sand, white, water at 300 (oil)	165	365
Shale	100	465
Sand, settling	20	485

Mississippian System.

	Thickness	Depth
Limestone, hard, shells	40	525
Limestone (Little Lime)	5	530
Shale (Pencil Cave)	10	540
Lime (Big Lime)	70	610
Shale, green	165	775
Limestone, hard, shell	55	830
Shale, dark	38	868
Sand (oil and gas)	2	870
Sand	10	880
Shale	27	907
Sand (oil)	23	930
Shale	9	939
Sand, gas showing	12	951
Sand	5	956
Total depth		956

AN EXCELLENT MISSISSIPPIAN EXPOSURE
The clifted strata above is the St. Louis Limestone (lower part of the "Big Lime"), and below occur the green, shaley and sandy Logan and Cuyahoga formations. Photo ¼ mile above Glencarin, Wolfe County, Kentucky.

Log No. 734

Buddie Blanton, lessor. L. S. Roberts, et. al., lessees. Lower No. 1. Location: ½ mi. from mouth of Panther's Lick. Elevation: 920 feet.

Strata.

Pennsylvanian System.	Thickness	Depth
Unrecorded sediments	600	600
Mississippian System.		
Limestone (Big Lime)	75	675
Unrecorded sediments	237	912
Sand (Wier) and shale	74	986
Sandy shale including Sunbury to		986
Total depth		986

NOTE—The last 74 feet of this well includes not only the Wier sand, but also the underlying Sunbury shale, and a small upper portion of the Berea. A nice show of oil in the Berea is reported. This well shot with 60 quarts.

Log No. 735

Milt Wheeler, No. 1, lessor. Bedrock Oil Co., lessee. Location: Litteral Fork near Wheelersburg. Production reported: 22 bbls. of oil.

Strata.

Pennsylvanian System.	Thickness	Depth
Unrecorded sediments	854	854
Mississippian System.		
Sand, 1st (Wier)	23	877
Shale	25	902
Sand (Wier)	30	932
Total depth		932

Log No. 736

D. B. Cooper, No. 1, lessor. Location: Head of Lick Creek. Drillers: Ben Creed, Algin Messer. Completed and shot April 9, 1921, with 30 quarts in first pay, and 40 quarts in second pay. Had 650 feet fluid in hole Monday, A. M., April 11th, 1921.

Strata.

Pennsylvanian System.	Thickness	Depth
Surface soil (conductor)	7	7
Shale, gritty	187	194

Pennsylvanian System.

	Thickness	Depth
Sand (50 ft. bottom settles)	220	414
Shale	75	489
Sand (3 breaks)	45	534
Shale	13	547
Limestone, sandy	8.	555
Shale	9	564

Mississippian System.

	Thickness	Depth
₰ Limestone, black	6	570
Shale	5	575
Limestone (Little Lime)	10	585
Shale (Pencil Cave)	20	605
Limestone (Big Lime), casing 613	83	688
Shale, light gray	160	848
Shale dark gray (shells)	125	973
Sand	4	977
Shale, black	16	993
Sandstone (Wier), (Top 993)	$\frac{1}{2}$	$993\frac{1}{2}$
First pay, Wier sand	25	$1,018\frac{1}{2}$
Break, Wier sand	3	$1,021\frac{1}{2}$
Second pay, Wier sand	22	$1,043\frac{1}{2}$
Total depth		$1,043\frac{1}{2}$

Log No. 737

Bud Gullet, No. 1' lessor. Location: State Road Fork. Elevation: 1,059.

Strata.

Pennsylvanian System.

	Thickness	Depth
Sand, gravel	20	20
Sandstone, hard	8	28
Shale	36	64
Sandstone	35	99
Shale	85	184
Sandstone	25	209
Shale	45	254
Shale, sandy, dark blue	25	279
Sand, gray	66	345
Shale	10	355
Sandstone	15	370
Sand, settling	70	440
Shale	100	540
Sand, dark	25	565
Shale, soft	10	575
Sand, hard, blue	5	580
Shale	..	...

NOTE—This record is all in the Coal Measures, and is incomplete.

Log No. 738

Jack Whittaker 'Well. Incomplete record, drilling Oct. 22, 1921. Drilling started, April 9, 1921. Location: Arnett Branch of Burning Fork 4½ miles (airline) southeast of Salyersville. Production: Oil and gas shows only; plugged and abandoned. Authority: S. L. Yunker.

Strata.

	Thickness	Depth
Pennsylvanian System.		
Soil, Pottsville	38	38
Shale, Pottsville	132	170
Sand, Pottsville	30	200
Shale, Pottsville	160	360
Sand, Pottsville	30	390
Shale, Pottsville	30	420
Sand, Pottsville	318	738
Shale, Pottsville	12	750
Mississippian System.		
Limestone (Little Lime)	10	760
Shale (Pencil Cave)	15	775
Limestone (Big Lime)	117	892
Shale (Waverly), (oil show 1,106-1,111)	215	1,107
Sand (Wier), (salt water 1,145-1,151)	131	1,238
Shale	54	1,292
Shale (Sunbury)	12	1,304
Sandstone (Berea)	53	1,357
Devonian System.		
Shale, black (Chattanooga)	378	1,735
Limestone and white shale	137	1,872
Limestone, black	28	1,900
Ordovician System.		
Limestone, brown	12	1,912
Limestone, gray	18	1,930
Limestone, flinty	20	1,950
Limestone, tight, (sulphur gas 1,952)	30	1,980
Limestone, white	30	2,010
Limestone, blue	50	2,060
Limestone, gray	43	2,103
Limestone, brown	17	2,120
Limestone, gray	42	2,162
Limestone, blue	20	2,182
Limestone, light gray	11	2,193
Limestone, black	17	2,210
Limestone, white	7	2,217
Limestone, gray	13	2,230

Ordovician System.

	Thickness	Depth
Limestone, brown, fine	62	2,292
Limestone, brown, coarse	26	2,318
Limestone, brown	14	2,332
Limestone, gray	56	2,388
Limestone, brown	15	2,403
Limestone, blue	19	2,422
Limestone, white, flaky	40	2,462
Shale, green	10	2,472
Shale (red rock)	68	2,540
Shale, green	50	2,590
Shale, white	55	2,645
Shale (red rock)	70	2,715
Shale, arenaceous	141	2,856
Shale (red rock)	24	2,880
Shale, arenaceous	70	2,950
Limestone, soft	387	3,337
Limestone, broken	613	3,950
Incomplete depth		3,950

NOTE—The Devonian-Silurian contact occurs toward the top of the 137 feet of limestone and shale above 1,872 feet, and was not noted by the driller.

Log No. 739

Sherman Rice, No. 1, lessor. Ivyton Oil & Gas Co., lessee. Location: Kelly Branch of Burning Fork, Ivyton. Date Drilled: July 9, 1921. Contractor: Gentry. Orig. Open Flow: 1,000,000 cu. ft. gas. Orig. Rock Press.: 390. lbs. Casinghead elevation: 965.5. Authority: Louisville Gas & Electric Co.

Strata.

Pennsylvanian System.

	Thickness	Depth
Soil	28	28
Shale	52	80
Lime shell	40	120
Shale	250	370
Sandstone	330	700
Shale	3	703
Sandstone	37	740
Shale	16	756

Mississippian System.

Limestone (Little Lime)	13	769
Cave	3	772

Mississippian System.	Thickness	Depth
Limestone, (Big Lime)	10	782
Limestone, (Big Lime)	58	840
Shale	225	1,065
Sandstone	20	1,085
Shale	11	1,096
Sandstone (Wier)	22	1,118
Total depth		1,118

Log No. 740

Cordelia Grace, No. 1, lessor. Ivyton Oil & Gas Co., lessee. Location: Ivyton. Contractor: Gentry. Orig. Open Flow: 350,000 cu. ft. gas. Shot. Orig. Rock Press.: 265 lbs. Casinghead elevation: 924.8. Authority: Louisville Gas & Electric Co.

Strata.

Pennsylvanian System.	Thickness	Depth
Soil	30	30
Shale, black	155	185
Shale, white	180	365
Sandstone	350	715
Shale	5	720
Sandstone	30	750

Mississippian System.		
Limestone (Little Lime)	25	775
Cave	6	781
Limestone (Big Lime)	59	840
Shale, gray	210	1,050
Shale, black	8	1,058
Sandstone	80	1,138
Shale	2	1,140
Sandstone (Wier)	38	1,178
Shale (Sunbury)	22	1,200
Sandstone, brown	35	1,235
Sandstone (Berea)	33	1,268
Shale, brown	6	1,274
Total depth		1,274

Log No. 741

W. Spradlin, Nò. 1, lessor. Ivyton Oil & Gas Co., lessee. Location: Middle Creek, Ivyton, Ky. Contractor: Gentry. Date Drilled: Aug. 13, 1921. Production: Dry. Authority: Louisville Gas & Electric Co.

Strata.

Pennsylvanian System.	Thickness	Depth
Soil	10	10
Limestone	30	40
Shale	285	325
Sandstone	345	670
Mississippian System.		
Shale, sandy, red	2	672
Limestone (Little Lime)	22	694
Cave	5	699
Limestone (Big Lime)	141	840
Shale, gray	220	1,060
Shale, black	10	1,070
Shale, black	12	1,082
Shale, white	38	1,120
Shale, black	10	1,130
Shale	15	1,145
Shale, brown	6	1,151
Sandstone	24	1,175
Shale	10	1,185
Sandstone	20	1,205
Shale, brown	16	1,221
Total depth		1.221

Log No. 742

George Grace, No. 1, lessor. Ivyton Oil & Gas Co., lessee. Location: Grace Branch of Middle Creek, Ivyton. Contractor: Gentry. Date drilled: Aug. 30, 1921. Orig. Open Flow: 556,000 cu. ft. gas. Shot. Orig. Rock Press.: 390 lbs. Authority: Louisville Gas & Electric Co.

Strata.

Pennsylvanian System.	Thickness	Depth
Soil	19	19
Limestone and sandstone	106	125
Shale	425	550
Sandstone	165	715
Shale	2	717

Pennsylvanian System.	Thickness	Depth
Sandstone	163	880
Shale	5	885
Sandstone	25	910
Shale	4	914

Mississippian System.		
Limestone (Little Lime)	10	924
Cave	22	946
Limestone (Big Lime)	24	970
Shale (Waverly)	124	1,094
Sandstone	20	1,114
Shale (Waverly)	58	1,172
Sandstone (Wier)	28	1,200
Sandstone (Wier), hard	12	1,212
Total depth		1,212

Log No. 743

Elzo Dotson, No. 2, lessor. Ivyton Oil & Gas Co., lessee. Location: Mash Branch of Burning Fork. Contractor: Potts. Date drilled: Oct. 20, 1921. Orig. Open Flow: 750,000 cu. ft. gas. Shot. Orig. Rock Press.: 350 lbs. Authority: Louisville Gas & Electric Co.

Strata.

Pennsylvanian System.	Thickness	Depth
Soil	15	15
Shale	40	55
Sandstone	15	70
Shale	165	235
Sandstone	35	270
Shale	145	415
Sandstone	20	435
Shale	10	445
Sandstone	350	795
Shale	40	835

Mississippian System.		
Limestone (Little Lime)	20	855
Cave	2	857
Limestone (Big Lime)	63	920
Shale, gray	240	1,160
Shale, black	15	1,175
Sandstone	15	1,190

Mississippian System.

	Thickness	Depth
Sandstone, hard	13	1,203
Sandstone	8	1,211
Sandstone, hard	2	1,213
Total depth		1,213

NOTE—This well showed some oil, but shot ruined same.

Herewith are given a number of "sand" records. These logs are all incomplete, the thickness of the Pennsylvanian System and the uppermost beds of the Mississippian System having been omitted by the driller.

Log No. 744

R. B. Griffith, No. 1, lessor. Bedrock Oil Company, lessee. Location: Litteral Fork near Wheelersburg.

	Top	Bottom
Top, Big Lime, all Mississippian	467	...
First sand, all Mississippian	819	836
Shale, all Mississippian	836	858
Sand (Wier oil), all Mississippian	858	888
Shale, all Mississippian	888	893
Sand with gas, all Mississippian	893	906
Shale, all Mississippian	906	924
Total depth		924

Log No. 745

R. B. Griffith, No. 2, lessor.

	Top	Bottom
Top, Big Lime.............................	...	580
Sand	943	965
Shale	965	984½
Sand, pay	984½	1,013
Shale	1,013	1,020
Sand, gas	1,020	1,038
Total depth		1,038

Shot, 20 qts. in 1st, and 60 qts. in 2nd. Production: 10 bbls. oil.

Log No. 746

Milt Wheeler, No. 3

		Bottom
Top, Big Lime at		460
First sand, show oil & gas	800	823
Shale	823	
Sand (Wier oil)	846½	872
Shale	872	
Sand, good flow gas	881½	
Shale	898½	901
Total depth		901

Shot with 20 qts. in 1st pay and 60 qts. in 2nd pay. Production: 18 bbls oil.

Log No. 747

Vernon Kelley, lessor. **Myers & Turner**, lessee. Location: Two miles west of Ivyton.

	Bottom
Top of Big Lime at	718
Top of Wier	1,015
Gas at	1,040
Show of oil	1,081

Log No. 748

Dave Conley, lessor. Mid South Oil Co., lessee. Location: Litteral Branch. Elevation: 970. Completed: June 12, 1920. Production: 30 bbls. oil.

	Top	Bottom
Wier sand top	...	866
Wier sand	866	866
Dark shale	886	909
Sand	909	936
Dark shale	936	948
Sand	948	961
Good show oil	909	931

Log No. 749

M. Collins, lessor. Location: One mile west of Oil Springs on State Road Fork of Little Paint Creek. Elevation: 906. Production: 6 bbls. oil. Shot, 40 qts. 884 ft. gas.

	Top	Bottom
Gas	60	200
Settling sand		185
Big Lime	560	620
Cased		525
Pea green shale sand	620	824
1st pay sand	854	863
Brown shale, shells	863	900
Gas, gray brown sand	900	911
Soft mud	903	907
Blue shale	911	945
Coffee shale	945	960
Berea	960	980

Log No. 750

Pit Whitten, No. 1, lessor. Sidney Oil Co., lessee. Location: Painter Lick Fork of Little Paint Creek. Production: 12 bbls. oil natural.

	Top	Bottom
Gas at	...	250
Top of lime	...	723
1st oil show	...	1,010
Depth		1,068

Log No. 751

Bud Blanton, lessor. Sidney Oil Co., lessee. Location: Painter Lick Fork of Little Paint Creek. Production: 15 bbls. oil. Shot, 60 qts.

	Top	Bottom
Gas & 1 bbl. oil	250	278
Top of lime	...	810
First oil	...	1,045
Sand, best oil	1,045	1,057
Break	1,060	1,064
Sand (pay)	1,064	1,069

	Top	Bottom
Break	1,069	1,075
Sand (gas & oil in last screw)	1,075	1,100
Break	1,100	1,118
Sand, good quality	1,118	1,135
Sand shale	1,135	1,147
Bottom hole	...	1,147

Log No. 752

W. B. Bailey, No. 1, lessor. Location: On State Road Fork, two miles west of Oil Springs. Drilled in June 23, 1920.

	Top	Bottom
Top of lime at	...	564
Bottom of lime at	...	654
Sand	...	897
Pay	...	921
Blue shale	921	934
Sand loft oil gas 8 ft.	834	952
Bottom hole	...	954

MARTIN COUNTY.

Production: Oil and Gas. Producing sands: Maxton, Big Lime, Big Injun, Wier, and Berea (Mississippian).

Log No. 753

Malissa Ward, No. 1, lessor. Mayo Gas & Oil Co., lessee. Location: On Rockhouse Fork of Rockcastle Creek, 4 miles west of Inez. Commenced October 6, 1919. Shut down October 20, 1919. Drilling recommenced Nov. 10, 1919. Completed March 24, 1920. Conductor 22' 1" 13½ Casing 10" 201' 6' Casing 8" 1043' 7" Casing 6⅝ 1255' Tubing 1274' 10" 2". Casing pulled out 8¼" 1043' 7".

Strata.

Pennsylvanian System.	Thickness	Depth
Soil and sand	21	21
Coal bloom	1	22
Shale, black (fresh water)	53	75
Limestone	50	125
Shale, white	75	200
Limestone	10	210
Coal	5	215
Shale, black	60	275
Limestone	25	300
Shale, gray	75	375
Sand (First Dunkard)	25	400

Pennsylvanian System.   Thickness Depth

	Thickness	Depth
Shale	75	475
Shale (Second Dunkard)	25	500
Sand, salt, (salt water 540)	125	625
Shale, dark	5	630
Sand, salt	100	730
Shale, dark	10	740
Sand	60	800

Mississippian System.

	Thickness	Depth
Limestone	25	825
Shale	15	840
Sand	20	860
Limestone	25	885
Shale (salt. water)	15	900
"Sand" (Maxton)	15	915
Limestone	10	925
Shale (Red Rock), (salt water)	5	930
"Sand" (Maxton)	80	1,010
Shale, dark	5	1,015
Shale, red, sandy	5	1,020
Shale, dark	40	1,060
Limestone, dark	29	1,089
Shale (pencil cave)	2	1,091
Limestone, dark	6	1,097
Limestone (Big Lime), white	173	1,270
Sand, red (Big Injun)	20	1,290
Shale, white	60	1,350
Shale, black, and shells	287	1,637
Limestone shells, dark	10	1,647
Shale, dark brown	40	1,687
Sand, (gas)	6	1,693
Shale, white	19	1,712
Limestone, black	11	1,723
Shale, dark, and shells	20	1,743
Sand, hard	10	1,753
Shale, blue	14	1,767
Limestone, sandy	23	1,790

Devonian System.

	Thickness	Depth
Shale, black (Chattanooga)	125	1,915
"Sand," dark	12	1,927
Shale, dark (Chattanooga)	682	2,609
Shale, brown	26	2,635
Limestone (Corniferous, upper 50 ft.)	500	3,135
Total depth		3,135

Break of shale 2 ft. at 3,090. Break of shale 5 ft. at 3,125.
Last limestone showed all colors, no two screws alike. Oi' show in Big
Lime 1210-1215. Gas in last limestone 2849-2864.

NOTE—Neither the Berea or Weir sands shows characteristically
in this record. The base of the Devonian and top of the Silurian, as
well as the base of the Silurian and top of the Ordovician, are included
within the 682 feet of "dark shale," probably a succession of limestones just below 1,927 feet. The last 1,220 feet of this record was
very slovenly kept.

Log No. 754

Lewis Dempsey, No. 1, lessor. United Fuel & Gas Co., lessee. Location: Forks of Pipe Mud & Holty Branches of Wolf Creek. Authority: Adkins, Supt. Completed: Dec., 1918. Driller. Lohman. Elevation: 620. (aneroid). 5 bailers of salt water per hour from Injun
sand.

Strata.

Pennsylvanian System.	Thickness	Depth
Conductor	16	16
Sand	34	50
Coal	5	55
Sand	60	115
Coal	6	121
Sand	97	218
Coal, cannel	4	222
Sand	13	235
Coal	3	238
Sand, (salt water 221-436)	198	436
Coal	1½	437½
Sand	½	438
Gray and broken sand and shells	507	945
Shale	5	950
Sand	60	1,010
Break	8	1,018
Sand	37	1,055
Shale	20	1,075
Sand, (base of Pottsville)	130	1,205 -

Mississippian. System.

	Thickness	Depth
Sand	80	1,285
Sand (Maxon), (light oil show 1,312)	40	1,325
Red rock	80	1,405
Pencil cave (6 in. casing to 1,455)	15	1,420
Limestone	150	1,570
Red rock and lime shells	30	1,600
Sand, Injun, (gas show)	60	1,660
Shale and lime shells, (more gas)	60	1,720
Shale and shells	235	1,955

Devonian System.

	Thickness	Depth
Shale, very black at bottom	260	2,215

Silurian System.

	Thickness	Depth
Sand, Niagara, (light oil show 2,215)	45	2,260
Shale, black	35	2,295
Total depth		2,295

Log No. 755

Lewis Dempsey, No. 1, (Elk Creek Tract) lessor. United Fuel & Gas Co., lessee. Location: Head of Big Elk Creek. Production: 559,-000 cu. ft. gas. Rock pressure: 275 lbs. Authority: C. M. Goodwill, driller.

Strata.

Pennsylvanian System.

	Thickness	Depth
Clay	11	11
Sand	59	70
Shale and shell	54	124
Coal	3	127
Shale	13	140
Limestone	20	160
Shale	10	170
Coal	4	174
Sand	16	190
Shale	40	230
Sand	20	250
Shale	10	260
Limestone	52	312
Limestone	10	322
Shale and shell	178	500
Sand	45	545
Shale, black	50	595

Pennsylvanian System.	Thickness	Depth
Sand	55	650
Shale	40	690
Sand (1st salt sand)	65	755
Shale	30	785
Sand (2nd salt sand)	205	990
Shale	40	1,030
Mississippian System.		
Sand	57	1,087
Shale	81	1,168
Sand (Maxon)	11	1,179
Shale shells	34	1,213
Red rock	2	1,215
Total depth		1,215

Log No. 756

Lewis Dempsey, No. 2, lessor. United Fuel & Gas Co., lessee. Location: Buck Creek, right fork. Drilled: June 15, 1916. Production: 7,500 cu. ft. gas per day. Rock pressure: 350 lbs. Authority: W. F. Taylor & R. N. Dunbar, drillers.

Strata.

Pennsylvanian System.	Thickness	Depth
Clay	20	20
Quicksand	7	27
Clay, blue	7	34
Sand	26	60
Shale	30	90
Limestone?	5	95
Coal	4	99
Shale	6	105
Limestone?	15	120
Shale	20	140
Limestone?	25	165
Shale	20	185
Limestone?	20	205
Shale	40	245
Sand, (show of gas 340)	105	350
Shale, brown	30	380
Shale, hard	20	400
Limestone?	15	415
Shale	40	455
Sand (1st salt sand)	120	575

Pennsylvanian System.	Thickness	Depth
Break	2	577
Sand (2nd salt sand), (2 bailers per hour at 695)	118	695
Sand	17	712
Sand	41	753
Break	1	754
Sand, black, (base Pottsville?)	2	756
Mississippian System.		
Sand, white	34	790
Shale	15	805
Limestone	10	815
Shale	45	860
Fire clay	8	868
Red rock	22	890
Limestone, hard	16	906
Shale, white	7	913
Limestone shells	1	914
Red rock	8	922
Limestone, black	20	942
Shale	8	950
Limestone	15	965
Red rock	7	972
Sand	3	975
Shale	8	983
Limestone shells	5	988
Red rock	37	1,025
Limestone shells	3	1,028
Shale	5	1,033
Limestone (Little Lime)	27	1,060
Shale	15	1,075
Shale (Pencil Cave)	10	1,085
Limestone (Big Lime)	230	1,315
Red rock	14	1,329
Sand, (Big Injun)	44	1,373
Limestone, sandy	52	1,425
Shale	45	1,470
Shale, black	25	1,495
Shale	35	1,530
Sand	10	1,540
Shale, black	135	1,675
Limestone shells	5	1,680
Shale, black (Sunbury)	120	1,800
Sand (Berea)	14	1,814
Limestone shells	25	1,839
Total depth		1,839

Log No. 757

Lewis Dempsey, No. 1, (Warfield Tract), lessor.   United Fuel &
Gas Co., lessee.   Location: Martha Boone Hollow of Right Fork of Buck
Creek.   Production: 96,000 cu. ft. gas.   Rock pressure: 310 lbs.
Authority: J. R. McCleary, driller.

Strata.

Pennsylvanian System.	Thickness	Depth
Conductor	16	16
Sand	242	258
Shale and limestone	257	515
Sand	45	560
Shale and limestone	30	590
Sand, salt	290	880

Mississippian System.		
Shale and limestone shells	35	915
Sand (Maxon)	25	940
Limestone shells	40	980
Shale, black (pencil cave)	40	1,020
Sand (Maxon), 2nd, (gas 1,040)	40	1,060
Shale and red rock	83	1,143
Sand	10	1,153
Red rock	27	1,180
Shale (Pencil Cave)	75	1,255
Limestone (Big Lime), (gas 1,313)	175	1,430
Red rock	15	1,445
Limestone shell	150	1,595
Shale	353	1,948
Shale	12	1,960
Sandstone (Berea grit), (gas 1,950)	75	2,035
Limestone shells	15	2,050
Total depth		2,050

Log No. 758

Lewis Dempsey, (Tract No. 1, well No. 1), lessor.   United Fuel
& Gas Co., lessee.   Location: Head of Big Elk Creek.   Production: Dry
hole.   Authority: D. S. Osborne & R. M. Dunbar, drillers.

Strata.

Pennsylvanian System.	Thickness	Depth
Clay	9	9
Shale	66	75
Sand	45	120

Pennsylvanian System.   Thickness Depth

	Thickness	Depth
Shale	42	162
Coal	5	167
Shale	18	185
Sand	35	220
Shale	35	255
Sand	20	275
Shale	45	320
Limestone	25	345
Shale	70	415
Limestone	30	445
Shale	25	470
Limestone	10	480
Shale	25	505
Sand	55	560
Limestone	40	600
Sand, salt	190	790
Shale	50	840
Limestone, black	25	865
Shale	5	870
Limestone, black	10	880
Shale	84	964
Sand	14	978
Shale	5	983

Mississippian System.

	Thickness	Depth
Red rock	15	998
Shale	27	1,025
Red rock	60	1,085
Sand (Maxon)	8	1,093
Shale	22	1,115
Red rock	10	1,125
Shale	25	1,150
Limestone	10	1,160
Shale	5	1,165
Limestone (Little Lime)	21	1,186
Shale (Pencil Cave)	4	1,190
Limestone (Big Lime)	190	1,380
Sand, Injun	15	1,395
Shale	75	1,470
Total depth		1,470

NOTE—Fresh water at 65 ft.; hole full. Salt water at 740 ft.; hole full.

McCRACKEN COUNTY.

Production: Neither oil or gas to date. Producing sands; none recognized to date.

Log No. 759

Paducah Well. Lessor unknown. Lessee unknown. Location: Within the City of Paducah. Drilling completed in 1888. Production: Dry. Drilling samples collected by J. C. Farley and W. L. Bradshaw. Authority: R. H. Loughridge, Ass't Geologist, Jackson Purchase Report of Kentucky Geological Survey, Series II, p. 321-326, pub. 1888.

Strata.

	Thickness	Depth
Quaternary System.		
Loam, brown, micaceous	40	40
Gravel, rounded chert and quartz	20	60
Tertiary System.		
Clay, black, and sand	90	150
Cretaceous System.		
Clay and sand, micaceous interlaminated	114	264
Chert, quart, and pyrite debris	71	335
Mississippian System.		
Limestone, shaly white, fossils, Chester Group	90	425
Limestone, dark, impure, cavernous, Chester Group	45	470
Limestone, silicious, cavernous, Chester Group	48	518
Shale, dark, limy, fossils, Chester Group	32	550
Shale, white, limy, fossils, Chester Group	185	735
Limestone, blue, Pentremital, Chester Group	400	1,135
Limestone, blue, fractured, loose sand, (St. Louis)	115	1,250
Total depth		1,250

NOTE—This record has been slightly revised from the original, chiefly to show the Tertiary representative which is regarded as present in this locality beneath the surface. Loughridge considered this record important as a proof of down throw faulting of 1,350 feet on the Kentucky side of the Ohio River as compared to the geologic section on the Illinois side of the Ohio River. This amount of faulting, though large, is indicated as altogether probable by recent detailed work done in Livingston, Crittenden and Caldwell Counties. In Livingston County the elongated areal outcrop of Pottsville sediments, extending in a northeast-southwest direction, is in reality a dropped fault block bordered on the northwest by a fault and on the southeast by another fault, each of which may be regarded as major faults of the region. In Livingston County the down throw attains a measured maximum of

—feet. This great fault block if it were to extend to the southwest as the two faults when last seen in Kentucky would indicate, would pass directly under the City of Paducah, though the thick recent deposits of unconsolidated sand, gravels and clays would obliterate any surface indication or proof of the great deformation below.

McCREARY COUNTY.

Production: Oil and gas. Producing sand: ''Beaver'' (Mississippian).

Log No. 760

Rock Creek Property Co., No. 16, lessee. Completed: March 5, 1914. Production: First day, 5 bbls. Authority: New Domain Oil & Gas Co.

Strata.

Pennsylvanian System.	Thickness	Depth
Shale, soft	15	15
Sandstone, yellow	105	120
Shale, blue, soft	180	300
Shale, red, sandy	50	350
Shale, blue, soft	114	464
Mississippian System.		
Limestone, variable in color	586	1,050
Shale, hard, blue, white	78	1,128
Limestone ''sand'' (Beaver), white	14	1,142
Shale, hard, blue (New Providence)	6	1,148
Total depth		1,148

Log No. 761

J. L. and J. A. Dobbs, No. 1, lessors. New Domain Oil & Gas Co., lessee. Completed: June 17, 1914. Production: After shot, 10 bbls. Authority: New Domain Oil & Gas Co.

Strata.

Pennsylvanian System.	Thickness	Depth
Clay (soil)	11	11
Sandstone	90	101
Clay, blue, red	345	446
Mississippian System.		
Limestone, gray, white	630	1,076
Shale, hard	50	1,126
Limestone ''sand'' (Beaver), brown	19	1,145
Shale, hard, blue (New Providence)	11	1,156
Total depth		1,156

Log No. 762

J. L. and J. A. Dobbs, No. 2, lessors. New Domain Oil & Gas Co., lessee. Completed: July 10, 1914. Production: 5 bbls. Authority: New Domain Oil & Gas Co.

Strata.

Pennsylvanian System.	Thickness	Depth
Clay	8	8
Sandstone, yellow	92	100
Shale, blue	150	250
Shale, red and blue	205	455

Mississippian System.		
Limestone, gray, white	405	860
Limestone, black	150	1,010
Shale, hard, mixed	120	1,130
Limestone "sand" (Beaver), brown	13	1,143
Shale, hard, blue (New Providence)	22	1,155
Total depth		1,155

Log No. 763

J. L. and J. A. Dobbs, No. 3, lessors. New Domain Oil & Gas Co., lessee. Completed: Dec. 12, 1914. Production: 15 bbls. Authority: New Domain Oil & Gas Co.

Strata.

Pennsylvanian System.	Thickness	Depth
Sandstone	125	125
Clay, shale, blue and red	355.	480

Mississippian System.		
Limestone, gray and white	390	870
Limestone, black	175	1,045
Shale, hard, mixed	125	1,170
Limestone "sand" (Beaver)	12	1,182
Shale, hard, blue (New Providence)	10	1,192
Total depth		1,192

Log No. 764

: Cephas Rice, No. 1, lessor. . New Domain Oil & Gas Co., lessee.
Completed: Oct. 8, 1915. Production: Dry. Authority: New Domain
Oil & Gas Co.
Strata.

Pennsylvanian System.	Thickness	Depth
Clay	10	10
Sandstone	150	160
Shale	300	460
Mississippian System.		
Limestone, gray, white	390	850
Limestone, black	175	1,025
Shale, hard, mixed	103	1,128
Limestone "sand" (Beaver), brown	12	1,140
Shale, hard, blue (New Providence)	20	1,160
Total depth		1,160

Log No. 765

Ephram Phipps, No. 1, lessor. New Domain Oil & Gas Co., lessee.
Completed: Dec. 9, 1919. Production: Dry. Well abandoned. Author-
ity: New Domain Oil & Gas Co.
Strata.

Mississippian System.	Thickness	Depth
Clay	20	20
Limestone, white	380	400
Limestone, gray	50	450
Limestone, black	175	625
Limestone and flint, black	107	732
Limestone "sand" (Beaver), white	5	737
Shale, hard, blue (New Providence)	11	748
Total depth		748

Log No. 766

Hoffman Bros., No. 1, lessees. Location: South of Silerville P. O.
Strata.

Pennsylvanian System.	Thickness	Depth
Soil	7	7
Sand	43	50
Shale, dark	28	78
Fire clay, sandy	4	82
Shale, sandy	25	107
Shale, dark	34	141
Fire clay, sandy	1 : 10	142 : 10

Pennsylvanian System.	Thickness	Depth
Sand	20	**162 : 10**
Shale, dark	26 : 2	189
Shale, black	25 : 9	214 : 9
Coal	0 : 3	215
Shale, dark, sandy	20 : 4	235 : 4
Sand	49 : 6	284 : 10
Shale, dark	0 : 8	285 : 6
Sand	23 : 6	309
Shale, dark	48	357
Sand	164 : 6	521 : 6
Coal	0 : 7	522 : 1
Fire clay, sandy	1 : 6	523 : 7
Shale, sandy	16 : 3	539 : 10
Shale, black	22	561 : 10
Fire clay, sandy	2 : 10	564 : 8
Shale, black	15 : 7	580 : 3
Coal	0 : 7	580 : 10
Shale, sandy	5 : 6	586 : 4
Sand	10	596 : 4
Shale, dark	33 : 2	629 : 6
Sand	8 : 10	638 : 4
Coal	0 : 2	638 : ½
Sandy rock binder	0 : 3½	638 : 4
Coal	0 : 10	639 : 2
Sand	25 : 2½	664 : 4½
Coal	0 : 5	664 : 9½
Fire clay	4 : 6	669 : 3½
Sand	10	679 : 3½
Coal	0 : 2	679 : 5½
Fire clay, sandy	3 : 9	683 : 2½
Sand	12	695 : 2½
Coal	0 : ½	695 : 3
Shale, limy	3	698 : 3
Sand	11 : 6	709 : 9
Shale, sandy	4 : 8	714 : 5
Sand, dark, limy	0 : 10	715 : 3
Shale, limy	4 : 4½	719 : 7½
Shale, dark	37	756 : 7½
Fire clay, sandy	4	760 : 7½
Shale, light	3	763 : 7½
Shale, gray	6	769 : 7½
Fire clay	1 : 6	771 : 1½
Sand rock	79 : 6	850 : 7½
Mississippian System.		
Shale, green, Mauch Chunk	11	861 : 7½
Shale, red, Mauch Chunk	16	877 : 7½

Mississippian System.   Thickness Depth
      Shale, gray, Mauch Chunk ....   2 : 11   880 : 6½
      Limestone, gray, Mauch Chunk   1   881 : 6½
           Total depth ....   881 : 6½

CHAPTER VIII.

McLEAN COUNTY.

Production: Oil and gas. Producing sands: "Beech Grove" and Sebree Sandstone (Alleghany-Pennsylvanian).

Log No. 767

J. L. Ford, No. 1, lessor. B. A. Kinney, Bradford, Pa., and Henry O'Hara, St. Louis, Mo., lessees. Location: Glennsville, 6 miles N. E. of Calhoun. Contractor: Clarence Shadwick, Owensboro, Ky. Authority: C. Shadwick and J. G. Stuart.

Strata.

Pennsylvanian System.

	Thickness	Depth
Soil	2	2
Loess	20	22
Sandstone, brown, shale, laminated	5	27
Sandstone, brown, shale laminated	13	40
Clay, gray, shale, laminated soapstone	3	43
Shale, black	2	45
Coal	2	47
Shale (fire clay)	3	50
Clay, gray, shale, slaty	30	80
Limestone, clayey, blue, hard	9	89
Shale, blue	10	99
Shale, black	3	102
Blue clay limestone "Bastard"	25	127
Shale, dark blue, petroliferous, very plastic fat water copious sulphate of iron	10	137
Trace, pronounced, taste of oil in all 50' blue shale	20	157
Total depth		157

Log No. 768

J. L. Ford, No. 2, lessor. B. A. Kinney and Henry O'Hara, lessees. Location: 6 miles N. E. of Calhoun. Authority: C. Shadwick and J. G. Stuart.

Strata.

Pennsylvanian System.

	Thickness	Depth
Shale and slate	5	5
Coal stain	0	5
Shale (fire clay), white	5	10
Unrecorded	156	166
Shale, black, slaty	19	185
Blue limestone shale	15	200
Shale, black, slaty	6	206

Pennsylvanian System.	Thickness	Depth
Coal stain	0	206
Shale (fire clay)	2	208
Limestone, gray, micaceous, coarse grained & porous, white grit, sandstone	36	244
Shale, black	20	264
White limestone "carbonate"	6	270
Limestone, white, clayey	9	279
Shale, black, slaty	6	285
Shale, black	12	297
Oil sand, coarse, grayish white	3	300
Sand, micaceous, fine grained, grayish white..	31	331
Total depth		331

Log No. 769

McLean County (Poor Farm), lessor. Drayton Drilling Syndicate, Decatur, Ill., lessee. Location: Waters of Pond and Cypress Creeks. Commenced: Aug. 30, 1920. Completed: Aug. 7, 1921. Casing head elev.: 402 feet. Geologist in charge: Dr. C. N. Gould. Stratigraphic determinations made from the cuttings by J. L. Ferguson.

Strata.

Pennsylvanian System.	Thickness	Depth
Soil	25	25
Coal, No. 14, (according to Hutchison K. G. S.) —	5	30
Shale, shelly, very hard	10	40
Shale, gray	60	100
Shale, dark gray, sandy	30	130
Sandstone	35	165
Shale, dark gray, sandy	12	177
Coal	3	180
Shale, shelly	6	186
Shale, dark gray, sandy	44	230
Shale, shelly	10	240
Shale	20	260
Limestone, light gray	2	262
Shale	8	270
Shale, dark gray, sandy	25	295
Shale, dark gray, sandy	25	320
Shale, dark gray, sandy	10	330
Sandstone, fine, white, calcareous	30	360
Shale	30	390
Sandstone, gray, fine	5	395

Pennsylvanian System.

	Thickness	Depth
Shale, gray, sandy	5	400
Shale	25	425
Shale, dark gray, sandy	45	470
Sandstone	10	480
Shale, dark gray, sandy	10	490
Sandstone, light gray	5	495
Shale, dark gray, sandy	65	560
Sandstone	5	565
Shale, dark gray, sandy	5	570
Shale, dark gray, sandy	10	580
Limestone, light gray, hard	10	590
Shale, dark gray, sandy	20	610
Shale	30	640
Sandstone, gray, coarse, calcareous	10	650
Limestone, gray, massive	5	655
Sandstone	5	660
Shale, dark gray, sandy	13	673
Sandstone, dark gray, shaly	16	689
Shale, dark gray, sandy	51	740
Shale, black	30	770
Limestone, white, massive	3	773
Coal	3	776
Shale, gray, sandy	16	792
Coal	7	799
Sandstone, water bearing	11	810
Sandstone, light gray, calcareous	10	820
Sandstone, light gray, medium grained	45	865
Shale, dark gray, sandy	5	870
Shale	20	890
Limestone	15	905
Shale, white	10	915
Shale, dark gray, sandy	60	975
Shale, black	15	990
Shale, white	5	995
Shale, dark gray, sandy	50	1,045
Shale, shelly	3	1,048
Sandstone, gray, fine grained, shaly	7	1,055
Shale, white	5	1,060
Shale, gray, sandy	45	1,105
Shale, shelly	5	1,110
Limestone, light gray, hard	10	1,120
Shale, gray, sandy	45	1,165
Sandstone, gray, fine grained	10	1,175
Shale, black	10	1,185
Sandstone, coarse, white, hard	48	1,233

Pennsylvanian System.

	Thickness	Depth
Shale, gray, sandy	7	1,240
Sandstone, gray, fine grained, ferruginous	10	1,250
Shale, (coal streak at top)	46	1,296
Shale, green-gray, sandy	25	1,321
Sandstone, green-white, soft, fine grained ferruginous	19	1,340
Sandstone, yellow, soft, fine grained, ferruginous	10	1,350
Sandstone, yellow-white, fine, ferruginous, calcareous	35	1,385
Shale, dark gray, hard, sandy	15	1,400
Sandstone, yellow-white, fine, ferruginous	25	1,425
Shale, dark gray, sandy	25	1,450
Shale, green-gray, very sandy	20	1,470
Sandstone, yellow-white, fine, ferruginous	5	1,475
Shale, gray, medium grained, very sandy	21	1,496
Sandstone	4	1,500
Shale	40	1,540
Sandstone, light gray, fairly hard	6	1,546
Shale (coal streak at top)	32	1,578
Shale, dark gray, hard, slightly sandy	18	1,596
Sandstone, dirty white, fairly hard	4	1,600
Shale, dark gray, hard, slightly sandy	25	1,625
Shale, gray, hard, sandy	20	1,645
Shale, light, and shells	20	1,665
Sandstone, light gray, fine, (water bearing) ..	10	1,675
Sandstone, white, soft, fine	23	1,698

Mississippian System.

	Thickness	Depth
Limestone	16	1,714
Shale, blue-gray, soft, sandy, ferruginous, very calcareous	4	1,718
Limestone, dirty white, hard, ferruginous	22	1,740
Shale, dark gray, fairly hard, ferruginous, calcareous	15	1,755
Limestone	11	1,766
Sandstone	5	1,771
Sandstone, dirty white, fine grained, hard, ferruginous, calcareous	19	1,790
Sandstone, gray, fine grained, hard, ferruginous, calcareous	5	1,795
Shale, dark gray, sandy, non-calcareous	5	1,800
Limestone	27	1,827
Sandstone, dirty white, friable, medium grained, calcareous	33	1,860
Sandstone, white, fine grained, ferruginous ..	25	1,885

Mississippian System.

	Thickness	Depth
Limestone, dirty gray, hard, ferruginous	37	1,922
Shale, reddish gray, brittle, sandy, non-calcareous	8	1,930
Limestone	56	1,986
Shale, dark gray, sandy, non-calcareous	24	2,010
Limestone, dark, greenish gray, very calcareous eous, ferruginous	44	2,054
Shale, dark red and green, very soft, calcareons, ferruginous, pyritic, shell frags....	4	2,058
Limestone	28	2,086
Shale, dark gray, soft, pyritic, non-calcareous	21	2,105
Limestone	55	2,160
Shale, dark gray, non-calcareous	50	2,210
Shale, dark gray, and limestone, dirty, white (Golconda)	160	2,370
Limestone, broken	25	2,395
Shale	23	2,418
Sandstone. dirty white, fine grained, friable, ferruginous, micaceous, non-calcareous..	5	2,423
Total depth		2,423

Log No. 770

John Smith, No. 1, lessor. Location: 4 miles northwest of Calhoun and 4 miles southwest of Glennville. Contractor, G. G. Billman. Authority: J. G. Stuart.

Strata.

Pennsylvanian System.

	Thickness	Depth
Clay, gravel and silt loam	39	39
Sandstone	4	43
Shale, gray, soft	15	58
Limestone, blue, very hard	4	62
Shale, blue, dries out	30	92
Shale, black	13	105
Limestone, white, (clay?)	2	107
Shale, blue, dries gray-white	40	147
Limestone, flinty	2	149
Sandstone, brown, greasy and oily, brown flakes like rust floating	5	154
Sandstone, white, water copious, shale with partings	18	172
Oil sand, gas pronounced, good showing of oil, 21 ft. of oil sand, depth 175, Sea Level		

Pennsylvanian System.	Thickness	Depth
400, stopped on account of water over casing head, oil show in the water	3	175
Sand, white, 51 ft. sand in all	25	200
Shale, soft, clay	3	203
Sand and limestone	12	215
Limestone, broken, shale	16	231
Limestone, gray, solid	15	246
Shale, black	3	249
Coal trace	0	249
Limestone	1	250
Total depth		250

NOTE—The drillers of this well were inexperienced, and probably by poor methods lost a good pay.

Log No. 771

Bess Oil & Gas Co., lessee. Location: At Beech Grove P. O. Production: Oil at 136 feet in depth. Authority: Kenney Bryce, Owensboro, Ky.

Strata.

Pennsylvanian System.	Thickness	Depth
Soils, etc	56	56
Quicksand	80	136
Sandstone, (pay) (excellent)	4	140
Total depth		140

NOTE—Wells Nos. 2 and 3 show same formation.

Log No. 772

Louis Iglehart, No. 1, lessor. McDoe Oil & Gas Co., lessee. Location: 14 miles southwest of Owensboro, Ky. Production: Heavy grade black oil.

Strata.

Pennsylvanian System.	Thickness	Depth
Sandstone and shale	200	200
Sandstone, (pay)	40	240
Total depth		240

NOTE—Wells Nos. 2 and 3 similar in their records.

MENIFEE COUNTY.

Production: Oil and Gas.   Producing sand: Corniferous (Devonian) limestone.

Log No. 773

John Fox, No. 1, lessor. Commenced: Mar. 30, 1920. Completed: April 15, 1920. Production: 40 bbls. oil after shot. Contractor, L. C. Imgrens. Driller, Tom Ingrens.

Strata.

Pennsylvanian System.	Thickness	Depth
Soil and shale	169	169
Sandstone	150	319
Mississippian System.		
Shale, green	55	374
Limestone (Big Lime)	55	429
Shale, soft	50	479
Sandstone	150	629
Shale, green, sandy	365	994
Devonian System.		
Shale, brown (Chattanooga)	185	1,179
Shale (fire clay)	12	1,191
Limestone (Ragland "sand")	3	1,194
Total depth		1,194

Log No. 774

Wells' Heirs, No. 1, lessor. Commenced: Mar. 24, 1920. Completed: Apr. 14, 1920. Production: 10 bbls. oil natural. Contractor, R. A. Lyons. Driller, Louis Cupper.

Strata.

Pennsylvanian System.	Thickness	Depth
Soil and shale	85	85
Sand	150	235
Mississippian System.		
Shale, green	60	295
Limestone (Big Lime)	50	345
Shale, soft (soapstone)	50	395
Freestone	150	545
Shale, green, sandy	365	910
Devonian System.		
Shale, brown (Chattanooga)	185	1,095
Shale (fire clay)	15	1,110
Limestone "sand" (Irvine), (oil)	1	1,111
Limestone	10	1,121
Total depth		1,121

Log No. 775

Dorsey Ratliff, No. 3, lessor. Contractors: Menifee Drilling Co. $8\frac{1}{4}$ inch Drive pipe at 14 feet. 6-5/8 inch National Casing at 327 feet. Authority: L. Beckner.

Strata.

Pennsylvanian System.	Thickness	Depth
Soil	5	5
Sandstone	60	65
Coal bloom and sandstone, broken	75	140
Shale, soft, blue	20	160
Sandstone	40	200
Coal bloom	20	220
Coal	5	225
Sandstone (Pottsville)	25	250
Mississippian System.		
Limestone	30	280
Shale, blue	10	290
Limestone, white, hard, (water)	35	325
Shale, blue	65	390
Sandstone	60	450
Rock, chalk	40	490
Sandstone, free	40	530
Shale, sandy, soft	230	760
Shale, shelly	15	775
Sandstone, hard	110	885
Devonian System.		
Shale, black	195	1,080
Shale (fire clay)	15	1,095
Limestone "sand" (Corniferous), (oil)	14	1,109
Total depth		1,109

Log No. 776

W. C. Taylor, No. 1, lessor. New Domain Oil & Gas Co., lessee. Completed: Oct. 1, 1904. Production: Dry. Authority: New Domain Oil & Gas Co.

Strata.

Mississippian System.	Thickness	Depth
Soil, dark, soft	5	5
Shale, blue, soft	8	14

Mississippian System.   Thickness Depth

    Shale, blue, soft, and sandstone, blue, hard .. 434   448
    Shale, hard, light, soft   90   538
    Limestone, gray, hard   5   543

Devonian System.

    Shale, black, firm (Chattanooga)   170   713
    Shale, white, soft   12   725
    Limestone (Corniferous)   20   745
    Shale, blue, soft   4   749
    Limestone, gray, hard   5   754
    Limestone, light, hard   10   764

Silurian System.

    Shale, red and green, soft   158   922
    Limestone, gray, hard   1   923
    Shale, light, soft, and limestone, gray, hard .. 80 1,003
        Total depth   1,003

Log No. 777

A. C. Skidmore, No. 1, lessor. Completed: Oct. 8, 1904. Production: 1,200,000 cu. ft. gas. Authority: New Domain Oil & Gas Co.

    Strata.

Mississippian System.   Thickness Depth

    Soil, brown, soft   5   5
    Sandstone, blue, hard, and shale, hard, blue,
      soft   240   245
    Sandstone, gray, hard   9   254
    Shale, hard, blue, soft   46   300
    Sandstone, gray, hard   5   305
    Shale, hard, blue, soft   15   320
    Shale, hard, pink, soft   5   325
    Shale, hard, blue, soft   41   366
    Sandstone, blue, hard   4   370
    Shale, hard, blue, soft   10   380

Devonian System.

    Shale, black, hard, Chattanooga   23   403
    Shale, brown, soft, Chattanooga   12   415
    Shale, black, hard and soft, Chattanooga   20   435
    Shale, black, hard, Chattanooga   22   457

Devonian System.	Thickness	Depth
Shale, brown, soft, Chattanooga	23	480
Shale, black, hard, Chattanooga	30	510
Shale, brown blue, soft Chattanooga......	18	528
Limestone (Corniferous) (gas)	43	· 571

Silurian System.		
Shale, blue, soft	3	574
Total depth		574

Log No. 778

G. W. Pitts, No. 1, lessor. Completed: March 14, 1905. Production: Dry. Pocket of gas at 315 feet; salt water at 511 feet. Authority: New Domain Oil & Gas Co.

Strata.

Mississippian System.	Thickness	Depth
Soil and gravel, yellow and loose	10	10
Sandstone, light, medium	33	43
Shale, light, medium ...·...................	66	109
Limestone, gray, medium	8	117
Shale, blue, soft	13	130
Sandstone, light, hard	9	139
Shale, light, soft	25	164
Limestone, blue, hard	56	220
Shale, blue, soft	5	225
Limestone, blue, hard	38	263
Shale, blue, soft	27	290
Shale (red rock), hard	8	298
Shale, light, soft	40	338

Devonian System.		
Shale, black medium (Chattanooga)	159	497
Shale (fire clay), white, soft	8	505
Limestone, hard (gas)	84	589
Shale, blue, medium	8	597
Total depth		597

NOTE—The Devonian-Silurian contact is probably about midway within the 84 feet of limestone above 589 feet.

Log No. 779

J. B. Phillips, No. 1, lessor. Completed: April 15, 1905. Production: Gas. Well was tubed and packed. Authority: New Domain Oil & Gas Co.

Strata.

Mississippian System.	Thickness	Depth
Soil and gravel, soft	6	6
Shale, blue, soft	10	16
Sandstone, blue, hard	15	31
Shale, blue, soft	9	40
Sandstone, blue, hard	27	67
Limestone, light, hard	30	97
Shale, blue, soft	8	105
Limestone, light, hard	8	113
Sandstone, blue, hard, firm	59	172
Shale, blue, soft	48	220
Limestone, light, hard	20	240
Shale, blue, soft, limestone, blue, hard	90	330

Devonian System.		
Shale, black, soft (Chattanooga)	146	476
Shale (fire clay), blue, hard	9	485
Limestone "sand," gray, soft and hard (gas)	23	508
Shale, blue, soft	1	509
Total depth		509

Log No. 780

Jefferson Brewer No. 1, lessor. Completed: April 29, 1905. Production: A little gas at 803 feet. Authority: New Domain Oil & Gas Co.

Strata.

Mississippian System.	Thickness	Depth
Soil and gravel, yellow, soft	19	19
Limestone, red, hard	4	23
Limestone, white, hard	8	31
Shale, blue, soft	1	32
Limestone, gray, hard	13	45
Flint, brown, hard	18	63
Limestone, white, hard	27	90
Shale, blue, soft	2	92
Limestone, white, hard	2	94
Shale, blue, red, soft	42	136

Mississippian System.	Thickness	Depth
Sandstone, blue, firm, hard	284	420
Limestone, yellow, hard	2	422
Sandstone, blue, soft, hard	195	617
Limestone, blue, hard	5	622
Shale, blue, soft , ...	7	629
Limestone, blue, hard	5	634

Devonian System.		
Shale, black, soft (Chattanooga)	154	788
Shale (fire clay), blue, soft	14	802
Limestone, "sand," brown, medium, (gas) ..	26	828
Shale, blue, soft	1	829
Total depth		829

Log No. 781

J. J. Dennis, No. 1, lessor. Completed: May 11, 1905. Production: The well was dry. Authority: New Domain Oil & Gas Co.

Strata.

Mississippian System.	Thickness	Depth
Gravel, yellow, coarse	5	5
Sandstone, blue, hard, soft	233	238
Shale, blue, soft	16	254
Sandstone, blue, hard	1	255
Shale, blue, soft	84	339
Sandstone, blue, firm	14	353
Shale, blue, soft	66	419
Limestone, gray, hard	2	421
Shale, blue, soft	10	431

Devonian System.		
Shale, black (Chattanooga)	170	601
Limestone "sand," blue, (gas)	26	627
Shale, blue, hard	10	637
Total depth		637

Log No. 782

E. M. Yocum, No. 1 lessor. Completed: Sept. 30, 1919. Production: The well was dry; was plugged and abandoned. Authority: New Domain Oil & Gas Co.

Strata.

Pennsylvanian System.	Thickness	Depth
Clay, black	170	170
Sandstone, yellow	80	250
Shale	80	330

Mississippian System.		
Limestone	20	350
Shale	20	370
Limestone	60	430
Shale, blue	50	480
Shale, white	500	·980

Devonian System.		
Shale, black (Chattanooga)	200	1,180
Shale (fire clay)	20	1,200
Limestone "sand"	75	1,275
Limestone, white	20	1,295
Shale, hard, blue	12½	1,307½
Total depth		1,307½

NOTE—The Devonian-Silurian contact is within the lower half of the 75 feet of limestone above 1275 feet.

Log No. 783

George Downing, No. 2, lessor. Completed: Sept. 2, 1919. Production: Dry. Casing pulled and well abandoned. Authority: New Domain Oil & Gas Co.

Strata.

Pennsylvanian System.	Thickness	Depth
Clay, red	6	6
Shale, black	139	145
Sandstone, white	90	235
Shale, dark	70	305

Mississippian System.		
Limestone, white	18	323
Shale, dark	18	341
Limestone, white	60	401

Mississippian System.	Thickness	Depth
Sandstone, light, shaly	164	565
Shale, blue, sandy	270	835
Limestone	43	878
Shale (soapstone), blue	92	970

Devonian System.		
Shale, black, Chattanooga	20	990
Shale brown, Chattanooga	15	1,005
Shale, black, Chattanooga	158	1,163
Shale, (fire clay), white	8	1,171
Shale, black, hard	5	1,176
Limestone "sand," blue	34	1,210
Limestone, white	20	1,230
Total depth		1,230

Log No. 784

George B. Downing, No. 3, lessor. Completed Sept 19, 1919. Production: The well was dry; plugged and abandoned. Authority: New Domain Oil & Gas Co.

Strata.

Pennsylvanian System.	Thickness	Depth
Clay	5	5
Sandstone, white	10	15
Shale, blue	55	70
Sandstone, white	85	155
Shale, blue	71	226

Mississippian System.		
Limestone, gray	10	236
Shale, blue	25	261
Limestone, gray	57	318
Sandstone, blue, shaly	15	333
Sandstone, blue, shaly	412	745
Sandstone, blue, shaly	140	885
Shale, black	20	905
Shale, brown	20	925

Devonian System.		
Shale, black (Chattanooga)	149	1,074
Shale (fire clay), white	15	1,089
Shale, black	2½	1,091½
Limestone "sand," gray, (oil)	40	1,131½

Silurian System.

	Thickness	Depth
Limestone, white	30	1,161½
Shale, hard, gray	6	1,167½
Total depth		1,167½

Log No. 785

H. F. Osborn, No. 1, lessor. Location:——— Commenced: Dec. 1, 1919. Completed: Dec. 15, 1919. Shot Dec. 18, 1919, between 1,165 and 1,175 feet. Production: 48 hours after shot, 12 bbls. Authority: Ohio Oil Co.

Strata.

Pennsylvanian System.

	Thickness	Depth
Soil, red	120	120
Sandstone, red, medium	80	200
Shale, blue, soft	40	240
Sandstone, red, medium	80	320

Mississippian System.

	Thickness	Depth
Limestone, hard, white, Big Lime	40	360
Limestone, hard, blue, Big Lime	40	400
Sandstone, hard, dark, (little gas)	200	600
Shale, hard, and limestone shells	325	925

Devonian System.

	Thickness	Depth
Shale, brown, soft, Chattanooga	200	1,125
Shale (fire clay), light, soft, Chattanooga	30	1,155
Shale, black, soft, Chattanooga	8	1,163
Limestone (cap rock), hard, black	1	1,164
Limestone "sand," dark, soft	14	1,178
Total depth		1,178

Log No. 786

H. F. Osborn, No. 2, lessor. Location:——— Commenced: Feb. 6, 1920. Completed: Feb. 19, 1920. Shot Feb. 22, 1920, between 1,141 and 1,152 feet. Production: First 24 hours after shot, 115 bbls. Authority: The Ohio Oil Co.

Strata.

Pennsylvanian System.

	Thickness	Depth
Clay, soft	8	8
Shale, blue, soft	20	28
Shale, white, soft	42	70

Pennsylvanian System.	Thickness	Depth
Sandstone, gray	35	105
Shale, hard, white	15	120
Sandstone, gray	75	195
Shale, blue, soft	75	270

Mississippian System.		
Limestone, hard, white, Little and Big Lime..	15	285
Limestone, shaly, blue, soft, Little and Big Lime	35	320
Limestone, hard, blue, Little and Big Lime ..	54	374
Shale, red, soft	3	377
Shale, green, soft	13	390
Sandstone, light, soft	175	565
Sandstone, light	160	725
Soapstone, blue, soft	180	905
Limestone, hard, blue	3	908
Sandstone, light, soft	22	930

Devonian System.		
Shale, black, medium, Chattanooga	25	955
Shale, hard, white, soft, Chattanooga	15	970
Shale, brown, soft, Chattanooga	153	1,123
Fire clay, white, soft, Chattanooga	15	1,138
Shale, black, hard, Chattanooga	3	1,141
Limestone "sand," hard, brown	121/2	1,1531/2
Total depth		1,1531/2

Log No. 787

H. F. Osborn, No. 3, lessor. Location:——— Commenced: March 3, 1920. Completed: March 26, 1920. Shot March 26, 1920, between 1,160 and 1,170 feet. Production: 24 hours after shot, 100 bbls. black oil. Authority: The Ohio Oil Co.

Strata.

Pennsylvanian System.	Thickness	Depth
Clay	5	5
Sandstone, gray, soft	10	15
Shale, gray, soft	50	65
Sandstone, yellow, soft	65	130
Shale, gray, soft	110	240

Mississippian System.

	Thickness	Depth
Limestone, hard, white, Big Lime	15	255
Shale, blue, soft, Big Lime	10	265
.Limestone, hard, blue, Big Lime	40	305
Shale, (red rock), soft	5	310
Shale, hard, gray	20	330
Sandstone, blue, soft	530	860

Devonian System.

	Thickness	Depth
Shale, black, soft (Chattanooga)	182	1,042
Shale (fire clay), white, soft	18	1,069
Limestone "sand," brown, medium	12	1,072
Total depth		1,072

Log No. 788

H. F. Osborn, No. 4, lessor. Location:—— Commenced: March 5, 1920. Completed: April 23, 1920. Commenced producing April 29, 1920. Production: 24 hours after shot, 50 bbls., green oil. Authority: The Ohio Oil Co.

Strata.

Pennsylvanian System.

	Thickness	Depth
Clay, soft	10	10
Shale, hard, dark	45	55
Sand, yellow, soft	15	70
Shale, hard	15	85
Shale, red, soft	10	95
Shale, hard, and sand, soft	45	140
Sand, hard, white	45	185
Shale, hard	100	285

Mississippian System.

	Thickness	Depth
Shale (red rock), soft	10	295
Shale, hard	17	312
Limestone (Big Lime)	46	358
Shale (red rock), soft	10	368
Shale (Waverly), soft	540	908

Devonian System.

	Thickness	Depth
Shale, brown, soft (Chattanooga)	211	1,119
Shale (fire clay)	15	1,134
Limestone "sand"	13	1,147
Total depth		1,147

Log No. 789

H. F. Osborn, No. 5, lessor. Location:——— Commenced: May 3, 1920. Completed: May 17, 1920. Production: 48 hours after shot, 60 bbls., oil. Authority: The Ohio Oil Co.

Strata.

Pennsylvanian System.	Thickness	Depth
Clay	10	10
Shale, hard, gray	70	80
Sandstone, gray	30	110
Shale, hard, white	16	126
Sandstone, brown, soft	75	201
Shale, blue, soft	80	281
Mississippian System.		
Shale (red rock), soft	10	291
Shale, light, soft	20	311
Limestone, hard, light	35	346
Shale, hard, blue	20	366
Limestone, hard, white	14	380
Shale, red, soft, sandy	13	393
Shale, light, soft	177	570
Sandstone, light, soft	60	630
Sandstone, light, soft	73	703
Devonian System.		
Shale, brown, soft (Chattanooga)	215	918
Shale (fire clay), white, soft	156	1,074
Limestone "sand," brown, medium	85	1,159
Total depth		1,159

Log No. 790

H. F. Osborn, No. 6, lessor. Location:——— Commenced: May 2, 1920. Completed: May 16, 1920. Production: 48 hours after shot, 25 bbls., oil. Authority: The Ohio Oil Co.

Strata.

Pennsylvanian System.	Thickness	Depth
Clay, soft	8	8
Shale, blue, soft	40	48
Shale, white, soft	44	92
Sandstone, gray, soft	35	127
Shale, hard, white	20	147
Sandstone, gray, soft	80	227

Mississippian System.

	Thickness	Depth
Shale, blue, soft	78	305
Limestone, hard, white	17	322
Shale, blue, soft	35	357
Limestone, gray, hard	60	417
Shale, red, soft, sandy	4	421
Shale, green, soft	13	434
Sandstone, light, soft, fine	175	609
Sandstone, light, fine	60	669
Shale, hard, gray	281	950

Devonian System.

	Thickness	Depth
Shale, black, soft (Chattanooga)	181½	1,131½
Shale (fire clay), white, soft	15	1,146½
Limestone "sand," brown, medium	13½	1,160
Total depth		1,160

Log No. 791

H. F. Osborn, No. 7, lessor. Location——: Commenced: May 1, 1920. Completed: May 12, 1920. Production: 24 hours after shot, 50 bbls., oil. Authority: The Ohio Oil Co.

Strata.

Pennsylvanian System.

	Thickness	Depth
Clay, soft	10	10
Shale, hard, blue	90	100
Sandstone, gray, soft	30	130
Shale, hard, blue	10	140
Sandstone, gray, soft	60	200

Mississippian System.

Shale, hard, blue	70	270
Limestone, hard, blue	35	305
Shale, hard, blue	6	311
Limestone, hard, blue	7	318
Shale (red rock), soft	11	329
Shale, gray, soft	175	504
Sandstone, light, soft, fine	135	639
Sandstone, light, soft, fine	250	889

Devonian System.

Shale, brown, soft (Chattanooga)	180	1,069
Shale, (fire clay), light, soft	14	1,083
Limestone "sand," brown	12	1,095
Total depth		1,095

Log No. 792

H. F. Osborn, No. 8, lessor. Location:——— Commenced: May 26, 1920. Completed: June 7, 1920. Production: 48 hours after shot, 10 bbls., oil. Authority: The Ohio Oil Co.

Strata.

Pennsylvanian System.	Thickness	Depth
Clay	10	10
Shale, hard, blue	93	103
Sand, white, soft	70	173
Shale, hard, dark	110	283

Mississippian System.		
Limestone, light	22	305
Shale, hard, blue	12	317
Limestone, hard, white	23	340
Shale, red, soft	10	350
Waverly Shale, light, soft	533	883

Devonian System.		
Shale, brown, soft (Chattanooga)	195	1,078
Shale (fire clay), white, soft	141½	1,092½
Limestone "sand," brown, hard	11	1,103½
Total depth		1,103½

Log No. 793

H. F. Osborn, No. 9, lessor. Location:——— Commenced: May 25, 1920. Completed: June 5, 1920. Production: 48 hours after shot, 10 bbls., oil. Authority: The Ohio Oil Co.

Strata.

Pennsylvanian System.	Thickness	Depth
Clay, soft	10	10
Shale, hard, blue	60	70
Sand, white	65	135
Shale, hard, dark	115	250

Mississippian System.		
Limestone, light	20	270
Shale, hard, blue	10	280
Limestone, hard, white	25	305
Shale, red, soft	10	315
Shale (Waverly), light, soft	535	850

Devonian System.	Thickness	Depth
Shale, black (Chattanooga)	200	1,050
Shale (fire clay), white, soft	14	1,064
Limestone "sand," brown, hard	11	1,075
Total depth		1,075

Log No. 794

H. Osborn, No. 10, lessor. Location:——— Commenced: June 17, 1920. Completed: July 5, 1920. Shot July 5, 1920, between 1,217 and 1,227 feet. Authority: The Ohio Oil Co.

Strata.

Pennsylvanian System.	Thickness	Depth
Soil	20	20
Shale, hard, blue	185	205
Sand, light, medium	55	260
Shale, hard, light	95	355

Mississippian System.		
Shale (red rock), soft	10	365
Limestone, light	30	395
Shale, hard, blue	10	405
Limestone, hard, gray	50	455
Shale (red rock) soft	10	465
Shale (Waverly), light, soft	530	995

.Devonian System.		
Shale, brown (Chattanooga)	205	1,200
Shale (fire clay), light, soft	15	1,215
Limestone "sand," brown	15	1,230
Total depth		1,230

Log No. 795

H. Osborn, No. 11, lessor. Location:——— Commenced: June 23, 1920. Completed: July 3, 1920. Shot July 4, 1920. Authority: The Ohio Oil Co.

Strata.

Pennsylvanian System.	Thickness	Depth
Clay	14	14
Shale, hard, blue	20	34
Sand, light	56	90
Shale, hard, light	90	180

Mississippian System.	Thickness	Depth
Shale (red rock), soft	10	190
Limestone, light	30	220
Shale, hard, blue	10	230
Limestone, gray, hard	45	275
Shale (red rock), soft	8	283
Shale (Waverly), light, soft	539	822

Devonian System.		
Shale, brown (Chattanooga)	209	1,031
Shale (fire clay), light, soft	20½	1,051½
Limestone "sand," brown	11½	1,063
Total depth		1,063

Log No. 796

John Becraft, No. 1, lessor. Location: Near Rothwell. Completed: June 2, 1904. Production: The well was dry. Authority: New Domain Oil & Gas Co.

Strata.

Mississippian System.	Thickness	Depth
Clay, yellow, soft	7	7
Sandstone, dark, soft	23	30
Sandstone, dark, hard	2	32
Sandstone, dark, soft	1	33
Sandstone, dark, hard and soft	35	68
Shale, blue, soft	4	72
Sandstone, dark, hard	8	80
Sandstone, dark, soft	13	93
Shale, blue, soft	1	94
Sandstone, dark, hard	6	100
Shale, blue, hard	45	145
Sandstone, hard, dark	3	148
Shale, blue, soft	12	160
Sandstone, hard, dark	10	170
Shale, blue, soft	13	183
Sandstone, hard, dark	11	194
Shale, blue, hard	46	240
Shale, blue, soft	272	512
Limestone, gray, hard	2	514
Shale, blue, hard	6	520
Limestone, gray, very hard	2	522
Shale, blue, soft and hard	23	545

Devonian System.   Thickness Depth

    Shale, black, hard Chattanooga   98   643
    Shale, brown, soft, Chattanooga   48   691
    Shale, blue, soft, Chattanooga   9   700
    Limestone, hard, dark (Corniferous)   36   736
    Shale, blue, soft   5   741
    Limestone, gray, hard   5   746

Silurian System.

    Shale, soft, blue and pink   111   857
    Shale, light, soft   38   895
    Limestone, gray, hard   8   903
    Shale, light, soft ·........................   27   930
    Limestone, gray, hard   20   950
    Shale, blue, soft   40   990
    Limestone, blue, soft   480   1,470
    Limestone, white, soft   12   1,482
    Limestone, gray, soft   33   1,515
    Limestone, light, soft   10   1,525
    Limestone, blue, soft   40   1,565
    ·Limestone, gray, hard   165   1,730
    Limestone, brown, hard   70   1,800

        Total depth   1,800

NOTE—The Silurian-Ordovician contact is within the upper part
of the 480 feet of limestone above 1,470 feet.

Log No. 797

    J. J. Chambers, No. 2, lessor. Completed: Sept. 15, 1904. Production: The well was dry. Authority: New Domain Oil & Gas Co.

    Strata.

Mississippian System.   Thickness Depth

    Clay, yellow, soft   7   7
    Sandstone, blue, hard   113   120
    Shale, blue, hard   180   300
    Shale, blue, soft 153½   453½
    Limestone, gray, hard 3½   457

Devonian System.

    Shale, black, hard, (Chattanooga)   156   613
    Shale, white, soft   8   621
    Limestone "sand," dark, hard, open, (gas) ..   15   636
    Limestone "sand," dark, close, (gas)   25   661

Silurian System.	Thickness	Depth
Limestone "sand," gray, close, (gas)	15	676
Limestone "sand," gray, hard, (salt water) ..	12	688
Limestone, gray, soft	7	695
Shale, blue, soft	13	708
Total depth		708

Log No. 798

John P. Crockett, No. 1, lessor. Location: Near Rothwell. Completed: July 29, 1904. Authority: New Domain Oil & Gas Co.

Strata.

Mississippian System.	Thickness	Depth
Clay yellow, soft	3	3
Sandstone, blue, hard	5	8
Shale, blue, soft	7	15
Sandstone, blue, hard	3	18
Shale, blue, soft	7	25
Sandstone, blue, hard	10	35
Shale, blue, soft	60	95
Sandstone, blue, hard	11	106
Shale, blue, soft	254	360
Limestone, gray, hard	2	362
Shale, blue, soft	53	415
Limestone, gray, hard	5	420

Devonian System.		
Shale, black, hard, (Chattanooga)	159	579
Shale, blue, soft, (Chattanooga)	8	587
Limestone "sand," dark, hard, open, (gas) ..	16	603
Limestone "sand," light, hard, open, (gas) ..	15	618
Limestone "sand," light, hard, close, (gas..	24	642
Total depth		642

Log No. 799

W. F. Fitzpatrick, No. 1, lessor. Completed: June 28, 1904.
Authority: New Domain Oil & Gas Co.

Strata.

Mississippian System.

	Thickness	Depth
Clay, yellow, soft	5	5
Shale, dark, soft	15	20
Sandstone, light, hard	10	30
Sandstone, dark, soft	10	40
Sandstone, dark, hard	10	50
Shale, dark, soft	120	170
Shale, dark, hard	10	180
Shale, dark, soft	137	317
Shale, light, hard	9	326

Devonian System.

Shale, black, hard, (Chattanooga)	40	366
Shale, dark brown, soft, (Chattanooga)	102	468
Shale, blue, soft	5	473
Limestone "sand," dark, hard, open, (gas)	6	479
Limestone "sand," dark, soft, close, (gas)	4	483
Limestone "sand," light, soft, close, (gas)	8	491
Limestone "sand," dark, soft, close, (gas)	4	495
Limestone "sand," light, soft, (gas)	4	499
Shale, blue, soft	4	503
Total depth		503

Skeleton "Sand" Records

These wells were drilled in the Alexander Pool on the waters of Meiers Creek, Menifee County, Ky. The elevations were run by Y-level, hand level, and barometer, by Louis Panyitti, Geologist for the Ohio Cities Gas Co., and W. S. Peck. The surficial rocks in this pool are Pennsylvanian in the hills and Mississippian in the bottoms.

G. H. Alexander, lessor. Location: 10 acre tract.

Log No. 800

No. 1. (3)	Feet	
Elevation	1,173	A. T.
Cap	1,079	
Cap above tide	94	

Log No. 801

No. 2.   Feet

Elevation .... 1,220 . 78 A. T.
Cap 1,132
Cap above tide 89

Log No. 802

No. 3.

Elevation .... 1,181 . 81 A. T.
Cap 1,092
Cap above tide 90

G. H. Alexander, lessor.   13 acre tract.

Log No. 803

No. 1.   (1)

Elevation .... 1,174 . 18 A. T.
Cap 1,076
Cap above tide 98

Log No. 804

No. 2.   (2)

Elevation .... 1,176 . 37 A. T.
Cap 1,079
Cap above tide 97

Log No. 805

No. 3

Elevation .... 1,194 . 01 A. T.
Cap 1,098½
Cap above tide 96

Log No. 806

No. 4.

Elevation .... 1,227 . 95 A. T.
Cap 1,132
Cap above tide 96

Log No. 807

No. 5.   Feet
   Elevation .... 1,182 . 38 A. T.
   ·Cap 1,084
   Cap above tide .... 98`

Log No. 808

No. 6.
   Elevation .... 1,187 . 15 A. T.
   Cap 1,093
   Cap above tide 94

G. H. Alexander, lessor, Big Side.
Log No. 809

No. 1. (4)
   Elevation .... 1,182 . 71 A. T.
   Cap 1,101
   Cap above tide 82

Log No. 810

No. 2.
   Elevation 1,187   A. T.
   Cap 1,100
   Cap above tide 87

Log No. 811

No. 3.
   Elevation 1,190   A. T.
   Cap 1,102½
   Cap above tide 88

Log No. 812

No. 4.
   Elevation .... 1,186 . 5 A. T.
   Cap 1,112
   Cap above tide 75

Log No. 813

No. 5.   Feet

    Elevation 1,105   A. T.
    Cap 1,049
    Cap above tide   56

Dorsey Ratliff, lessor.
Log No. 814

No. 1.

    Elevation .... 1,226.43 A. T.
    Cap 1,117
    Cap above tide   109

Log No. 815

No. 2.

    Elevation .... 1,199.32 A. T.
    Cap 1,094
    Cap above tide   105

Log No. 816

No. 3.

    Elevation .... 1,190.03 A. T.
    Cap 1,093
    Cap above tide   97

Log No. 817

No. 4.

    Elevation ....
    Cap 1,064   A. T.
    Cap above tide ....

Log No. 818.

No. 5.

    Elevation .... 1,246.83 A. T.
    Cap 1,152
    Cap above tide   95

Log No. 819

No. 6.   Feet

Elevation ....

Cap .... 1,083 . 'A. T.

Cap above tide ....

Log No. 820

No. 7.

Elevation .... 1,193 . 39 A. T.

Cap 1,103

Cap above tide 90

Log No. 821

No. 8.

Elevation .... 1,223 . 91 A. T.

Cap 1,133

Cap above tide 91

Log No. 822

No. 9.

Elevation .... 1,239 . 78 A. T.

Cap 1,135

Cap above tide 105

Log No. 823

No. 10.

Elevation .... 1,211 . 43 A. T.

Cap 1,113

Cap above tide 98

Log No. 824

No. 11.

Elevation .... 1,232 . 51 A. T.

Cap 1,124

Cap above tide 109

Log No. 825

    No. 12.   Feet

        Elevation ....

        Cap   1,089   A. T.

        Cap above tide ....

    Dorsey Ratliff, lessor, (Hog Lot).

Log No. 826

    No. 1.

        Elevation ....   1,185.69 A. T.

        Cap   1,095

        Cap above tide   91

    Pete Brown.

Log No. 827

    No. 1.

        Elevation ....   1,191.91 A. T.

        Cap   1,089

        Cap above tide   103

Log No. 828

    No. 2.

        Elevation ....   1,198.06 A. T.

        Cap   1,095

        Cap above tide   103

Log No. 829

    No. 3.

        Elevation ....   1,228.16 A. T.

        Cap   1,131

        Cap above tide   97

Log No. 830

    No. 4.

        Elevation ....   1,233.37 A. T.

        Cap   1,122

        Cap above tide   111

Log No. 831

No. 5.   Feet
   Elevation .... 1,307.60 A. T.
   Cap .... 1,184 .
   Cap above tide 124

Log No. 832

No. 6.
   Elevation .... 1,285.96 A. T
   Cap 1,172
   Cap above tide 114

Log No. 833

No. 7.
   Elevation ....
   Cap 1,179   A. T.
   Cap above tide .... 107 (?)

Log No. 834

No. 8.
   Elevation .... 1,246.81 A. T.
   Cap ....
   Cap above tide ....

Log No. 835

No. 9.   .
   Elevation .... 1,260.39 A. T.
   Cap ....
   Cap above tide ....

Log No. 836

No. 10.
   Elevation .... 1,251.03 A. T.
   Cap 1,145
   Cap above tide 106
                                          .

Log No. 837

No. 11.   Feet

Elevation .... 1,275..71 A. T.
Cap 1,171
Cap above tide 105

Log No. 838

No. 12.

Elevation .... 1,264.29 A. T.
Cap 1,157
Cap above tide 107

Tilford Back, lessor.

Log No. 839

No. 1.

Elevation .... 1,195.07 A. T.
Cap 1,094
Cap above tide 102

Log No. 840

No. 2.

Elevation .... 1,260.69 A. T.
Cap 1,155
Cap above tide 106

Log No. 841

No. 3.

Elevation .... 1,205.07 A. T.
Cap 1,103
Cap above tide 102

Log No. 842

No. 4.

Elevation .... 1,256.81 A. T.
Cap 1,143
Cap above tide 114

Log No. 843

No. 5.   Feet

    Elevation .... 1,241.99 A. T.
    Cap 1,145
    Cap above tide 97

W. K. Wells, lessor, South Half.

Log No. 844

No. 1.

    Elevation .... 1,159.89 A. T.
    Cap 1,071
    Cap above tide 89

Log No. 845

No. 2.

    Elevation .... 1,183.89 A. T.
    Cap 1,095
    Cap above tide 89

Log No. 846

No. 3.

    Elevation .... 1,184.63 A. T.
    Cap 1,098
    Cap at tide 87

W. K. Wells, lessor, North Half.

Log No. 847

No. 1.

    Elevation .... 1,195.59 A. T.
    Cap 1,103
    Cap above tide 93

Log No. 848

No. 2.

    Elevation .... 1,226.56 A. T.
    Cap 1,132
    Cap above tide 95

George O. Downing, lessor.
Log No. 849

No. 1.   Feet
    Elevation 1,208   A. T.
    Cap 1,119½
    Cap above tide 89

Log No. 850

No. 2.
    Elevation 1,261   A. T.
    Cap 1,176
    Cap above tide 85

Log No. 851

No. 3.
    Elevation 1,176   A. T.
    Cap 1,131
    Cap above tide 45

Francis Bowhn, lessor.
Log No. 852

No. 1.
    Elevation .... 1,216.42 A. T.
    Cap 1,123
    Cap above tide 93

Log No. 853

No. 2.
    Elevation .... 1,245.45 A. T.
    Cap 1,145
    Cap above tide 100

Log No. 854

No. 3.
    Elevation .... 1,275.89 A. T.
    Cap 1,173
    Cap above tide 103

Log No. 855

John Fox, lessor.
No. 1.   Feet
    Elevation .... 1,321.26 A. T.
    Cap 1,196
    Cap above tide 125

Log No. 856

No. 2.
    Elevation .... 1.282.86 A. T.
    Cap 1,153
    Cap above tide 130

Martha Botts, lessor.
Log No. 857

No. 1.
    Elevation .... 1,290.81 A. T.
    Cap 1.193
    Cap above tide 89

Log No. 858

No. 2.
    Elevation .... 1,329.95 A. T.
    Cap 1,217½
    Cap above tide 112

H. F. Osborn, lessor.
Log No. 859

No. 1.
    Elevation .... 1,294 A. T.
    Cap 1,163
    Cap above tide 132

Log No. 860

No. 2.
    Elevation .... 1,257.82 A. T.
    Cap 1,141
    Cap above tide 117

Log No. 861

No. 3.   Feet
        Elevation .... 1,179.17 A. T.
        Cap 1,060
        Cap above tide 119

Log No. 862

No. 4.
        Elevation .... 1,254.17 A. T.
        Cap 1,134
        Cap above tide 120

Log No. 863

No. 5.
        Elevation .... 1,270.84 A. T.
        Cap 1,148
        Cap above tide 123

Log No. 864

No. 6.
        Elevation .... 1,269.18 A. T.
        Cap 1,146½
        Cap above tide 123

Log No. 865

No. 7.
        Elevation .... 1,206.35 A. T.
        Cap 1,089
        Cap above tide 117

Log No. 866

No. 8.
        Elevation .... 1,212.45 A. T.
        Cap 1,092½
        Cap above tide 120

Log No. 867

No. 9.   Feet

Elevation .... 1,177 . 28 A. T.
Cap 1,064
Cap above tide .... 113 .

Log No. 868

No. 10.

Elevation 1,330   A. T.
Cap 1,215½
Cap above tide 114

Log No. 869

No. 11.

Elevation .... 1,160 . 28 A. T.
Cap .... 1,046-8 in.
Cap above tide 114
Cased at 280

Log No. 870

No. 12.

Elevation .... 1,211 . 35 A. T.
Cap 1,100
Cap above tide 111

Log No. 871

No. 13.

Elevation .... 1,229 . 82 A. T.
Cap 1,119
Cap above tide 111

Log No. 872

No. 14.

Elevation 1,210   A. T.
Cap 1,108
Cap above tide 147
Cased at 340

Log No. 873

No. 15.   Feet
    Elevation 1,147   A. T.
    Cap 1,038
    Cap above tide 109

Martha Botts, lessor.
Log No. 874

No. 1.
    Elevation .... 1,290 81 A. T.
    Cap 1,193
    Cap above tide   98

Log No. 875

No. 2.
    Elevation .... 1,329 95 A.
    Cap 1,217½
    Cap above tide 112
       Lock level from Pete Brown No. 3 to Martha Botts No.
       109 ft. higher. (Pete Brown No. 3, 1,228.16.)

Scott Ledford, lessor.
Log No. 876

No. 1.
    Elevation .... 1,166 .01 A. T.
    Cap 1,093
    Cap above tide   73
    Cased at 324

Martin Ledford, lessor.
Log No. 877

No. 1.
    Elevation 1,190   T.
    Cap 1,134
    Cap above tide   56

G. W. Denniston Heirs, lessors.
Log No. 878

No. 1.
    Elevation .... 1,187 27 A. T.
    Cap 1,109
    Cap above tide   78

Log No. 879

No. 2.   Feet

Elevation .... 1,148 . 98 A. T.
Cap 1,068
Cap above tide 82

Phil Denniston Heirs, lessors.
Log No. 880

No. 1.

Elevation 1,206   A. T.
Cap 1,114
Cap above tide 93
Cased at 355

W. J. Dennis, lessor.
Log No. 881

Elevation 1,110   A. T.
Cap 1,066
Cap above tide 44

Hattie Sallie, lessor.
Log No. 882

No. 1.

Elevation .... 1,190 . 54 A. T.
Cap 1,110
Cap above tide 81

W. E. Little, lessor.
Log No. 883

No. 1.

Elevation .... 1,210 . 84 A. T.
Cap 1,140
Cap above tide 80

L. N. Sexton, lessor.
Log No. 884

No. 1.

Elevation .... 1,145 . 02 A. T.
Cap 1,065
Cap above tide 80

.William Trimble, lessor.

Log No. 885

No. 1.   Feet
    Elevation .... 1,202 .51 A. T.
    Cap 1,043
    Cap above tide 160

Rebecca Dennis, lessor.

Log No. 886

No. 1.
    Elevation 1,242   A. T.
    Cap 1,194
    Cap above tide 48

E. M. Yokum, lessor.

Log No. 887

No. 1.
    Elevation 1,271   A. T.
    Cap 1,203
    Cap above tide 68

James Wilson, lessor.

Log No. 888

No. 1.
    Elevation ....
    Cap 1,080   A. T.
    Cap above tide ....

B. Swango, lessor.

Log No. 889

No. 1. ·
    Elevation .... 1.084   A. T.
    Cap 1099
    Cap above tide .... —15

Jos. Collingsworth, lessor.

Log No. 890

No. 1.
    Elevation ....:............... 1,153   A. T.
    Cap ....
    Cap above tile ....

J. C. Ledford, lessor.
Log No. 891

No. 1.   Feet
    Elevation .... 1,185 . 70 A. T.
    Cap .... 1,032 .
    Cap above tide 154

Thos. Greenwald, lessor.
Log No. 892

No. 1.
    Elevation 914   A. T.
    Cap 626
    Cap above tide 288

Lon Barker, lessor.
Log No. 893

No. 1.
    Elevation .... 954 . 87 A. T.
    Cap 808
    Cap above tide 147

William Baty, lessor.
Log No. 894

No. 1.
    Elevation 1,223   A. T.
    Cap, tight 1,106
    Cap above tide 117

Brooks Tract, lessor.
Log No. 895

No. 1.
    Elevation ....
    Cap 1,120   A. T.
    Cap above tide ....

Powers Heirs, lessors.
Log No. 896

No. 1.
    Elevation .... 1,270 . 84 A. T.
    Cap 1,140
    Cap above tide 131

Log No. 897

No. 2.   Feet
    Elevation ....   1,263.34 A. T.
    Cap   1,142
    Cap above tide   121

Log No. 898

No. 3.
    Elevation ....   1,190 35 A. T.
    Cap   1,080
    Cap above tide   110

Log No. 899

No. 4.
    Elevation ....   1,193 35 A. T.
    Cap   1,082
    Cap above tide   111

O. D. Barker, lessor.
Log No. 900

No. 1.
    Elevation   1,234   A. T.
    Cap   1,120
    Cap above tide   114

Log No. 901

No. 2.
    Elevation   1,215   A. T.
    Cap   1,101
    Cap above tide   114

Mart Barker, lessor.
Log No. 902

No. 1.
    Elevation ....   867.87 A. T.
    Cap, ...(722?)   702
    Cap above tide ....(146?)   166
    Black Shale   602

Oscar Motley, (J. R. Lyon), lessor.
Log No. 903

No. 1.   Feet
    Elevation 1,099   A. T.
    Cap 1,134
    Cap above tide .... —35
    Total depth 3,131

W. M. Whitt, lessor.
Log No. 904

No. 1.
    Elevation 831   A. T.
    Cap 786
    Cap above tide 45

Jas. Collingsworth, lessor.
Log No. 905

    Elevation 1,080   A. T.
    Cap 1,153
    Cap above tide 73

Jim Phelps, lessor.
Log No. 906

So. of E.
    Elevation 1,263   A. T.
    Cap 1,090
    Cap above tide 173

Beaty Heirs, lessors.
Log No. 907

No. 1.
    Elevation 1,145   A. T.
    Cap .... 1,030 Dead Oil
    Cap above tide 115

Frank Lawson, lessor.
Log No. 908

No. 1.
    Elevation 882   A. T.
    Cap 707
    Cap above tide 175

J. T. Powers, lessor.
Log No. 909

No. 1.   Feet
    Elevation .... 1,257.85 A. T.
    Cap 1,149.50
    Cap above tide   108

Log No. 910

No. 2.
    Elevation .... 1,241.65 A. T.
    Cap 1,138½
    Cap above tide   103

Log No. 911

No. 3.   Feet
    Elevation .... 1,208,94 A. T.
    Cap .... 1,108½ plus 8½
                                          No Water.
    Cap above tide 100.4

Log No. 912

No. 4.
    Elevation .... 1,203.94 A. T.
    Cap .... 1,103 plus 6½·
                                      S. W. plus 2. plugged.
    Cap above tide ....   100.94 to 6 ft.
                                          shot 20 qts. Salt
                                          water shut off and
                                          365 feet of oil.

Silas Montgomery, lessor.
Log No. 913

No. 1.
    Elevation .... 1,198.65 A. T.
    Cap .... 1,093 plus 13½
    Cap above tide 105.65

Wells Heirs, lessors.
Log No. 914

No. 1.
    Elevation .... 1,223.85 A. T.
    Cap 1,110
    Cap above tide   114
        Frof Wells Heirs to J. T. Powers No. 1, 26 feet higher.
        (J. T. Powers No. 1 is 1,257.85.)

MONROE COUNTY.

Production: **Oil and Gas.** Producing "Sand": Sunnybrook (Ordovician).

Log No. 915

Dux Oil Co. Location: About 6 miles west of Thompkinsville. Commenced: Oct. 31, 1919. Completed: Nov. 27, 1919. Authority: Dux Oil Co., through L. Beckner.

Strata.

Mississippian System.	Thickness	Depth
Soil	10	10
Limestone, flinty	70	80
Sand and limestone, grayish, (gas)	3	83
Limestone, dark gray and flinty	37	120
Limestone, blue	21	141
Limestone, white	3	144
Shale, hard, green, (sulphur gas)	5	149
Devonian System.		
Shale, black (Chattanooga)	25	174
Silurian System.		
Limestone, brown, sandy	15	189
Shale, hard, gray	5	194
Ordovician System.		
Limestone, light blue	50	244
Limestone, purplish	3	247
Limestone, brownish	5	252
Shale, greenish, hard	4	256
Limestone, gray	4	260
Limestone "sand," (oil)	2½	262½
Total depth		262½

Log No. 915-A.

W. L. Douglas, No. 1, lessor. Location: Near Fountain Run, Monroe County, Ky. Commenced:— Completed:— (Partial Record).

Strata.

Mississippian System.	Thickness	Depth.
Shale and limestone	170	170
Devonian System.		
Shale, black (Chattanooga)	35	205
Limestone (Corniferous)	15	220
Silurian System.		
Limestone (Niagara)	30	250

Ordovican System.	Thickness	Depth
Limestone and shaly limestone (includes Trenton)	1,150	1,400
Limestone, dark, hard	47	1,447
Limestone, dark gray, Knoxville Dolomite ...	73	1,520
Limestone, dark, shaly, Knoxville Dolomite ..	60	1,580
Limestone, dark, compact	22	1,602
Incomplete depth (April 1, 1922)		1,602

NOTE—The Trenton and Calciferous is found within the lower half of the 1150 feet above 1447 feet in depth. The Knoxville Dolomite is regarded as the producing sand of the new Beech Bottom wells of Clinton County, Ky. These wells produced oil at a depth of 1365 feet below the black shale (Devonian).

MORGAN COUNTY.

Production: Oil and Gas. Producing "Sands": Big Lime, Big Injun, Wier, Berea (Mississippian), Corniferous (Devonian).

Log No. 916

E. H. Oldfield, No. 1, lessor. Location: At Mize P. O. Production: 2,000,000 cu. ft. gas.

Strata.

Pennsylvanian System.	Thickness	Depth
Shale and shells	100	100
Sand, white	215	315
Shale and shells	25	340
Mississippian System.		
Limestone (Little lime)	30	370
Shale (pencil cave)	22	392
Limestone (Big Lime)	110	502
Shale (Waverly)	575	1,077
Devonian System.		
Shale, black (Chattanooga)	204	1,281
Limestone "sand" (Irvine)	8	1,289
Total depth		1,289

Log No. 917

Clearfield Lumber Co., lessor. Northwestern Oil Co., No. 1, lessee. Location: Head of Yocum Creek, near Blaze P. O. Completed: Feb. 6, 1920. Driller: Andrew Shearard. Authority: Sam Shearard, contractor.

THE "BIG LIME" OF EASTERN KENTUCKY

This is a characteristic, though not a complete exposure, of the sequence of Mississippian Limestones of Eastern Kentucky. View in the quarry at Limestone, Carter County, Kentucky.

Strata.

Pennsylvanian System.	Thickness	Depth
Soil	7	7
Gravel	6	13
Limestone, black	9	22
Shale, blue	5	27
Mississippian System.		
Limestone (Big Lime)	163	190
Shale, green	15	205
Shale, (red rock)	15	220
Shale, blue	305	525
Limestone, black	25	550
Limestone, white	115	665
Sandstone (Berea grit)	60	725
Shale, blue	20	745
Devonian System.		
Shale, black (Chattanooga)	195	940
Limestone, black (Chattanooga)	20	960
Shale, white (Chattanooga)	20	980
Shale, brown (Chattanooga)	20	1,000
Shale, white (Chattanooga)	40	1,040
Limestone ''sand,'' (Corniferous)	10	1,050
White water sand (Corniferous)	5	1,055
Sand, hard, brown (Corniferous)	10	1,065
White water sand· (Corniferous)	15	1,080
Total depth		1,080

20 feet 10 inch casing.
95 feet 8 inch casing.
520 feet 6¼ inch casing.

Log No. 918

J. T. Fugett, No. 1, lessor. Iron City Oil Co., No. 1, lessee. Location: Brushy Fork of Caney Creek. Completed: Oct. 21, 1917. Authority: L. Beckner.

Strata.

Pennsylvanian System.	Thickness	Depth
Drift	18	18
Shells, lime	42	60
Shale, hard	290	350
Sand	85	435
Shale, hard, sandy	40	475
Sand	140	615
Shale, hard	60	675
Sand	65	740

Mississippian System.

	Thickness	Depth
Shale, hard	10	750
Limestone (Little Lime)	5	755
Shale, hard	5	760
Limestone (Big Lime)	105	865
Shale (Waverly)	485	1,350
Sandstone (Berea)	40	1,390
Limestone, sandy	50	1,440

Devonian System.

Shale, brown (Chattanooga)	319	1,759
Shale, hard, white	30	1,789
Limestone "sand," (oil & gas shows)	22	1,811
Total depth		1.811

A little gas at 1 foot in sand.

A show of oil at 4 feet in sand.

Second show of oil at 12 feet in sand.

Size of hole at mouth was 10 inches, and at bottom 6-5/8 inches.

Log No. 919

A. J. Linden, No. 1, lessor. Location: About 3 miles east of Adele, Ky. Commenced: July 15, 1917. Completed: Aug. 31, 1917. Production: Dry. Authority: The Eastern Gulf Oil Co.

Strata.

Pennsylvanian System.

	Thickness	Depth
Drift	10	10
Shale, hard, shelly	30	40
Lime, shell	35	75
Shale, hard, (coal at 175)	100	175
Sand	25	200
Shale, hard	100	300
Sand	15	315
Shale, hard	35	350
Sand	5	355
Shale, hard	5	360
Sand	110	470
Shale, hard	105	575
Lime shells	20	595
Sand	75	670
Shale, hard	60	730
Sand	33	763

Mississippian System.

Limestone (Little Lime)	5	768
Shale, hard	10	778
Limestone (Big Lime)	114	892

Mississippian System.	Thickness	Depth
Shale (Waverly	434	1,326
Shale, black (Sunbury)	5	1,331
Sandstone (Berea)	20	1,351
Shale, hard, white	25	1,376
Devonian System.		
Shale, brown (Chattanooga)	319	1,695
Shale, hard, white	30	1,725
Limestone	54	1,779
Total depth		1,779

Log No. 920

V. P. Haney, No. 1, lessor. Location: Upper Tract No. 2. Commenced: July 28, 1913. Completed: Aug. 18, 1913. Drillers: Harry Creel, Grover Barnes and W. R. Forman. Authority: L. Beckner.
Strata.

Pennsylvanian System.	Thickness	Depth
Soil	9	9
Coal, bituminous and shale	266	275
Sand	45	320
Sand, soft	210	530
Sand, black	200	730
Sand, settling	45	775
Mississippian System.		
Limestone (Little Lime)	105	880
Limestone (Big Lime)	20	900
Shale (Waverly)	185	1,085
Shale	419	1,504
Sandstone (Berea)	16	1,520
Shale, hard	24	1,544
Devonian System.		
Shale, black (Chattanooga)	36	1,580
Shale (Chattanooga)	248	1,828
Limestone ''sand''	31'8"	1,859'8"
Limestone ''sand''	20'4"	1,880
Total depth		1,880

First pay at 31¼ feet in sand and runs to 13 feet.

Log No. 921

V. P. Haney, No. 3, lessor. Location: Upper Tract. Commenced: Oct. 14, 1913. Completed: Nov. 12, 1913. Drillers: J. Dennis, H. R. Newland, G. Barnes and W. R. Forman. Authority: L. Beckner.

Strata.

Pennsylvanian System.	Thickness	Depth
Sandstone, shale and cannel coal	470	470
Sand	152	622
Sand, settling	98	720
Mississippian System.		
Limestone (Little Lime)	100	820
Shale, hard	8	828
Limestone (Big Lime)	179	1,007
Shale (Waverly)	458	1,465
Sandstone (Berea)	15	1,480
Shale, hard	35	1,515
Devonian System.		
Shale, black (Chattanooga)	255	1,770
Shale	40	1,810
Limestone "sand"	21	1,831
Total depth		1,831

Pay from 1,812 to 1,823½ feet.

Log No. 922

Mason Jones, No. 1, lessor. Location: Cannel City. Commenced: May 2, 1913. Completed: June 3, 1913. Authority: L. Beckner.

Strata.

Pennsylvanian System.	Thickness	Depth
Soil	12	12
Shale and sandstone	117	129
Cannel coal	6	135
Shells and shale, hard	265	400
Sand	200	600
Shale, hard	50	650
Sand	100	750
Mississippian System.		
Limestone (Little Lime)	10	760
Shale	5	765
Limestone (Big Lime)	185	950
Shale (Waverly)	440	1,390
Sandstone (Berea)	25	1,415
Shale, hard	50	1,465
Devonian System.		
Shale, hard, black (Chattanooga)	251	1,716
Shale	30	1,746
Limestone "sand"	16½	1,762½
Total depth		1,762½

No pay until 7 feet below cap.

Log No. 923

Jim Little, No. 1, lessor. Mullins & Mullins Oil & Gas Co., lessee.
Location: Near Mize P. O., about 200 yards above post office on Mur-
phy Fork of Grassy Creek. Commenced: July 10, 1917. Completed:
Aug. 9, 1917. Initial production: 900,000 cu. ft. gas. Authority: C.
E. Bales.

Strata.

Pennsylvanian System.	Thickness	Depth
Soil	6	6
Shale and shells	94	100
Sandstone, white	210	310
Shale	50	360
Mississippian System.		
Limestone (Little Lime)	20	380
Limestone (Big Lime)	120	500
Shale (Waverly)	540	1,040
Devonian System.		
Shale, brown (Chattanooga)	200	1,240
Shale, white (Chattanooga)	20	1,260
Shale, brown (Chattanooga)	21	1,281
Limestone ''sand,'' (gas)	11	1,292
Total depth		1,292

Log No. 924

Jim Little, No. 2, lessor. Location: 1 mile southwest of Mize
P. O. Authority: L. V. Mullen.

Strata.

Pennsylvanian, Mississippian and Devonian Systems.	Thickness	Depth
Sandstone, limestone and shale	1,034	1,034
Limestone ''sand'' (Corniferous)	32	1,066
Silurian and Ordovician Systems.		
Limestone (gas at 1,306)	240	1,306
Total depth		1,306

Log No. 925

Clay Murphy, No. 1, lessor. Forman Oil & Gas Co., lessee. Lo-
cation: Near Mize P. O., about 1 mile up the Murphy Fork on Grassy
Creek from the post office. Commenced: May, 1917. Completed:
June, 1917. Production: Dry.

Strata.

Pennsylvanian System.	Thickness	Depth
Soil	18	18
Limestone and shells	12	30
Limestone, blue	10	40
Sandstone	125	165
Shale	3	168
Sandstone	57	225
Shale	20	245
Sandstone	5	250
Shale	10	260
Sandstone	60	320
Shale	6	326
Sandstone	6	332
Shale	18	350
Mississippian System.		
Limestone (Little Lime)	20	370
Shale	22	392
Limestone (Big Lime)	108	500
Shale (Waverly)	461	961
Shale, black	19	980
Shale, white	10	990
Sandstone (Berea)	5	995
Shale, white	15	1,010
Devonian System.		
Shale, brown (Chattanooga)	250	1,260
Shale, white (Chattanooga)	20	1,280
Shale, black	3	1,283
Limestone "sand"	37	1,320
Total depth		1,320

Log No. 926

Hurt Dowery, No. 1, lessor. Murphy Fork Oil & Gas Co., lessee. Location: Near Mize P. O.. about 2½ miles from Mize P. O., on the left hand fork of Murphy Fork of Grassy Fork. Commenced: April, 1917. Completed: May, 1917. Production: Dry. Authority: C. E. Bales.

Strata.

Pennsylvanian System	Thickness	Depth
Soil	17	17
Shale	23	40
Sandstone	185	225
Shale	2	227
Sandstone	48	275
Shale, (show of gas)	60	335

Mississippian System.	Thickness	Depth
Limestone and shells	35	370
Shale, red	18	388
Limestone and shells	22	410
Limestone (Big Lime)	150	560
Shale (Waverly)	440	1,000
Sandstone (Berea)	5	1,005
Shale, white	5	1,010
Devonian System.		
Shale, black (Chattanooga)	25	1,035
Shale, white (Chattanooga)	15	1,050
Shale, brown (Chattanooga)	205	1,255
Shale, white (Chattanooga)	15	1,270
Shale, black (Chattanooga)	19	1,289
Limestone ''sand''	136	1,425
Shale, white	15	1,440
Shale, red	6	1,446
Total depth	.	1,446

Log No. 927

Charles Coffee, No. 1, lessor. Kentucky Oil Land Investment Co., lessee. Location: White Oaks Creek, near Williams P. O. Authority: L. Beckner.

Strata.

Pennsylvanian System.	Thickness	Depth
Soil	10	10
Sand	30	40
Shale, hard	25	65
Sand	15	80
Shale, hard	90	170
Sand	125	295
Shale, hard	5	300
Sand	48	348
Shale, hard	52	400
Sand	35	435
Shale, hard	30	465
Mississippian System.		
Limestone (Little Lime)	35	500
Shale, hard	17	517
Limestone (Big Lime)	123	640
Shale (Waverly)	410	1,050
Shells, gritty	10	1,060
Shale, hard, white	45	1,105
Sandstone	20	1,125

Mississippian System.

	Thickness	Depth
Shale, brown (Sunbury)	5	1,130
Sandstone (Berea)	35	1,165
Shale, hard, white	30	1,195
Sandstone (Berea), gray	35.	1,230
Shale, hard, white	110	1,340

Devonian System.

	Thickness	Depth
Shale, brown (Chattanooga)	220	1,560
Shale, hard, white	37	1,597
Total depth		1,597

NOTE—This record is irregular in the lower part of the Mississippian System. A white shale of 110 feet is quite out of place above the Chattanooga Shale, and indicates faulty recordation. The Sunburst is also very thin.

Log No. 928

Andy Gose, No. 2, lessor. Location:—— Commenced: September 8, 1913. Completed: Oct. 15, 1913. Drillers: J. A. Frentz and S. E. Ewing. Production: Pay oil from 1918' 9" to 1,925' 9". Authority: L. Beckner.

Strata.

Pennsylvanian System.

	Thickness	Depth
Soil	13	13
Shale, hard, black	122	135
Sand	50	185
Shale, hard	25	210
Cannel coal	5	215
Shale	395	610
Sand	160	770
Shale, hard	30	800
Sand, white	80	880
Shale, hard	30	910

Mississippian System.

	Thickness	Depth
Limestone, black	10	920
Limestone (Big Lime), white	190	1,110
Shale, light gray	25	1,135
Shells, shale, hard	440	1,575
Shale, hard, black	10	1,585
Sandstone (Berea)	20	1,605
Shale, hard, white	20	1,625

Devonian System.

	Thickness	Depth
Shale, black, hard (Chattanooga)	284'9"	1,909'9"
Limestone "sand," (Corniferous)	20'3"	1,930
Total depth		1,930

Log No. 929

A. A. Gose, No. 3, lessor.   Commenced: Nov. 13, 1913.   Completed: Dec. 22, 1913.   Authority: L. Beckner.

Strata.

Pennsylvanian System.	Thickness	Depth
Soil	18½	18½
Cannel coal and shale	8½	100
Shale, hard	402	502
Sand, soft	118	620
Shale, hard	160	780
Sand, settling	60	840
Shale, hard	90	930

Mississippian System.		
Limestone (Little Lime)	10	940
Shale	10	950
Limestone (Big Lime)	20	970
Shale (Waverly)	155	1,125
Shale, hard	460	1,585
Sandstone (Berea)	10	1,595
Shale, hard	20	1,615
Shale, hard, black	25	1,640

Devonian System.		
Shale (Chattanooga)	264	1,904
Limestone ''sand'' (Corniferous)	22 '2"	1,926 '2"
Limestone ''sand'' (Corniferous)	19	1,945 '2"
Total depth		1,945 2"

Log No. 930

L. M. Haney, No. 1, lessor.   Completed: Aug. 5, 1913.   Drillers: W. R. Forman and H. R. Newland.   Production: First pay 3'6" from top of sand; second pay 9 to 14 feet in sand.   Casinghead alt.: 982.1 feet.   Authority: L. Beckner.

Strata.

Pennsylvanian System.	Thickness	Depth
Soil	5	5
Shale, etc.	49	54
Cannel coal	5	59
Shale, hard	81	140
Sand, soft top	187	327
Sandstone (Pottsville)	253	580

Mississippian System.   Thickness Depth

    Limestone (Little Lime) 15   595
    Limestone (Big Lime) 110   705
    Shale (Waverly) 165   870
    Shale, brown 430 1,300
    Sandstone (Berea) 30   1,330
    Shale, hard 65   1,395

Devonian System.

    Shale, black (Chattanooga) 256 1,651
    Limestone ''sand'' (Corniferous) 20 1,671
        Total depth 1,671

Log No. 931

L. M. Haney, No. 2, lessor. Commenced: July 24, 1913. Completed: Sept. 1, 1913. Drillers: T. Christie, E. Guignon, G. Barnes and W. R. Forman. Casinghead alt.: 1,136.38 feet. Authority: L. Beckner.

   Strata.

Pennsylvanian System.   Thickness Depth
    Soil 13   13
    Cannel coal and shale 117   130
    Coal, bituminous and shale 91   221
    Sand 259   480
    Shale 205   685

Mississippian System.

    Limestone (Little Lime) 20   705
    Limestone (Big Lime) 155   860
    Shale, brown 610 1,470
    Sandstone (Berea) 20 1,490
    Shale, hard 5   1,495

Devonian System.

    Shale, black (Chattanooga) .... 318'9" 1,813'9"
    Limestone (Corniferous), (pay 1·9) .... 20'8" 1,834'5"
        Total depth .... 1,834'5"

Log No. 932

L. M. Haney, No. 3, lessor. Commenced: Aug. 27, 1913. Completed: Sept 15, 1913. Drillers: W. R. Forman and H. Creel. Production: Pay oil from 1,746′6 to 1,756′6″. Authority: L. Beckner.

Strata.

Pennsylvanian System.	Thickness	Depth
Soil	18	18
Shale and sand	72	90
Shale, hard	60	150
Soft coal	0	150
Cannel coal and shale	35	185
Shale	5	190
Sand	230	420
Shale, hard	200	620
Sand, white	50	670
Shale, hard, black	95	765
Sand	31	796
Mississippian System.		
Limestone (Little Lime)	34	830
Limestone (Big Lime), light gray, hard	154	984
Sandstone	166	1,150
Sandstone (Berea in part)	265	1,415
Shale, brown, sandy	34	1,449
Devonian System.		
Shale, black (Chattanooga)	285	1,744
Limestone "sand"	20	1,764
Total depth		1,764

Log No. 933

I. N. Caskey, No. 1, lessor. Completed: Feb. 6, 1918. Driller G. Barnes. Authority: L. Beckner.

Strata.

Pennsylvanian System.	Thickness	Depth
Soil	18	18
Sand	12	30
Coal (cannel)	3	33
Shale, hard, sandy, (dark, heavy oil)	60	93
Sand and shale, hard	389	482

Mississippian System.   Thickness Depth

    Limestone (Big Lime), (cased) ....:.   15   497
    Limestone and shale (Big Lime in part)   250   747
    Sand (Big Injun), (gas)   5   752
    Shale ...:................................   250   1,002
    Sand (Berea)   45   1,047
    Shale, hard   50   1,097

Devonian System.

    Shale, black (Chattanooga)   300   1,397
    Shale and fire clay   53   1,450
    Limestone "sand" (Corniferous), (salt
       water) .... $121\frac{1}{2}$ $1,462\frac{1}{2}$
       Total depth .... $1,462\frac{1}{2}$

Log No. 934

    Mattie Burton, No. 1, lessor. Completed: Dec. 31, 1913. Driller: C. E. Stalker. Authority: L. Beckner.

    Strata.

Pennsylvanian System.   Thickness Depth

    To salt sand   580   580
    Sand, salt   10   590
    Sand   155   745
    Shale, hard   35   780
    Sand, salt   126   906

Mississippian System.

    Limestone (Big Lime)   25   930
    Shale (Waverly)   160   1,090
    Limestone   300   1,390
    Shale, brown   130   1,520
    Sandstone (Berea)   10   1,530
    Shale, hard, white   25   1,555

Devonian System.

    Shale, brown (Chattanooga)   20   1,575
    Shale, hard (Chattanooga)   247   1,822
    Limestone "sand"   33   1,855
    Limestone "sand," (gas 1,860) (oil 1867-71)   25   1,880
       Total depth   1,880

Log No. 935

Home Oil Co., No. 1, lessee. Location: Cannel City. Commenced: Jan. 15, 1913. Completed: Feb. 4, 1913. Casinghead elevation: 930 feet. Authority: L. Beckner.

Strata.

Pennsylvanian System.	Thickness	Depth
Soil	15	15
Sand	10	25
Shale, hard	55	80
Coal	3	83
Sandstone	77	160
Shale	40	200
Sand	70	270
Shale, hard, white	20	290
Shale	20	310
Sand, white	95	405

Mississippian System.		
Limestone (Big Lime)	180	585
Limestone and shale, hard	40	625
Sandstone, shaly, hard	175	800
Shale, hard, light, gray	480	1,280
Sandstone (Berea)	22	1,302
Shale, hard	38	1,340

Devonian System.		
Shale, hard, black (Chattanooga)	260	1,600
Limestone ''sand'' (Corniferous)	36	1,636
Limestone ''sand''	42	1,678
Total depth		1,678

Log No. 936

Buck Jones, No. 1, lessor. Commenced: Sept. 24, 1913. Completed: Oct. 17, 1913. Casinghead elevation: 1,175.26 feet. Authority: L. Beckner.

Strata.

Pennsylvanian System.	Thickness	Depth
Soil	10	10
Sand and shale, hard	110	120
Shale, hard	390	510

Mississippian System.

	Thickness	Depth
Limestone and sandstone	420	930
Shale, hard, light gray	130	1,060
Sandstone (Berea in part)	460	1,520
Shale, hard	25	1,545

Devonian System.

	Thickness	Depth
Shale, hard, black (Chattanooga)	31	1,576
Shale (Chattanooga)	250	1,826
Limestone ''sand''	56	1,882
Limestone	27½	1,909½
Total depth		1,909½

Log No. 937

A. E. Sebastian, No. 1, lessor. Commenced: May 18, 1913. Completed: June 5, 1913. Drillers: C. E. Musser and Mike Dolan. Production: Oil 6-11 feet in ''sand.'' Authority: L. Beckner.

Strata.

Pennsylvanian System.

	Thickness	Depth
Soil	10	10
Cannel coal and shale	95	105
Sand	275	380
Shale, hard	204	584
Sand?	56	640
Shale, hard	85	725

Mississippian System.

	Thickness	Depth
Limestone (Little Lime)	10	735
Shale	15	750
Limestone (Big Lime), (cased)	165	915
Shale, hard	440	1,355
Sandstone (Berea)	15	1,370
Shale, hard	30	1,400

Devonian System.

	Thickness	Depth
Shale, hard, black (Chattanooga)	30	1,430
Shale, black (Chattanooga)	286½	1,716½
Limestone ''sand,'' (first oil 1,724)	12½	1,729
Total depth		1,729

Log No. 938

Daniel Gullet, No. 1, lessor. Ohio Fuel Oil & Gas Co., lessee. Location: Cannel City District. Commenced: Oct. 27, 1913. Completed: Dec. 22, 1913.

Strata.

Pennsylvanian System.	Thickness	Depth
Soil	12	12
Shale	84	96
Coal	2	98
Shale and shells	272	370
Sand	270	640
Shale	30	670
Mississippian System.		
Limestone (Little Lime)	20	690
Shale and shells	30	720
Limestone (Big Lime)	120	840
Shale (Waverly)	440	1,280
Sandstone (Berea)	25	1,305
Shale and shells	60	1,365
Devonian System.		
Shale, black (Chattanooga)	294	1,659
Shale	40	1,699
Limestone (Corniferous)	48	1,747
Limestone ''sand,'' brown	2	1,749
Total depth		1,749

Log No. 939

J. B. Whitt, No. 1, lessor. Ohio Fuel Oil & Gas Co., lessee. Location: Cannel City District. Commenced: July 7, 1913. Completed: July 27, 1913. Shot, 50 qts.

Strata.

Pennsylvanian System.	Thickness	Depth
Soil	14	14
Shale, (1 bailer water 100)	86	100
Sand and shells	105	205
Native coal	4	209
Shale and shells	81	290
Shale and shells	290	580
Sand, salt, (first)	160	740

Pennsylvanian System.

	Thickness	Depth
Shale	50	790
Sand, salt	90	880
Shale	25	905

Mississippian System.

Limestone (Little Lime)	17	922
Shale (Pencil Cave)	4	926
Limestone (Big Lime)	154	1,080
Shale	40	1,120
Sandstone (Big Injun)	165	1,285
Shale and shells	235	1,520
Sand	20	1,540
Shale, brown	10	1,550
Sandstone (Berea)	20	1,570
Shale	40	1,610

Devonian System.

Shale, brown (Chattanooga)	260	1,870
Shale	20	1,890
Limestone ''sand'' (Corniferous)	19	1,909
Total depth		1,909

Pay 1897-1905.

Did not drill through sand.

Last 4 feet brown, sandy lime.

Log No. 940

H. C. Keeton, No. 3, lessor. Ohio Fuel Oil & Gas Co., lessee. Location: Cannel City District. Commenced: July 31, 1913. Completed: Aug. 31, 1913. Shot Sept. 1, 1913, 50 qts.; 2nd shot, 50 qts.

Strata.

Pennsylvanian System.

	Thickness	Depth
Soil and clay	40	40
Shale	220	260
Cannel coal	2	262
Shale	288	550
Sand, salt	165	715
Shale	60	775
Sand, salt	105	880
Shale	5	885

Mississippian System.   Thickness Depth

	Thickness	Depth
Limestone (Little Lime)	13	898
Limestone (Big Lime)	174	1,072
Shale (Waverly)	28	1,100
Sandstone (Big Injun)	30	1,130
Shale, shelly	70	1,200
Shale	30	1,230
Shale, shelly	70	1,300
Shale	50	1,350
Shale, shelly	75	1,425
Shale white	75	1,500
Limestone and shale	20	1,520
Sandstone (Berea)	20	1,540
Shale	20	1,560

Devonian System.

	Thickness	Depth
Shale, black (Chattanooga)	280	1,840
Shale	27	1,867
Limestone (Corniferous)	10	1,877
Limestone	8½	1,885½
Total depth		1,885½

1st oil 1870; more gas and oil 1872; bottom sand 1877.

Log No. 941

H. C. Keeton, No. 4, lessor. Ohio Fuel Oil & Gas Co., lessee. Location: Cannel City District. Commenced: Nov. 21, 1913. Completed: Dec. 20, 1913.

Strata.

Pennsylvanian System.   Thickness Depth

	Thickness	Depth
Shale	30	30
Sand	40	70
Coal	3	73
Shale and shells	100	173
Limestone	60	233 ·
Shale and shells	307	540
Sand, salt	190	730
Shale	25	755
Sand, salt	120	875
Shale	5	880

Mississippian System.	Thickness	Depth
Limestone (Little Lime)	20	900
Limestone (Big Lime)	190	1,090
Shale and shells	20	1,110
Sandstone (Big Injun)	3.0	1,140
Shale, shelly	380	1,520
Shale, copper	10	1,530
Sandstone (Berea)	50	1,580

Devonian System.		
Shale, brown (Chattanooga)	260	1,840
Shale, white	25	1,865
Limestone (Corniferous), (oil 1871)	17	1,882
Total depth		1,882

Log No. 942

H. C. Keeton, No. 6, lessor. Ohio Fuel Oil & Gas Co., lessee. Location: Cannel City District. Commenced: Sept. 12, 1913. Completed: Oct. 18, 1913. Shot 50 qts.

Strata.

Pennsylvanian System.	Thickness	Depth
Sand and mud	11	11
Stone	5	16
Shale and shells	114	130
Sand	20	150
Shale	50	200
Sand	30	230
Shale	112	342
Sand and shells	178	520
Shale	28	548
Sand	12	560
Shale	74	634
Sand, salt	126	760
Shale	5	765
Sand	30	795
Shale	70	865
Sand	90	955
Shale	5	960

Mississippian System.		
Limestone (Little Lime)	12	972
Shale	8	980
Sand	15	995

Mississippian System.   Thickness Depth
    Limestone (Big Lime) 167   1,162
    Shale, white   30   1,192
    Sandstone (Big Injun)   28   1,220
    Shale and shells 396   1,616
    Shale, brown   9   1,625
    Sandstone (Berea)   15   1,640
    Shale, white   30   1,670

Devonian System.
    Shale, brown (Chattanooga) 267   1,937
    Shale, white   30   1,967
    Limestone "sand" (Corniferous), (oil 1974)   16   1,983
        Total depth   1,983

Log No. 943

H. C. Keeton, No. 7, lessor. Ohio Fuel Oil & Gas Co., lessee. Location: Cannel City District. Shot Oct. 8, 1913, 50 qts.

    Strata.
Pennsylvanian System.   Thickness Depth
    Soil   25   25
    Coal   2   27
    Shale and shells   511   538
    Sand Salt   192   730
    Break   70   800
    Sand   50   850
    Break   15   865
Mississippian System.
    Limestone   20   885
    Shale   10   895
    Sand   12   907
    Limestone (Big Lime) 160 1,067
    Shale and shells 30 1,097
    Sand 23 1,120
    Shale and shells 405 1,525
    Shale, brown 10 1,535
    Sandstone (Berea) 40 1,575
Devonian System.
    Shale, brown (Chattanooga) 271 1,846
    Shale, white .... $23\frac{1}{2}$ $1,869\frac{1}{2}$
    Limestone "sand," (Corniferous) .... $4\frac{1}{2}$ 1,874
    Limestone, (oil and gas) (Corniferous) 8 1,882
    Limestone (Corniferous) 7 1,889
        Total depth   1,889

Log No. 944

J. B. Whitt, No. 3, lessor. Ohio Fuel Oil & Gas Co., lessee. Location: Cannel City. Commenced: Aug. 27, 1913. Completed: Sept. 11, 1913. Shot Oct. 10, 1913, 60 qts.

Strata.

Pennsylvanian System.

	Thickness	Depth
Clay	158	158
Coal	3	161
Shale	57	218
Coal	2	220
Shale and shells	297	517
Sand, salt	201	718
Shale	12	730
Sand, salt	112	842
Shale, black	5	847
Sand, black	26	873
Shale, white	5	878

Mississippian System.

Limestone (Little Lime)	9	887
Shale	2	889
Limestone (Big Lime)	170	1,059
Shale, white	27	1,086
Sandstone (Big Injun)	30	1,116
Shale and shells	71	1,187
Shale	18	1,205
Shale and shells, (water at 542)	245	1,450
Shale, white	42	1,492
Shale, brown (Sunbury)	8	1,500
Sandstone (Berea)	23	1,523
Shale, white	17	1,540
Shale, shelly	8	1,548

Devonian System.

Shale, brown (Chattanooga)	276	1,824
Shale, white	25	1,849
Limestone "sand," (Corniferous)	18	1,867
Total depth		1,867

Pay 1854-1863; gas 1858.

Log No. 945

J. B. Whitt, No. 4, lessor.   Ohio Fuel Oil & Gas Co., lessee.   Location: Cannel City District.   Commenced: Sept. 27, 1913.   Completed: Oct. 28, 1913.   Shot, 50 qts.

Strata.

Pennsylvanian System.	Thickness	Depth
Soil	250	250
Cannel coal?	10	260
Sand and shells	90	350
Limestone	50	400
Sand and shells	50	450
Shale, black	50	500
Limestone and shell	100	600
Shale, brown	60	660
Sand, salt	180	840
Shale, black	40	880
Sand, salt	100	980
Shale, white	15	995
Mississippian System.		
Limestone (Little Lime)	11	1,006
Shale	2	1,008
Limestone (Big Lime)	150	1,158
Shells and shale	42	1,200
Sand shells	80	1,280
Limestone	20	1,300
Shale and shells	100	1,400
Sand, hard	50	1,450
Shale	190	1,640
Sandstone (Berea)	20	1,660
Devonian System.		
Shale, brown (Chattanooga)	300	1,960
Shale, white	21	1,981
Limestone "sand" (Corniferous) (oil 1900).	15½	1,996½
Total depth		1,996½

Log No. 946

J. B. Whitt, No. 11, lessor.   Ohio Fuel Oil & Gas Co., lessee.   Location: Cannel City District.

Strata.

Pennsylvanian System.	Thickness	Depth
Clay	10	10
Sand	80	90

Pennsylvanian System.   Thickness Depth

Shale, soft	120	210
Sand	65	275
Shale	15	290
Limestone, gritty?	30	320
Shale	90	410
Limestone	120	530
Shale and shells	35	565
Limestone?	35	600
Limestone, white?	87	687
Sand, salt, (water 745)	143	830
Shale, black	5	835
Sand	25	860
Shale	40	900
Sand	105	1,005
Shale	25	1,030
Sand	10	1,040

Mississippian System.

Limestone (Little Lime)	7	1,047
Muck, black	4	1,051
Limestone (Big Lime)	157	1,208
Shale, white	42	1,250
Limestone	25	1,275
Shale and shells	30	1,305
Limestone	20	1,325
Shale and shells	155	1,480
Limestone, hard	90	1,570
Shale, white	45	1,615
Limestone, white	25	1,640
Sand, hard	30	1,670
Shale, black	15	1,685
Sandstone (Berea)	15	1,700
Shale and shells	25	1,725

Devonian System.

Shale, brown (Chattanooga)	275	2,000
Shale, white	13	2,013
Limestone ''sand'' (Corniferous), (oil 2019)..	16	2,029
Total depth		2,029

Log No. 947

Riley Benton, No. 2, lessor. Dreadnaught Oil & Refining Co., lessee. Location: Brush Creek, 1½ miles from Cannel City, Morgan Co., Ky.

Strata.

Pennsylvanian System.	Thickness	Depth
Conductor 8¼" pipe	21	21
Shale, dark	50	71
Shale, white	177	248
Sand	171	419
Shale	12	431
Sand	20	451
Shale	49	500
Limestone?	25	525
Shale	14	539
Sand	50	589
Shale, sandy	6	595
Shale and limestone shells?	23	618
Sand	7	625
Shale	3	628

Mississippian System.		
Limestone (Little Lime)	14	642
Shale (break)	12	654
Limestone (Big Lime), (cased 6-5/8)	150	794
Shale, green	50	844
Sandstone (Big Injun), (oil and gas)	36	880
Shale (Waverly), (gas)	354	1,234
Shale, brown (Sunbury)	10	1,244
Sandstone (Berea), (showing oil)	39	1,283
Shale, white	30	1,313

Devonian System.		
Shale, brown (Chattanooga)	290	1,603
Fire clay	42	1,645
Limestone (cap rock)	8	1,653
Limestone, (pay oil at 5, 7, 11 and 17)	11	1,664
Total depth		1,664

Log No. 948

Lewis Williams, No. 1, lessor. Mid South Gas Co., lessee. Location: 1 mile north of fault on Mine Fork. Casinghead elevation: 740 feet.

Strata.

Pennsylvanian System.	Thickness	Depth
Surface gravel	16	16
Sand	70	86
Shale, soft	9	95

Mississippian System.		
Limestone (Little Lime)	10	105
Shale (Pencil Cave)	5	110
Limestone (Big Lime), (little gas 130)	50	160
Sand, (oil show 207)	200	360
Limestone shale (oil show 390)	122	482
Sandstone	37	519
Shale, blue	30	549
Sandstone (Squaw)	14	563
Shale	20	583
Shale, brown (Sunbury)	19	602
Sandstone (Berea)	87	689

Devonian System.		
Shale, brown (Chattanooga)	23	712
Total depth		712

Little gas 415. Dry hole.

Log No. 949

J. C. Hill, No. 1, lessor. Location: Open Fork, near Johnson County line. Casinghead elevation: 725 feet.

Strata.

Pennsylvanian System.	Thickness	Depth
Soil	3	3
Sand, (water 50)	150	153
Sand, loose, (large pebbles)	4	157
Shale, soft	16	173
Shale	52	225
Shale, red and blue	14	239
Limestone, white	3	242
Sand and shale	24	266

Pennsylvanian System.   Thick

    Sand, black, and limestone   18
    Shale   18
    Sand   3

Mississippian System.

    Limestone (Little Lime), gray   30
    Limestone (Big Lime), white   90
    Sand (break 268-372)   17
    Shale (Waverly), (gas and salt water 392-5)..   108
    Shale, (little oil 425)   45
    Limestone   10
    Shale   59
    Limestone and sand, (show oil 590)   21
    Shale, (show oil 666-670)   7
    Sand   31
    Shale and sand   22
    Sand   7
    Shale and limestone   21
    Shale   17
    Sandstone (Berea)   98

Devonian System.

    Shale   16

        Total depth ....

CHAPTER IX.

MUHLENBERG COUNTY.

Production: Oil and Gas. Producing sands: Pottsville Sandstone (Pennsylvanian), and Penrod (Chester age) (Mississippian).

Log No. 950

Cox, No. 1, lessor. Location: 3 miles north of Dunmor. Production: 800,000 cu. ft. Gas. Gray Sand Oil and Gas Co,. Central City, lessee. Authority: H. F. Storer, Central City.
Strata.

Pennsylvanian System.	Thickness	Depth
Shale	30	30
Sandstone (water) ..,....................	300	330
Shale	45	375
Mississippian System.		
Limestone	130	505
Sandstone	20	525
Shale ;...................................	19	544
Sandstone, (oil and gas show)	12	556
Limestone	73	629
Sandstone, (gas)	7	636
Shale	7	643
Sandstone, (gas)	22	665
Limestone	286	951
Total depth		951

Casing record:
    71 ft. of 8¼ casing.
    365 ft. of 6¼ casing.
    397 ft. of 2 in. tubing on Packer.
NOTE—This well probably finished in the Chester.

Log No. 951

Poole, No. 1, lessor. Location: Twin Tunnels. Production: Dry.
Strata.

Pennsylvanian System.	Thickness	Depth
Sandstone	41	41
Coal	2	43
Shale	71	114
Coal	4	118
Sandstone	6	124

Pennsylvanian System.	Thickness	Depth
Shale	96	220
Coal	2	222
Sandstone, (water)	8	230
Shale	140	370
Sandstone, (water)	300	670
Shale (Pencil Cave)	30	700
Mississippian System.		
Limestone, sandy, hard	15	715
Shale	35	750
Limestone, hard	20	770
Shale	6	776
Limestone	6	782
Shale	8	790
Limestone	15	805
Shale	20	825
Limestone, sandy, (water)	25	850
Shale	35	885
Limestone, cherty	15	900
Limestone, cherty, very hard	30	930
Shale	10	940
Limestone	5	945
Sandstone	15	960
Shale	10	970
Sandstone	30	1,000
Sandstone	12	1,012
Total depth		1,012

NOTE—This well probably finished in the Chester.

Log No. 952

Oakes Heirs, No. 1, lessor. Commenced: Oct. 15, 1918. Production: Dry; casing pulled, well plugged and abandoned. Authority: The Ohio Oil Co.

Strata.

Pennsylvanian System.	Thickness	Depth
Soil, yellow, soft	9	9
Rock, gray, soft	10	19
Shale, hard, gray, soft	4	23
Coal	3½	26½
Shale, hard	3½	30
Sand, hard, dry	2	32
Shale, hard, blue, medium	74	106

Mississippian System.

	Thickness	Depth
Limestone, hard, white	20	126
Shale, hard, blue	60	186
Shale, hard, white	80	266
Sand, gray, soft	6	272
Shale, hard, black, soft	4	276
Limestone, hard, gray	3	279
Limestone, hard, brown	3	282
Shale, hard, blue, soft	19	301
Shale, hard, sandy	15	316
Shale, hard, gray, sandy	50	366
Shale, hard, black	10	376
Shale, hard, white	30	406
Limestone, hard, white	10	416
Shale, hard, blue, soft	50	466
Sand, soft, brown	10	476
Shale, hard	15	491
Sand, gray, soft	45	536
Shale, hard, blue	50	586
Limestone, hard, white	5	591
Shale, hard	15	606
Shale, hard, brown, soft	30	636
Shale, hard, blue, soft	50	686
Sand, hard, white, (water 720)	85	771
Shale, brown, soft	35	806
Shale, hard, blue	20	826
Shale, hard, sandy	65	891
Sand, hard, gray	15	906
Shale, hard, blue	15	921
Sand, gray, soft	25	946
Sand, fine, white, hard, (hole full fresh water)	100	1,046
Shale, hard, blue, soft	20	1,066
Shale, hard, black, soft	25	1,091
Limestone, sandy, hard, brown	15	1,106
Limestone, dark, extra hard	22	1,128
Shale, hard, gray, soft	33	1,161
Limestone, hard, dark, sandy	3	1,164
Shale, hard, blue, soft	15	1,179
Limestone, hard, brown	12	1,191
Limestone, hard, white	23	1,214
Shale, hard, green, soft	15	1,229
Shale, hard, blue, soft	55	1,284
Limestone, hard, brown	20	1,304
Shale, hard, blue	8	1,312
Shale, hard	10	1,322
Shale, hard, dark, extra soft	32	1,354
Limestone, gray, extra hard	3	1,357

	Thickness	Depth
Mississippian System.		
Shale, hard, blue	7	1,364
Limestone, hard, brown	3	1,367
Shale, hard, blue, soft	5·	1.372
Limestone, hard, brown, sandy	13	1,385
Shale, hard, blue	7	1,392
Limestone, hard, brown	5	1,397
Shale, hard, green	10	1,407
Limestone, hard, gray	8	1,415
Shale, hard, gray	4	1,419
Shale, hard, black, soft	18	1,437
Shale, hard, blue, soft	85	1,522
Limestone, hard, brown	5	1,527
Shale, hard, blue, soft	20	1,547
Sand, white, soft	34	1,581
Shale, hard, blue	5	1,586
Sand, white, extremely hard	31	1,617
Limestone, white, extra hard	13	1,630
Limestone, yellow, soft	5	1,635
Shale, hard, blue	2	1,637
Limestone, yellow, soft	3	1,640
Limestone, gray, extra hard	26	1,666
Shale, hard, blue	17	1,683
Shale, brown, soft	7	1,690
Shale, blue, soft	13	1,703
Sand, white	15	1,718
Shale, hard, green	17	1,735
Limestone, gray	20	1,755
Shale,·hard, blue	20	1,775
Limestone, brown	13	1,788
Shale (red rock)	3	1,791
Sand, gray, green	19	1,810
Shale, hard, blue	25	1,835
Sand, gray, green, (New Providence)	5	1,840
Shale, hard, blue, (New Providence)	25	1,865
Limestone, brown, (New Providence)	5 ,	1,870
Shale, hard, blue, (New Providence)	30	1,900
Devonian System.		
Shale, brown (Chattanooga)	35	1,935
Limestone, hard, brown	5	1,940
Shale, hard, blue	2	1,942
Limestone, brown	4	1,946
Limestone, white	16	1,962
Shale, hard, blue	3	1,965
Limestone, hard, dark	15 · ·1,980	
Limestone, sandy	7	1,987

Mississippian System.

	Thickness	Depth
Shale, hard, blue	3	1,990
Limestone ''sand,'' green	15	2,005

Silurian System.

Limestone, white	19	2,024
Limestone, hard, white	21	2,045
Shale, hard, blue	40	2,085
Limestone, hard, gray	20	2,105
Shale, hard, blue	25	2,130

Ordovician System.

Limestone, white	80	2,210
Limestone, gray and brown	10	2,220
Limestone ''sand,'' green	38	2,258
Limestone, soft, white, (salt water 2,245)....	7	2,265
Total depth		2,265

Log No. 953

Lacy well, No. 1, lessor. Casinghead elevation: 450 feet. Bar. Authority: L. Beckner.

Strata.

Pennsylvanian System.

	Thickness	Depth
Soil and clay	10	10
Sandstone	33	43
Shale	11	54
Coal	1	55
Clay	3	58
Sandstone	16	74
Shale	34	108
Coal No. 12	5	113
Fire clay	1	114
Limestone, black	1	115
Coal, No. 11	6	121
Fire clay	5	126
Sandstone	7	133
Shale	5	138
Sandstone	23	161
Shale	32	193
Shale, hard, black	3	196
Coal No. 9	5	201
Total depth		201

Log No. 954

St. Bernard Mining Co.   Location: 1¼ miles northeast of White Plains. Casinghead elevation: 400 feet, Bar. Authority: St. Bernard Mining Co., and L. Beckner.

Strata.

Pennsylvanian System.	Thickness	Depth
Clay	15	15
Shale, soft, and limestone	17	32
Shale, hard, black, and coal	2	34
Sandstone	10	44
Shale	5	49
Sandstone	14	63
Shale and coal	17	80
Fire clay	1	81
Limestone	2	83
Gob	1	84
Coal	3	87
Fire clay	5	92
Sandstone	9	101
Shale	3	104
Limestone	5	109
Shale	2	141
Sandstone	30	201
Limestone	1	202
Shale	8	210
Sandstone	20	230
Shale	6	236
Sandstone	26	262
Total depth		262

Log No. 955

Pond Creek Bottom Well.   Location: ½ mile north of Rochester Road. Authority: L. Beckner.

Strata.

Pennsylvanian System.	Thickness	Depth
Clay	30	30
Quicksand, blue	8	38
Gravel bed	8	46
Sandstone, blue	8	54
Shale, soft	8	62
Shale, hard, gray	17	79
Coal	½	79½
Shale (Kidney)	1½	81

Pennsylvanian System.

	Thickness	Depth
Shale, hard, black	3	84
Fire clay	2	86
Shale, soft	11	97
Shale, hard, gray	33	130
Shale, hard, black	2	132
Coal	1	133
Fire clay	2	135
Shale, soft	6	141
Total depth		141

Log No. 956

Concord Well. Location: At Concord Schoolhouse. Casinghead elevation: 220 feet. Authority: L. Beckner.

Strata.

Pennsylvanian System.

	Thickness	Depth
Soil and clay	4	4
Sandstone	35	39
Shale, sandy, hard, brown	1	40
Shale	34	74
Sandstone and shale	109	183
Shale	5	188
Coal	1	189
Fireclay	7	196
Limestone and sandstone	15	211
Shale	24	235
Coal No. 12	6	241
Limestone	4	245
Coal No. 11	6	251
Fireclay	3	254
Shale	8	262
Sandstone	15	277
Shale	44	321
Shale, hard, black	2	323
Coal No. 9	4	327
Total depth		327

Log No. 957

Location: ½ mile west of White Plains. Casinghead elevation: 465 feet, Bar. Authority: L. E. Littlepage and L. Beckner.

Strata.

Pennsylvanian System.	Thickness	Depth
Soil	5	5
Sandstone	13	18
Shale	23	41
Shale, hard, and coal, rotten	2	43
Sandstone, blue	13	56
Shale, hard, gray	10	66
Shale	10	76
Sandstone, soft, blue	13	89
Shale	8	97
Sandstone, white	39	136
Shale, hard, gray	22	158
Limestone, hard	4	162
Shale, hard and gray, black	29	191
Coal, (clay parting)	1	192
Coal, (clay parting)	2	194
Coal, (Bone coal)	1	195
Fireclay, hard	2	197
Shale, sandy	11	208
Sandstone	52	260
Shale, hard, gray	25	285
Sandstone, white	3	288
Shale, blue	4	292
Fireclay	1	293
Shale, gray	11	304
Shale, hard and gray	26	330
Fireclay	2	332
Sandstone or hard rock	9	341
Shale, hard, sandy	4	345
Shale, hard and gray	19	364
Sandstone, white	2	366
Total depth		366

Log No. 957-A.

Lucy Garrett, No. 1, lessor. Gray Sand Oil & Gas Co., Central City, lessee. Location: 800 feet north and west of Cox No. 2. Production: 12 bbls. oil and 2,500,000 cu. ft. gas approx. Authority: H. F. Storer, Central City.

Strata.

Pennsylvanian System.	Thickness	Depth
Soil	20	20
Slate	80	100
Sand (water)	47	147

Pennsylvanian System.	Thickness	Depth
Slate	15	162
Sand (gas 294)	204	366
Slate	10	376
Lime	14	390
Slate	28	418
Lime	22	440
Slate	40	480
Lime	30	510
Slate	15	525
Lime	30	555
Lime, broken	39	594
Sand, (gas-show oil)	5	599
Lime	36	635
Slate	15	650
Lime	18	668
Slate	20	688
Sand (gas)	12	700
Slate	14	714
Sand (oil)	13	727
Total depth		727

366 ft. 6¼ casing.
727 ft. 2 in. tubing.
Elevation about 50 ft. higher than Cox No. 1.

Log No. 957-B

Cox, No. 2, lessor. Gray Sand Oil & Gas Co., Central City, lessee.
Location: 400 feet N. W. of Cox No. 1. Elevation: About 30 feet
higher than Cox No. 1. Production: 5,000,000 cu. ft. gas approx.
This well blew wide open for 3 months, due to accident attending
measurement. Finally caught fire and burned for 17 days, destroying
rig, etc. Extinguished by steam. Authority: H. F. Storer, Central
City, Ky.
Strata.

Pennsylvanian System.	Thickness	Depth
Soil	12	12
Shale	43	55
Sandstone (water)	65	120
Shale	25	145
Limestone, broken and shale	20	165
Sandstone	183	348
Shale	10	358
Limestone	25	383
Shale	·27	410
Limestone	27	437
Shale	66	503

Pennsylvanian System.	Thickness	Depth
Limestone	65	568
Sandstone (gas-oil show)	4	572
Limestone	38	610
Shale, green, broken	15	625
Shale	19	644
Sandstone, (large gas)	6	650
Total depth		650

Finished in sand at 650.

348 ft. 6¼" casing.

650 ft. 2" tubing.

NOTE—Well only drilled to second sand.

OHIO COUNTY.

Production: Oil and Gas. Producing Sands: Major and other Mississippian Sands; Corniferous (Devonian).

Log No. 958

Patterson Well No. 1, lessor. Location: Near Olaton, Ky.

Strata.

Mississippian System.	Thickness	Depth
Shale	12	12
Limeston, white, hard	15	27
Limestone "sand," (oil)	5	32
Shale, blue	16	48
Limestone, white, hard	5	53
Shale, blue	11	64
Limestone, white, hard	31	95
Limestone, blue, broken	9	104
Limestone, sandy	10	114
Limestone, white	36	150
Limestone, white	60	210
Limestone, brown	55	265
Limestone, white	32	297
Limestone "sand," (oil)	6	303
Limestone, gray	32	335
Blue Lick formation	61	396
Limestone, brown, (cased 8" at 400)	4	400
Limestone, white	2	402
Shale lime	2	404
Limestone, white, hard	11	415
Limestone, gray	5	420
Limestone, brown	6	426

Mississippian System.

	Thickness	Depth
Limestone, brown and gray	5	431
Limestone, light brown, hard	5	436
Gas sand	10	446
Limestone, light brown	19	465
Limestone, gray, hard	5	470
Limestone, dark gray	44	514
Limestone, gray brown	8	522
Limestone, dark brown	23	545
Limestone, dark brown	37	582
Limestone, gray and brown, hard	8	590
Limestone, gray, hard	10	600
Limestone, dark gray	35	635
Limestone, blue and white	15	650
Limestone, dark gray, sandy	5	655
Limestone, brown, hard	35	690
Limestone, dark gray, hard	45	735
Limestone, black, soft	29	764
Limestone, dark gray, soft	71	835
Limestone, black, soft	90	925
Limestone, gray, soft	15	940
Limestone "sand," (oil)	6	946
Limestone, gray	11	957
Limestone "sand," (oil)	10	967
Limestone "sand," (oil)	9	976
Limestone, gray	59	1,035
Limestone, gray, sandy	20	1,055
Limestone, blue shell	5	1,060
Limestone, blue, and shale	5	1,065
Shale, blue	23	1,088
Shale, black	184	1,272
Limestone, black, hard	4	1,276
Limestone, black, dark	4	1,280
Limestone, gray black	4	1,284
Limestone, black, soft	6	1,290
Limestone, black and gray	6	1,296
Limestone, gray	4	1,300
Sand, light brown, hard, (show of gas)	14	1,314
Sand, brown	20	1,334
Sand, brown, soft	10	1,344
Limestone, black	6	1,350
Limestone, black, soft	15	1,365
Limestone, black, hard	15	1,380
Limestone, gray	7	1,387
Limestone, white, soft	5	1,392
Total depth		1,392

NOTE—This well is located near the Grayson County line in Ohio

County. It was first published in Ser. V, Bull. I, under the Grayson County records. To correct that error it is herewith published as an Ohio County record. It is all in the Mississippian Series, but finished probably close to the Devonian.

OWSLEY COUNTY.

Production: Gas, oil show.   Producing Sand: Corniferous (Devonian).

Log No. 959

John G. White Oil & Gas Co., No. 1.   Location: On Meadow Creek. Commenced: Feb. 10, 1909.   Completed: April 12, 1909.   Production: Dry hole.   Authority: C. E. Bales.

Strata.

Pennsylvanian System.	Thickness	Depth
Soil	8	8
Shale	60	68
Sandstone	32	100
Shale, blue	100	200
Sandstone	175	375
Shale, blue	50	425
Sandstone, (salt water)	193	618
Shale	6	624
Mississippian System.		
Limestone (Little Lime)	8	632
Shale, blue	20	652
Limestone (Big Lime)	184	836
Shale (Waverly)	438	1,274
Devonian System.		
Shale, brown (Chattanooga)	173	1,447
Shale (fire clay)	15	1,462
Shale, brown	10	1,472
Limestone (Corniferous)	31	1,503
Total depth		1,503

Log No. 960

Rufus Barker, No. 1, lessor.   Location: At Traveler's Rest P. O. Production: No oil, gas under cap.

Strata.

Pennsylvanian System.	Thickness	Depth
Sand (Mountain)	469	469

Mississippian System.

	Thickness	Depth
Limestone (Big Lime)	101	570
Limestone, white	14	584
Shale, green (Waverly)	398	982

Devonian System.

	Thickness	Depth
Shale, blus, Chattanooga	130	1,112
Shale, black, Chattanooga	16	1,128
Fire clay	4	1,132
Limestone ''sand,'' (gas)	20	1,152
Shale, black	15	1,167
Limestone ''sand''	34	1,201
Shale, blue	1	1,202
Total depth ....·......		1,202

Casing record: 32 ft. 8¼ in. casing; 584 ft. 6¼ in. casing.

PENDLETON COUNTY.

Production: Small gas. Producing Sands: unnamed, possibly of Trenton age (Ordovician).

Log No. 961

Location: About 200 yards from the Campbell County line, near Morning View. Authority: L. Beckner.

Strata.

Ordovician System.

	Thickness	Depth
Clay and stone	10	10
Shale, blue (salt water)	80	90
Black sulphur lime, hard	35	125
Shale, blue, (s. w. 145)	27	152
Limestone, gray, hard	12	164
Shale, blue	48	212
Limestone, blue, very hard	8	220
Limestone, gray, hardest yet	16	236
Limestone, light gray	8	244
Limestone, black, (gas 248)	12	256
Shale, dark	4	260
Limestone, dark, very hard	8	268
Limestone, brown	8	276
Limestone, gray, (gas)	16	292
Limestone, dark gray	8	300
Limestone, blue, hard	28	328
Limestone, black, not so hard	20	348
Limestone, gray, very hard	36	384
Flint, brown	24	408

Ordovician System.	Thickness	Depth
Limestone, gray, flinty, very hard	100	508
Flint, brown	32	540
Limestone, light gray, not so hard	117	657
Flint, brown	15	672
Limestone, black	25	697
Shale, dark, and limestone	25	722
Limestone, brown, sandy, (blk. sul. s. w.)	68	790
Black sulphur lime, very hard	25	815
Limestone, blue	..	...

PIKE COUNTY.

Production: Small oil and gas. Producing Sands: Pottsville (Pennsylvanian); and Maxton (Mauch Chunk age) (Mississippian).

Log No. 962

Big Sandy Co., No. 1, owner and operator. Location: John Moore's Branch, ¼ mile from Elkhorn City, Elkhorn Creek, Pike County, Ky. Elevation of Lower Elkhorn Coal at this point, 1,500 feet approx. Drilling stopped at 918 feet, Nov. 21, 1912. Re-commenced and completed to 1,223 feet in 1920. Production: 10 gal. green crude oil daily. Casing head elevation: 1,000 A. . Approx. Authority: Big Sandy Co. and L. Beckner.

Strata.

Pennsylvanian System.	Thickness	Depth
Clay	6	6
Sand	4	10
Shale	4	14
Sand	4	18
Coal, (Auxier Seam)	2	20
Sand	20	40
Sand and shale	23	63
Shale, (ran core drill)	2	65
Sand, hard, (10 in. casing 70)	5	70
Limestone, sandy	20	90
Shale	11	101
Coal (8 in.) (Little Cedar Seam) and shale (core 101)	2	103

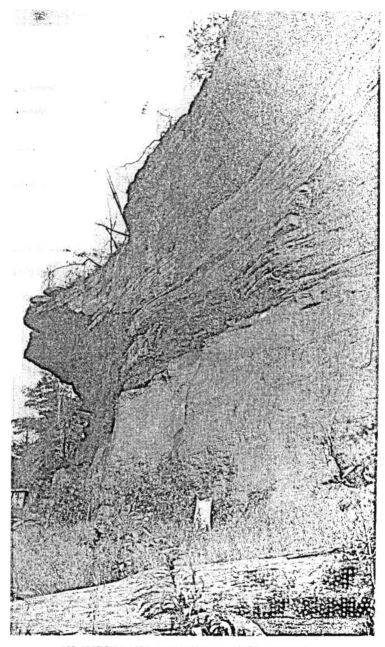

AN IMPORTANT KENTUCKY (OIL) "SAND"
The massive Pottsville Conglomerate, is not a large flush producer, but one of extremely long life, as evidenced by Floyd County wells drilled in 1891. These Pottsville cliffs are at Grahn, Carter County, Kentucky.

Pennsylvanian System.	Thickness	Depth
Shale, black		110
Sand, gray, hard, (total depth May 31, 1919)	5	115
Sand, gray	6	121
Limestone, sandy	5	126
Sand, gray, (50,000 cu. ft. gas at 169)	44	170
Limestone, sandy	5	175
Shale	15	190
Shale, (indication coal 228, ran core drill) ..	38	228
Coal, (Ellswick Seam)	2	230
Limestone, sandy	6	236
Shale	59	295
Sand	275	570
Coal, (Gilbert or Grundy Seam)	2½	572½
Shale	60	632½
Sand	233½	866
Sand, (oil show)	4	870
Shale and coal, (Jaegar Seam)	6	876
Sand and rotten shale	17	893
Sand, hard, and shale	2	895
Sand, hard	3	898
Limestone, very hard	2	900
Sand, rotten, (4 gals. oil)	2½	902½
Sand, hard and white	7½	910
Sand, white, hard	8	918
Sand, hard, white	5	923
Sand, hard, white	10	933
Sand, hard, changing to blue	15	948
Sand, bluish	17	965
Sand and shale, soft	17	982
Shale and rotten shaly sand	42	1,024
Sand, hard	36	1,060
Sand, shaly	20	1,080
Sand, hard, white, (show of oil 1,129)	55	1,135
Shale, black	14	1,149
Shale, black and gray, coal (Sewall Seam) ..	71	1,220
Sand, hard, gray, (gas and oil 1,223)	3	1,223
Total depth		1,223

DRILLERS' NOTE.—Broke pin off at 299 feet; crooked hole from 299 to 308 feet. The record is all in the Pottsville. Set: 8¼ casing 379 feet; 6¼ casing 794 feet; packed on bottom 6¼ casing.

Log No. 963

T. J. Williamson, No. 1, lessor. Location: Pikeville, Ky. Well completed: May 29, 1920. Drilled by A. B. Brode & Son. Tool Pusher: S. L. Anderson. Drillers: J. T. O'Laughlin and L. E. Smith.
Strata.

Pennsylvanian System.	Thickness	Depth
Drift, (12½ in. casing)	37	37
Sand	75	112
Shale, (10 in. casing 133)	28	140
Limestone, black	23	163
Shale and sand, broken	177	350
Sand, salt, sand	300	650
Shale	46	696
Sand, salt (2nd), hard	39	735
Coal and shale	2	737
Sand, salt, (gas 800 and 840)	115	852
Shale	2	854
Sand, salt	130	984
Shale, (8 in. casing 987½)	7	991
Limestone, dark	20	1,011
Sand, white	74	1,085
Shale, light	40	1,125
Sand, white	35	1,160
Shale	5	1,165
Sand, hard	10	1,175
Limestone, black	11	1,186
Shale, light	6	1,192
Mississippian System.		
Red rock	20	1,212
Limestone, sandy	10	1,222
Red rock	30	1,252
Sand, dark	23	1,275
Shale	5	1,280
Limestone, dark	20	1,300
Shale and sand	20	1,320
Sand, white	21	1,341
Shale, white	14	1,355
Sand, shells	5	1,360
Shale, white	7	1,367
Sand (Maxon)	87	1,454
Limestone, black	90	1,544
Shale and limestone	30	1,574
Limestone (Little Lime)	19	1,593
Limestone (Big Lime) (6⅝ casing 1,800)	200	1,793
Sandstone (Big Injun)	35	1,828
Sand (Squaw)	20	1,848

Mississippian System.	Thickness	Depth
Shale and shells	23	1,871
Red rock	15	1,886
Shale and shells	12	1,898
Sand, white, and limestone	100	1,998
Shale and shells	102	2,100
Sand, shelly	18	2,118
Shale	23	2,141
Shale and shells	60	2,201
Devonian System.		
Shale, brown, and slate	106	2,307
Sandstone	38	2,345
Shale and slate	20	2,365
Total depth		2,365

POWELL COUNTY.

Production: Oil and Gas. Producing Sand: Corniferous (Devonian), Niagaran (Silurian).

Log No. 964

Williams, No. 2, lessor. Location: Northeast edge of Stanton.
Production: 2 bbls. oil; gas at 156.
Strata.

Devonian System.	Thickness	Depth
Shale, black (Chattanooga) and soil	132	132
Fire clay and shale	18	150
Limestone (Irvine Sand)	7	157
Shale, light	58	215
Limestone ''sand,'' (oil)	8	223
Shale, light	17	240
Limestone, sandy (?)	5	245
Total depth		245

Log No. 965

Will Aiam, No. 1, lessor. Location: Near Xena P. O. Authority:
Lucien Beckner and Dr. I. T. Rogers.
Strata.

Pennsylvanian System.	Thickness	Depth
Sand and shale	400	400
Devonian System.		
Shale, black (Chattanooga)	150	550
Limestone (Corniferous)	60	610
Total depth		610

Log No. 966

Wix Day, No. 1, lessor. Completed: April 5, 1905. Production: Well was dry; show of gas at 134´feet; the casing was pulled and well plugged and abandoned. Authority: New Domain Oil & Gas Co.

Strata.

Mississippian System.	Thickness	Depth
Soil and gravel, red and loose	18	18
Sandstone, blue, soft	77	95
Sandstone, blue, firm	23	118
Shale, blue, soft	9	127
Shale, blue, hard	3	130
Sandstone, blue, hard	4	134
Sandstone, blue, firm	19	153
Lime shells, blue, hard	1	154
Shale, blue, hard	30	184
Shale, blue, soft	16	200
Limestone, blue, hard	6	206

Devonian System.		
Shale, black, hard, (Chattanooga)	50	256
Shale, brown, firm, (Chattanooga)	94	350
Shale (fire clay), light, soft	8	358
Limestone "sand," gray, hard, (gas)	39	397
Shale, hard, blue, soft	3	400
Total depth		400

Log No. 967

J. G. Skidmore, No. 1, lessor. Commenced: Feb. 3, 1905. Production: Dry. Authority: New Domain Oil & Gas Co.

Strata.

Mississippian System.	Thickness	Depth
Soil, soft	7	7
Gravel, blue, soft	1	8
Sandstone, blue, hard	18	26
Shale, blue, soft	50	76
Shale, light, soft	79	155
Limestone and shells, blue, hard	4	159
Shale, light, hard	10	169
Shale, blue, very hard	2	171

Mississippian System.   Thickness Depth

  Shale, light, soft   89   260
  Sandstone, red, hard   10   270
  Shale, light, soft   39   309
  Limestone, blue hard   3   312

Devonian System.

  Shale, black, soft (Chattanooga)   145   457
  Shale, hard, blue, soft   10   467
  Limestone ''sand,'' (gas) open   31   498
  Total depth   498

Log No. 968

J. S. Skidmore, No. 2. Completed: March 22, 1905. Production:
gas. Authority: New Domain Oil & Gas Co.

  Strata.

Mississippian System.   Thickness Depth

  Soil and gravel, blue, soft   9   9
  Shale, blue soft   30   39
  Limestone, light, hard   6   45
  Shale, blue, soft   37   82
  Limestone, light, hard   13   95
  Shale, light, soft   45   140
  Shale, blue, medium   30   170
  Shale, light, soft   44   214
  Shale (red rock), soft   8   222
  Shale, light, soft   35   257
  Sandstone, light, hard   3   260
  Shale, light, soft   2   262

Devonian System.

  Shale, black, soft (Chattanooga)   151   413
  Shale (fire clay), white, soft   8   421
  Limestone ''sand,'' light, open, (gas)   18   439
    Total depth   439

Log No. 969

Cornelia Wymore, No. 1, lessor. Completed: Sept. 28, 1904. Production: Dry; small show of oil at 338 feet. Authority: New Domain Oil & Gas Co.

Strata.

Mississippian System.	Thickness	Depth
Clay, soft	12	12
Shale, soft, blue	165	177
Shale, soft, pink	10	187
Limestone, blue, hard	8	195
Shale, soft, blue	10	205
Devonian System.		
Shale, black, soft (Chattanooga)	133	338
Limestone, gray, hard	20	358
Limestone, blue, soft, shaly	160	518
Limestone, blue, hard	45	563
Limestone, gray, hard	20	583
Limestone, blue, hard	245	828
Total depth		828

NOTE—The Devonian-Silurian contact is within the 160 feet above 518 feet in depth.

Log No. 970

Joseph Willoughby, No. 1, lessor. Completed: Oct. 22, 1904. Production. Dry; show of gas at 120 and 418 feet. Authority: New Domain Oil & Gas Co.

Strata.

Mississippian System.	Thickness	Depth
Clay and gravel	8	8
Shale, soft	40	48
Shale or sandstone, blue	85	133
Sandstone, blue, soft, shaly	117	250
Mississippian System.		
Shale, black (Chattanooga)	140	390
Fire clay	10	400
Limestone, brown, hard	18	418
Shale, soft, limy	157	575
Limestone, brown, medium	15	590
Shale, soft	10	600

Mississippian System.	Thickness	Depth
Limestone, brown, hard	10	610
Limestone, blue, hard	20	630
Shale, soft	10	640
Limestone, blue, hard	160	800
Total depth		800

NOTE—The Devonian-Silurian contact is within the 157 feet above 575 feet.

Log No. 971

M. D. Rogers, No. 1, lessor. Commenced: Dec. 10, 1919. Completed: Jan. 18, 1920. Production: Commenced Jan. 28, 1920; production 48 hours after shot, 15 bbls. oil. Shot Jan. 25, 1920, between 766 and 790 feet. Authority: The Ohio Oil Co.

Strata.

Mississippian System.	Thickness	Depth
Soil, brown, soft	20	20
Sandstone, red, medium	49	67
Limestone, hard, white	90	157
Shale, blue, medium	412	569
Shale, red, hard, sandy	18	587
Fire clay, white, soft	32	619
Devonian System.		
Shale, hard, brown (Chattanooga)	142	761
Fire clay, white, soft	4	765
Limestone ''sand,'' hard, dark, (little oil) ...	8	773
Limestone ''sand,'' brown, medium, (good pay)	16	789
Shale, hard, blue, soft	12	801
Total depth		801

Log No. 972

M. D. Rogers, No. 2, lessor. Commenced: Feb. 8, 1919. Completed: Feb. 24, 1919. Production: Commenced Feb. 28, 1919; production 48 hours after shot, 24 bbls. oil. Shot Feb. 26, 1919, between 768 and 790 feet. Authority: The Ohio Oil Co.

Strata.

Mississippian System.	Thickness	Depth
Soil, red, soft	20	20
Sandstone, red, hard	20	40

Mississippian System.

	Thickness	Depth
Limestone, hard, white	110	150
Shale, hard and soft, bluish, medium	460	610
Shale, red, soft, sandy	15	625
Fire clay, white, soft	27	652

Devonian System.

Shale, brown, medium (Chattanooga)	105	757
Fire clay, white, soft	3	760
Limestone "sand," hard, black	8	768
Limestone "sand," brown, soft, (oil)	22	790
Shale, hard, blue, soft	10	800
Total depth		800

Log No. 973

M. D. Rogers, No. 3, lessor. Commenced: March 8, 1919. Completed: April 5, 1919. Production: Commenced producing April 10, 1919; production 48 hours after shot, 15 bbls. oil. Shot April 8, 1919, between 755 and 771 feet. Authority: The Ohio Oil Co.

Strata.

Mississippian System.

	Thickness	Depth
Soil, red, soft	10	10
Sandstone, red, hard	20	30
Limestone, hard, white	102	132
Shale, hard and soft, bluish, medium	436	568
Shale, red, soft, sandy	12	580

Devonian System.

Shale, brown, soft (Chattanooga)	150	730
Fire clay, white, soft	17	747
Limestone "sand," hard, brown	30	777
Total depth		777

Log No. 974

Joe Mullins, No. 4, lessor. Commenced: Dec. 15, 1918. Completed: Feb. 26, 1919. Production: Commenced producing Feb. 28, 1919; natural production after 48 hours, 3 bbls. oil. No shot. Authority: The Ohio Oil Co.

Strata.

Mississippian System.

	Thickness	Depth
Soil, red, soft	6	6
Shale (red rock), hard	82	88
Limestone, hard, white	140	228

Mississippian System.	Thickness	Depth
Shale, hard and soft, blue, medium	520	748
Shale, red, soft, sandy	18	766
Fire clay, white, soft	14	780

Devonian System.		
Shale, brown, medium (Chattanooga)	80	860
Fire clay, white, soft	8	868
Limestone "sand," black, medium	10	878
Total depth		878

NOTE—The Devonian (Chattanooga) black shale is usually thin at 80 feet in this record.

Log No. 975

J. B. Rogers, No. 5, lessor. Commenced: April 19, 1919. Completed: May 6, 1919. Production: Commenced producing May 10, 1919; production 48 hours after shot, 8 bbls. oil. Shot May 7, 1919, between 707 and 731 feet. Authority: The Ohio Oil Co.

Strata.

Mississippian System.	Thickness	Depth
Soil, red, soft	8	8
Limestone, hard, white	80	88
Shale, hard, soft, bluish, medium	420	508
Shale, red, soft, sandy	12	520
Fire clay, white, soft	16	536

Devonian System.		
Shale, brown, medium (Chattanooga)	148	684
Fire clay, white, soft	12	696
Limestone "sand," hard, dark, (no oil)	11	707
Limestone "sand," gray, medium, (some oil).	24	731
Total depth		731

Log No. 976

J. N. Rogers, No. 1, lessor. Commenced: Dec. 20, 1918. Completed: Jan. 18, 1919. Production: 48 hours after shot, 12 bbls. oil. Shot Jan. 25, 1919, between 783 and 759 feet. Authority: The Ohio Oil Co.

Strata.

Mississippian System.	Thickness	Depth
Soil, brown, soft	20	20
Sandstone, red, soft	50	70
Shale, hard, white, blue, soft	85	155

Mississippian System.

	Thickness	Depth
Shale and shells, blue and soft	400	555
Shale, red, hard, sandy	12	567
Fire clay, white, soft	24	591

Devonian System.

	Thickness	Depth
Shale, brown, soft (Chattanooga)	136	727
Fire clay, white, soft	4	731
Limestone ''sand,'' hard, black	8	739
Limestone ''sand,'' brown, soft	24	763
Shale, hard, blue, soft	5	768
Total depth		768

Log No. 977

J. N. Rogers, No. 2, lessor. Commenced: Feb. 5, 1919. Completed: Feb. 22, 1919. Production: Commenced producing Feb. 28, 1919; production 48 hours after shot, 11 bbls. oil. Shot Feb. 25, 1919, between 740 and 746 feet. Authority: The Ohio Oil Co.

Strata.

Mississippian System.

	Thickness	Depth
Soil, light, soft	20	20
Sandstone, red	50	70
Limestone, hard, white	80	150
Shale and slate, blue, medium	450	600
Shale (red rock), soft	12	612
Fire clay, white, soft	23	635

Devonian System.

	Thickness	Depth
Shale, brown (Chattanooga)	95	730
Fire clay, white, soft	4	734
Limestone ''sand,'' brown	27	761
Shale, hard, blue, medium	9	770
Total depth		770

NOTE—The Devonian (Chattanooga) black shale at 95 feet is somewhat thin in this record.

Log No. 978

J. N. Rogers, No. 3, lessor. Commenced: March 6, 1919. Completed: March 26, 1919. Production: Commenced producing April 1, 1919; production 48 hours after shot, 10 bbls. oil. Shot March 27, 1919, between 817 and 841 feet. Authority: The Ohio Oil Co.

Strata.

Mississippian System.	Thickness	Depth
Soil, yellow, soft	20	20
Sandstone, hard, white	75	95
Limestone, white, very hard	80	175
Shale, hard, blue	450	625
Shale, red, soft, sandy	25	650
Devonian System.		
Shale, brown, soft (Chattanooga)	150	800
Fire clay, white, medium	10	810
Limestone "sand," hard, dark, (gas sand).	13	823
Limestone "sand," brown, medium, (oil sand)	18	841
Shale, hard, blue, soft	5	846
Total depth		846

Log No. 979

J. N. Rogers, No. 4, lessor. Commenced: April 9, 1919. Completed: April 19, 1919. Production: Dry. Authority: The Ohio Oil Co.

Strata.

Mississippian System.	Thickness	Depth
Soil, red, soft	12	12
Shale (red rock), medium	36	48
Limestone, hard, white	88	136
Shale, hard and soft	480	616
Shale, red, soft, sandy	12	628
Devonian System.		
Shale, brown, medium (Chattanooga)	140	768
Fire clay, white, soft	12	780
Limestone "sand," light, hard, (all salt water)	32	812
Total depth		812

Log No. 980

W. Adams, No. 9, lessor. Commenced: Dec. 20, 1918. Completed: Jan. 22, 1919. Production: Commenced producing Jan. 27, 1919; production 48 hours after shot, 25 bbls. oil. Authority: The Ohio Oil Co.

Strata.

Mississippian System.	Thickness	Depth
Sand, pink	40	40
Limestone, hard, white	106	146
Shale, hard, white	457	603
Shale, hard, pink	12	615
Shale, hard, white	28	643
Devonian System.		
Shale, hard, brown (Chattanooga)	132	775
Shale, white, soft	8	783
Limestone ''sand,'' white	32	815
Limestone ''sand,'' pink	6	821
Limestone ''shale,'' white, medium	6	827
Total depth		827

Log No. 981

W. Adams, No. 10, lessor. Commenced: Feb. 3, 1919. Completed: Feb. 18, 1919. Production: Commenced producing Feb. 21, 1919; production 48 hours after shot, 25 bbls. oil. Authority: The Ohio Oil Co.

Strata.

Mississippian System.	Thickness	Depth
Sand, pink	20	20
Limestone, hard, white	110	130
Shale, hard, white	451	581
Shale, hard, pink	12	593
Shale, hard, white	28	621
Devonian System.		
Shale, brown (Chattanooga)	132	753
Shale, white, soft	8	761
Limestone ''sand,'' white	33	794
Shale, white	4	798
Total depth		798

Log No. 982

W. Adams, No. 12, lessor. Commenced: June 12, 1919. Completed: June 21, 1919. Production: Commenced producing July 4,
1919; production 48 hours after shot, 12 bbls. oil. Authority: The
Ohio Oil Co.

Strata.

Mississippian System.	Thickness	Depth
Soil, red, soft	10	10
Limestone, hard, white	90	100
Shale and slate, blue	365	465
Shale (red rock), soft	28	493
Fire clay, white, soft	17	510
Devonian System.		
Shale, black, medium (Chattanooga)	143	653
Fire clay, white, soft	8	661
Limestone "sand," dark, medium	31	692
Total depth		692

Log No. 983

Dana Lumber Co., No. 1, lessor. Commenced: Jan. 28, 1918.
Completed: April 8, 1918. Production: Dry. Authority: The Wood
Oil Co.

Strata.

Mississippian System.	Thickness	Depth
Sandstone, shale, etc.	563	563
Devonian System.		
Shale, black (Chattanooga)	160	723
Fire clay	12	735
Limestone "sand," blue and brown, (no oil)	4	739
Limestone "sand," lighter and finer	4	743
Limestone "sand," (fine white water)	11	754
Limestone "sand," light and fine, (skim of oil)	6	760
Limestone "sand," yellow, muddy, (oil smell)	9	769
Limestone "sand," (filled 300 ft. with salt water)	9	778
Limestone "sand," fine, red	5	783
Shale, soft, blue	4	787
Total depth		787

Log No. 984

G. B. Caudill, No. 4, lessor. Location: On Hatton Creek, 3 miles south of Stanton. Casinghead Alt.: 720 feet, Bar. Top of Big Lime: 1,230 feet, Bar. Authority: F. W. Caldwell.

Strata.

Mississippian System.	Thickness	Depth
Soil	10	10
Shale	95	105
Shale, red, sandy	21	126
Limestone, white	58	184
Devonian System.		
Shale, black (Chattanooga)	95	279
Limestone (Irvine)	11	290
Silurian System.		
Shale, green	70	360
Limestone "sand," green, (pay)	½	360½
Shale, blue	36½	397
Limestone	43	440
Total depth		440

NOTE—The Devonian (Irvine-Corniferous) limestone at 11 feet is very thin in this record.

Log No. 985

Miller, Prewitt, Goff, No. 20, lessors. Petroleum Exploration Co., lessee. Location: Headwaters of South Fork of Red River. Completed: March 3, 1918. Authority: Petroleum Exploration Co.

Strata.

Mississippian System.	Thickness	Depth
Soil	15	15
Shale, soft	82	97
Sandstone and shale	520	617
Devonian System.		
Shale, black (Chattanooga)	126	743
Fire clay	15	758
Limestone "sand," (oil)	5	763
Limestone "sand"	8	771
Total depth		771

NOTE—The two Miller, Prewitt and Goff Land Co. tracts of 1,300 acres and 3,000 acres, totaling 4,300 acres, are located in Powell, Estill and Lee Counties on the headwaters of South Fork and Big Sinking Creeks. The location of the wells of the following eleven records is on the head of the South Fork of Red River in Powell County.

Log No. 986

Miller, Prewitt, Goff, No. 21, lessors. Petroleum Exploration Co., lessee. Commenced: Feb. 27, 1918. Completed: March 9, 1918. Authority: The Petroleum Exploration Co.

Strata.

	Thickness	Depth
Pennsylvanian System.		
Sandstone and shale	240	240
Mississippian System.		
Limestone (Big Lime)	155	395
Sandstone and shale	495	890
Devonian System.		
Shale, black (Chattanooga)	145	1,035
Fire clay	13	1,048
Limestone ''sand''	16	1,064
Total depth		1,064

Log No. 987

Miller, Prewitt, Goff, No. 22, lessors. Petroleum Exploration Co., lessee. Commenced: Feb. 16, 1918. Completed: Feb. 26, 1918. Authority: The Petroleum Exploration Co.

Strata.

	Thickness	Depth
Pennsylvanian System.		
Sandstone and shale	95	95
Mississippian System.		
Limestone (Big Lime)	159	254
Sandstone and shale	456	710
Devonian System.		
Shale, black (Chattanooga)	130	840
Fire clay	15	855
Limestone ''sand''	13	868
Total depth		868

Log No. 988

Miller, Prewitt, Goff, No. 23, lessors. Petroleum Exploration Co., lessee. Commenced: Feb. 25, 1918. Completed: March 14, 1918. Authority: Petroleum Exploration Co.

Strata.

	Thickness	Depth
Pennsylvanian System.		
Sandstone and shale	260	260
Mississippian System.		
Limestone (Big Lime)	140	400
Sandstone and shale	455	855
Shale, red, sandy	15	870
Sandstone and shale	40	910
Devonian System.		
Shale, black (Chattanooga)	130	1,040
Fire clay	16½	1,056½
Limestone ''sand''	17½	1,074
Total depth		1,074

Log No. 989

Miller, Prewitt, Goff, No. 24, lessors. Petroleum Exploration Co., lessee. Commenced: March 5, 1918. Completed: March 16, 1918. Authority: Petroleum Exploration Co.

Strata.

	Thickness	Depth
Pennsylvanian System.		
Sandstone and shale (Pottsville)	150	150
Mississippian System.		
Limestone (Big Lime)	100	250
Sandstone and shale	465	715
Shale, red, sandy	6	721
Sandstone and shale	24	745
Devonian System.		
Shale, black (Chattanooga)	135	880
Fire clay	20	900
Limestone (cap rock)	3	903
Limestone ''sand,'' (oil)	13½	916½
Total depth		916½

Log No. 990

Prewitt, Miller, Goff, No. 41, lessors. Petroleum Exploration Co., lessee. Commenced: Dec. 6, 1918. Completed: Feb. 30, 1919. Authority: Petroleum Exploration Co.

Strata.

Pennsylvanian System.	Thickness	Depth
Soil	7	7
Sandstone and shale (Pottsville)	428	435

Mississippian System.		
Limestone	15	450
Limestone, sandstone and shale	485	935

Devonian, Silurian Systems.		
Shale, brown (Chattanooga)	130	1,065
Fire clay	13	1,078
Limestone (cap rock)	4	1,082
Limestone "sand," (oil 1,088, salt water 1,114)	121	1,203
Shale, hard, white	22	1,225
Shale, pink, limy	75	1,300
Shale, hard, white	30	1,330
Shale, pink, limy	35	1,365

Ordovician System.		
Limestone, gray	7	1,372
Shale, hard, white	8	1,380
Shale, red, limy	10	1,390
Limestone, gray	20	1,410
Limestone and sand	25	1,435
Shale, white	55	1,490
Shale, gray	42	1,532
Shale, blue	68	1,600
Limestone	50	1,650
Shale, gray	65	1,715
Shale, hard, white	20	1,735
Limestone	283	2,018
Limestone, black	17	2,035
Shale, black	75	2,110
Total depth		2,110

NOTE—The Devonian-Silurian contact is within the upper half of the 121 feet of limestone above 1,203 feet in depth.

Log No. 991

Prewitt, Miller, Goff, No. 42, lessors. Petroleum Exploration Co., lessee. Commenced: Jan. 8, 1920. Completed: Feb. 13, 1920. Production: Show for about 8 bbls. oil. Authority: Petroleum Exploration Co.

Strata.

Pennsylvanian System.	Thickness	Depth
Soil and mud	35	35
Sand	25	60
Shale, soft	50	110
Mississippian System.		
Limestone (Big Lime)	120	230
Shale, hard, green	30	260
Shale, soft, and sandstone	465	725
Devonian System.		
Shale, black (Chattanooga)	140	865
Fire clay	10	875
Shale, hard	10	885
Fire clay	5	890
Shale, hard	3	893
Limestone (cap rock) and "sand," (oil pay, good)	97	990
Shale	13	1,003
Limestone, gray	8	1,011
Total depth		1,011

Log No. 992

Prewitt, Miller, Goff, No. 45, lessors. The Petroleum Exploration Co., lessee. Commenced: March 2, 1920. Completed: March 26, 1920. Estimated production: First 24 hours, 2 bbls. oil.

Strata.

Pennsylvanian System.	Thickness	Depth
Shale, soft	55	55
Sand	20	75
Shale, soft	35	110
Mississippian System.		
Limestone (Big Lime)	100	210
Shale and sandstone	501	711

Devonian System.	Thickness	Depth
Shale, black (Chattanooga)	135	846
Fire clay	20	866
Limestone "sand," (salt water)	15	881
Limestone, black	9	890
Shale, hard	10	900
Limestone, black	6	906
Limestone (water)	11	917

Silurian System.		
Limestone, black	28	945
Limestone "sand," (oil)	33	978
Total depth		978

NOTE—The lowest oil "pay" in this well is undoubtedly in the Silurian.

Log No. 993

Prewitt, Miller, Goff, No. 46, lessors. Petroleum Exploration Co., lessee. Commenced: June 11, 1920. Completed: July 30, 1920. Production: The hole was dry.

Strata.

Pennsylvanian System.	Thickness	Depth
Soil	40	40
Limestone (Big Lime)	140	180
Shale, blue	481	661

Devonian System.		
Shale, brown (Chattanooga)	140	801
Shale, red, sandy	20	821
Fire clay	15	836
Limestone "sand," (oil show 840)	9	845
Limestone, shelly, (oil show 886)	41	886
Limestone	79	965
Total depth		965

Log No. 994

Prewitt, Miller, Goff, No. 47, lessors. The Petroleum Exploration Co., lessee. Commenced: June 11, 1920. Completed: June 30, 1920. Estimated production: First 24 hours, 10 bbls. oil.

Strata.

Mississippian System.	Thickness	Depth
Soil and shale, hard and black	40	40
Limestone (Little Lime)	30	70

Mississippian System.	Thickness	Depth
Limestone (Big Lime)	100	·170
Shale, green	34	204
Sandstone and shale	456	660

Devonian System.		
Shale, brown (Chattanooga)	152	812
Fire clay	15	827
Limestone ''sand,'' (water)	13	840
Limestone	52	892
Limestone, (oil) (pay)	25	917
Limestone	13	930
Total depth		930

Log No. 995

Prewitt, Miller, Goff, No. 50, lessors. Petroleum Exploration Co., lessee. Commenced: July 6, 1920. Completed: July 23, 1920. Estimated production: First 24 hours, 10 bbls. oil. Authority: Petroleum Exploration Co.

Strata.

Pennsylvanian System.	Thickness	Depth
Shale and sandstone (Pottsville)	115	115

Mississippian System.		
Limestone (Big Lime)	130	245
Sandstone and shale	479	724

Devonian System.		
Shale (Chattanooga)	140	864
Fire clay	20	884
Limestone ''sand,'' (pay 964-983)	109	993
Total depth		993

Log No. 996

Miller, Prewitt, Goff, No. 71, lessors. Petroleum Exploration Co., lessee. Completed: Oct. 18, 1917.

Strata.

Mississippian System.	Thickness	Depth
Limestone, sandstone and shale	610	610

Devonian System.		
Shale, black (Chattanooga)	130	740
Fire clay	14½	754½
Limestone ''sand''	14	768½
Total depth		768½

Log No. 997

Thomas McCoy, No. 2, lessor. The Wood Oil Co., lessee. Commenced: Aug. 3, 1917. Completed: Aug. 18, 1917. Estimated capacity: 15 bbls. oil.

Strata.

Mississippian and Devonian Systems	Thickness	Depth
To top of "sand" (Irvine)	530	530
Limestone "sand" (Irvine)	37	567
Total depth		567

Log No. 998

Thomas McCoy, No. 3, lessor| The Wood Oil Co., lessee. Commenced: October 8, 1917. Completed: October 27, 1917. Estimated capacity: 15 bbls. oil.

Strata.

Mississippian and Devonian Systems.	Thickness	Depth
To top of "sand" (Irvine)	685	685
Limestone "sand" (Irvine)	33	718
Total depth		718

PULASKI COUNTY.

Production: Oil and Gas. Producing Sands: Pottsville (Pennsylvanian); Big Lime (Mississippian); Corniferous (Devonian); Niagaran (Silurian); Upper Sunnybrook, Maysville age (Ordovican)

Log No. 999

Newell, No. 1, lessor. Somerset Petroleum Corp., lessee. Location: Fishing Creek, 6 miles N. W. of Somerset. Driller: George Cox. Production: 1 barrel green oil. Authority: W. A. White, General Manager.

Strata.

Devonian System.	Thickness	Depth
Soil	10	10
Limestone, hard	18	28
Silurian System.		
Fire clay	10	38
Limestone, oil sand	2	40
Shale, brown	10	50
Limestone	10	60
Limestone and sand, (salt water)	5	65
Limestone, gray, (6¼ in. casing at 70)	5	70
Limestone	11	81
Oil sand	10	91
Limestone	3	94
Total depth		94

Log No. 1000

Newell, No. 1, lessor. Somerset Petroleum Corp., lessee. Location: Fishing Creek, 6 miles N. W. of Somerset. Driller: George Cox. Production: 1 barrel green oil. Authority: W. A. White, General Manager.

Strata.

Devonian System.	Thickness	Depth
Soil	9	9
Limestone, hard	18	27
Silurian System.		
Fire clay	10	37
Oil sand	2	39
Shale, brown	11	50
Limestone	5	55
Limestone, gray, and sand, (6¼" casing 65)	10	65
Limestone	15	80
Oil sand	10	90
Limestone	12	102
Total depth		102

Log No. 1001

A. J. Spaugh, No. 1, lessor. Somerset Petroleum Corp., lessee. Location: Fishing Creek, 6 miles N. W. of Somerset. Driller: George Cox. Production: 1 barrel green oil. Authority: W. A. White, General Manager.

Strata.

Devonian System.	Thickness	Depth
Soil	10	10
Limestone, hard	20	30
Silurian System.		
Fire clay	10	40
Oil sand	2	42
Shale, brown, (6¼ in. casing at 65)	23	65
Limestone	15	80
Oil sand, (first)	10	90
Limestone	32	122
Oil sand, (second)	6	128
Limestone	8	136
Oil and sand, (third)	16	152
Limestone	48	200
Total depth		200

Log No. 1002

Dun Bogle, No. 1, lessor. Location: 2½ miles southwest of Somerset. Completed: Nov. 28, 1921. Casing head elevation: 859 feet. Authority: Mr. Bee Whitis, Box 510, Somerset, Ky.

Strata.

Mississippian System.	Thickness	Depth
Soil	4	4
Limestone, hard, gray	46	50
Limestone, brown	40	90
Limestone, dark	28	118
Sandstone, brown grit, (some gas)	12	130
Limestone, gray, hard	35	165
Limestone, broken	35	200
Shale, blue	130	330
Devonian System.		
Shale, black (Chattanooga)	51	381
Limestone, (Irvine "sand"), (some oil)	8	389
Shale, blue, soft	19	408
Silurian System.		
Limestone, soft, blue, shaly	16	424
Total depth		424

ROCKCASTLE COUNTY.

Production: Oil and Gas shows. Producing Sands: Big Lime (Chester-Mississippian); Corniferous (Devonian).

Log No. 1003

Albert Albright, No. 1, lessor. New Domain Oil & Gas Co., lessee. Completed: Aug. 19, 1904. Production: Dry. Authority: New Domain Oil & Gas Co.

Strata.

Mississippian System.	Thickness	Depth
Gravel	15	15
Limestone, blue, open	91	106
Limestone, blue, hard (New Providence)	85	191
Devonian System.		
Shale, black, soft (Chattanooga)	50	241
Fire clay, soft, white	10	251
Limestone "sand" (Ragland), hard	20	271
Limestone, white, hard	220	491
Limestone, gray, hard	112	603
Total depth		603

NOTE—The Devonian-Silurian contact is within the upper quarter of the 220 feet of limestone above 491 feet in depth. The Silurian-Ordovician contact is toward the middle of the lower part of the 112 feet of limestone above 603 feet in depth.

Log No. 1004

William Hepinger, No. 1, lessor. New Domain Oil & Gas Co., lessee. Completed: Sept. 6, 1904. Production: Dry. Authority: New Domain Oil & Gas Co.

Strata.

Mississippian System.	Thickness	Depth
Gravel	5	5
Limestone, white, hard	20	25
Shale, blue, soft	50	75
Limestone, gray, hard	70	145
Limestone shells	155	300
Devonian System.		
Shale, black, soft (Chattanooga)	84	384
Limestone shells and shale, soft	126	510
Limestone, hard	97	607
Total depth		607

NOTE—The Devonian-Silurian contact is within the upper half of the 126 feet of limestone above 510 feet.

Log No. 1005

J. E. Tate & Co., No. 1, lessors. Completed: Aug. 8, 1904. Production: Dry; easing pulled, well plugged and abandoned. Authority: New Domain Oil & Gas Co.

Strata.

Mississippian System.	Thickness	Depth
Soil and gravel, yellow	10	10
Limestone, white, blue, hard	60	70
Shale, blue, soft	35	105
Limestone, gray, hard	60	165
Shale, blue, soft	105	270
Limestone, white, hard	8	278
Shale, blue, green, soft (New Providence)	82	360
Devonian System.		
Shale, black, soft, hard (Chattanooga)	94	454
Shale, blue, soft	45	499
Shale, pink, soft	15	514
Limestone shells, blue, white, soft	45	559
Silurian System.		
Limestone, gray, white, very hard	42	601
Sand, gray, very hard	12	613
Shale, blue, soft	12	625
Total depth		625

NOTE—The varicolored shales below the Chattanooga are probably in reality a part of same.

Log No. 1006

David Hysinger, No. 1, lessor.   New Domain Oil & Gas Co., lessee.
Completed: Oct. 1, 1904.   Production: Dry.   Authority: New Domain
Oil & Gas Co.

Strata.

Mississippian System.	Thickness	Depth
Sand, gravel and mud	65	65
Limestone, hard	63	128
Shale, soft, sandy	117	245

Devonian System.		
Shale, black, hard (Chattanooga)	70	315
Shale, limestone and shells	81	396
Limestone and shale, soft	89	485
Limestone, gray, hard	112	597
Total depth		597

NOTE—The Devonian-Silurian contact occurs toward the base of
the 81 feet of shale and limestone above 396 feet in depth.

Log No. 1007

C. L. Lear, No. 1, lessor.   Completed: Sept. 20, 1904.   Production: Dry.   Authority: New Domain Oil & Gas Co.

Strata.

Mississippian System.	Thickness	Depth
Soil and quicksand	10	10
Limestone, blue, hard	60	70
Shale and shells, soft	170	240

Devonian System.		
Shale, brown (Chattanooga)	93	333
Shale, blue, soft	65	398
Limestone, white, hard	100	498
Total depth		498

NOTE—The Devonian-Silurian contact occurs toward the base of
the upper half of the last 100 feet of this record.   The 65 feet of blue
shale is probably partly at least Chattanooga.

Log No. 1008

B. S. Devault, No. 1, lessor. New Domain Oil & Gas Co., lessee. Completed: Sept. 23, 1904. Production: Dry. Authority: New Domain Oil & Gas Co.

Strata.

Mississippian System.	Thickness	Depth
Soil, brown, soft	5	5
Limestone, white, hard	75	80
Shale, blue, soft	250	330
Devonian System.		
Shale, brown, hard (Chattanooga)	90	420
Shale, white, soft	12	432
Limestone, white, very hard	179	611
Total depth		611

NOTE—The Devonian-Silurian contact occurs toward the base of the upper one-third of the last 179 feet of limestone of this record.

ROWAN COUNTY.

Production: Oil. Producing Sand: Ragland (Corniferous) (Devonian) Niagaran (Silurian).

Log No. 1009

J. E. Johnson, No. 1, lessor. Location: 4½ miles northwest of Morehead. Commenced: January, 1920. Completed: February, 1920. Production: The hole was dry, and the casing was pulled and plugged. Authority: Mohney Bros. and Brown, drillers.

Strata.

Devonian System.	Thickness	Depth
Gravel	25	25
Shale, blue, and shale (Chattanooga)	55	80
Shale, blue (Chattanooga)	70	150
Shale (Chattanooga)	10	160
Shale, blue, and shale (Chattanooga)	28	188
Limestone "sand," (Corniferous)	5	193
Limestone, white (Corniferous)	32	225
Silurian System.		
Shale, blue	40	265
Shale, hard, red	30	295
Shale, hard, blue	55	350
Shale, hard, red	25	375
Limestone and shell	25	400
Shale, hard, red	10	410
Shale, hard, white	20	430
Limestone, white	20	450
Ordovician System.		
Limestone and shale, hard	150	600
Total depth		600

Log No. 1010

W. J. Fletcher, No. 1, lessor. Location: Near Morehead. Completed: April 27, 1904. Production: Well was dry. Water at 30 and 975 feet. Authority: New Domain Oil & Gas Co.

Strata.

Mississippian System.	Thickness	Depth
Sand and gravel, brown, soft	13	13
Shale, blue, soft	10	23
Freestone, blue, hard	167	190
Limestone, white, hard	50	240
Shale, hard, white, soft	45	285
Limestone, white, hard	105	390
Shale, white, soft	110	500
Shale, brown, soft	15	515
Shale, hard, white, soft	65	580
Devonian System.		
Shale, brown, soft (Chattanooga)	235	815
Fire clay, white, soft	10	825
Limestone "sand" (Ragland in part), white, hard	105	930
Shale, white, soft	30	960
Limestone "sand," white, hard, (gas 975) ..	15	975
Shale, hard, white, soft	35	1,010
Shale, red, hard, limy	50	1,060
Ordovician System.		
Limestone shells and shale, hard, white, soft	150	1,210
Limestone, white, very hard	291	1,501
Total depth		1,501

NOTE—The Devonian-Silurian contact is within the first half of the 105 feet of limestone above 930 feet in depth.

RUSSELL COUNTY.

Production: **Oil** and **Gas.** Producing **Sand: Sunnybrook** and **Trenton** (**Ordovician**).

Log No. 1011

E. G. Wilson, No. 1, lessor. Completed: Sept. 22, 1904. Production: Dry. Authority: New Domain Oil & Gas Co.

Strata.

Mississippian System.	Thickness	Depth
Limestone, light, hard	251	251
Devonian System.		
Shale, black, soft (Chattanooga)	35	286

Ordovician System.

	Thickness	Depth
Limestone ''sand'' light, hard	641	927
Limestone ''sand,'' dark, hard	18	945
Limestone ''sand,'' hard	5	950
Limestone ''sand,'' light, hard	10	960
Limestone ''sand,'' dark, soft	3	963
Limestone and shale, hard, dark, soft	4	967
Total depth		967

Log No. 1012

Simco Popplewell, No. 1, lessor. New Domain Oil & Gas Co., lessee. Completed: Aug. 23, 1904. Production· Well was dry: casing pulled, well plugged and abandoned. Authority: New Domain Oil & Gas Co.

Strata.

Ordovician System.

	Thickness	Depth
Limestone, gray, hard	175	175
Limestone ''sand,'' white, hard	10	185
Limestone, gray, medium	440	625
Limestone ''sand,'' gray, soft	5	630
Limestone, gray, soft	37	667
Limestone ''sand,'' white, hard	10	677
Limestone, gray, hard	458	1,135
Total depth		1,135

Log No. 1013

J. C. Wilson, No. 1, lessor. Location: Near Steubenville. Completed: Aug. 1, 1904. Production: Well was dry. Authority: New Domain Oil & Gas Co.

Strata.

Ordivician System.

	Thickness	Depth
Clay	5	5
Limestone, white, hard	40	45
Limestone, blue, soft	400	445
Sand, red, soft	5	450
Limestone, blue, soft	200	650
Limestone, blue, hard, (oil show 676)	26	676
Limestone ''sand,'' blue, gray, hard	16	692
Limestone, white, hard	106	798
Shale, caving, soft	2	800
Limestone, white, hard	51½	851
Total depth		851½

Log No. 1014

Kyle, No. 1, lessor.   T. A. Sheridan, lessee.   Location: Pumpkin Creek.   Commenced: In winter of 1920-21.   Elevation: About 640 feet A. T.   Starts 2 feet below Chattanooga Shale in the Ordovician Richmond Shale.

Strata.

Ordovician System.	Thickness	Depth
Limestone, blue	598	598
Shale (pencil cave)	3	601
Limestone, coarse, light brown	39	640
Limestone, coarse, light brown	16	656
Limestone, coffee-colored, harder	12	668
Shale, blue-black, limy, coarser	12	680
Limestone, coffee-colored, coarse	8	688
Limestone, dark blue	36	724
Limestone, dark brown	16	740
Limestone, lighter blue, (gas)	4	744
Limestone, light coffee-colored	4	748
Total depth		748

NOTE—Well unfinished, Jan 7, 1921.

Creelsboro Wells.   McMeade Co., lessee.   Drilled in 1920. Authority: L. Beckner.

Log No. 1015

No. 1.

Starts about 60 feet below base of Chattanooga Shale in the base of the Richmond or top of Maysville.   Limestone all the way.   Oil at 245 feet, dark blue to brown limestone to 275 feet in depth.   This well produces considerable gas.   Elevation, 597. Approx.

Log No. 1016

No. 2.

Same as above. Elevation, 595. Approx. Got oil at 246.

Log No. 1017

No. 3.

Same as above.   Elevation, 590.   Approx.   Got oil at 247 feet flowing.

Log No. **1018**

.

No. 4.

Same as above. Elevation, 485, Barometric. Got oil at 255 feet.

No. 2 was tubed for pumping and nothing about it could be learned.

No. 3 is flowing a small trickle of light gassy oil into a trough, about ½ bbl. a day. It has considerable gas.

No. 4 has oil about 200 feet down from which gas is rising, not as good as No. 1.

Log No. **1019**

Bacon No. 1. (Called Creelsboro, No. 5.) Elevation, 610. Approx. Got no oil or sand at same horizon as other wells. Got no oil at 605 feet in depth.

SIMPSON COUNTY.

Production: **Oil and Gas.** Producing Sands: "**Shallow**" (**St. Louis age**) (**Mississippian**); **Corniferous** (**Devonian**); "**Deep**" (**Niagaran age**) (**Silurian**).

Log No. **1020**

Henry Reeder, No. 2, lessor. Tidewater Oil Co., Norfolk, Va., lessee. Location: 4 miles northeast of Franklin, and east of Drakes Creek. Production: Considerable gas, and 25 bbls. oil, natural flow to tank 20 feet above casing head. Authority: Barney Calvert.

	Thickness	Depth
Strata.		
Mississippian System.		
Soil	10	10
Limestone	515	525
Devonian System.		
Shale, black (Chattanooga)	50	575
Sand, (gas)	4	579
Limestone (cap rock)	2	581
Limestone "sand," brown (oil)	12	593
Limestone, light	2	595
Limestone, blue	7	602
Total depth		602

Log No. 1021

Tom Lewis, No. 1, lessor. Location: Southwest of Rolands Mill in Drakes Creek bottom, 4½ miles northeast of Franklin. Production: Considerable sulphur gas, which flowed open for over a year.

Strata.

Mississippian System.	Thickness	Depth
Soil	18	18
Limestone, (first water)	107	125
Limestone, (gas)	245	370
Limestone	130	500
Shale, green	40	540

Devonian System.		
Shale, black (Chattanooga)	50	590
Limestone	10	600
Total depth		600

Log No. 1022

W. H. Lewis, No. 11, lessor. Location: 2 miles north of Franklin, left of Bowling Green Road. Production: This well was dry.

Strata.

Mississippian and Devonian Systems.	Thickness	Depth
Limestone and shale, (gas)	1,100	1,100
Total depth		1,100

Log No. 1023

Boyd, No. 1, lessor. McGlothlin, Moore & Co., lessees. Location: 3½ miles south of Franklin. Completed: June 25, 1920. 205 feet of casing set. Authority: Walter Moore.

Strata.

Mississippian System.	Thickness	Depth
Surface and limestone	503	503

Devonian System.		
Shale, black (Chattanooga)	57	560
Limestone (cap rock)	10	570
Limestone	48	618

Silurian System.		
Limestone "sand," (oil) (first pay)	32	650
Total depth		650

Log No. 1024

Boyd, No. 2, lessor. McGlothlin, Moore & Co, lessees. Location: 3½ miles south of Franklin. Completed: Aug. 1, 1920. 205 feet casing set. Authority: Walter Moore.

Strata.

Mississippian System.	Thickness	Depth
Surface and limestone	500	500
Devonian System.		
Shale, black (Chattanooga)	61	561
Limestone (cap rock)	10	571
Limestone	29	600
Limestone ''sand,'' (oil)	37	637
Total depth		637

Log No. 1025

W. M. McGlothlin, No. 2, lessor. Blue Goose Oil Co., lessee. Location: 3¾ miles south of Franklin. Completed: July 20, 1920. 190 feet casing set. Authority: Walter Moore.

Strata.

Mississippian System.	Thickness	Depth
Surface and limestone	492	492
Devonian System.		
Shale, black (Chattanooga)	58	550
Limestone (cap rock)	9	559
Limestone ''sand,'' (oil at 600)	68	627
Total depth		627

NOTE—The Devonian-Silurian contact occurs within the last **68** feet of this record.

Log No. 1026

J. E. Hagan, No. 1, lessor. Location: 7½ miles east of Franklin, off Gold City Road. Completed: Sept. 29, 1919. Casing set at **149** feet. Water struck at 78 feet. Authority: B. W. Lightburn.

Strata.

Mississippian System.	Thickness	Depth
Red mud and boulders	35	35
Limestone, white, hard	104	139
Shale, black, soft, limy	55	194

Mississippian System.	Thickness	Depth
Lime rock, dark	20	214
Shale, dark, soft, limy	10	224
Limestone, gray, hard	20	244
Limestone, black, soft	5	249
Limestone, white, hard	30	279
Limestone, gray, hard	55	334
Limestone, gray and white	20	354
Sand and limestone, grayish	21	375
Shale, green, hard (New Providence)	30	405
Devonian System.		
Shale, black (Chattanooga)	50	455
Limestone and ''sand''	75	530
Limestone ''sand,'' dark, (oil)	29	559
Total depth		559

NOTE—The Devonian-Silurian contact is within the 75 feet of limestone above 530 feet in depth. The oil ''sand'' in the last 29 feet is therefore Silurian.

Log No. 1027

J. E. Hagan, No. 2, lessor. Location: 7½ miles east of Franklin, off Gold City Road. Authority: B. W. Lightburn.

Strata.

Mississippian System.	Thickness	Depth
Clay boulders	50	50
Limestone, gray	20	70
Oil sand	5	75
Limestone, black	115	190
Limestone, gray	35	225
Limestone, flinty	45	270
Limestone, flinty	105	375
Shale, green (New Providence)	40	415
Devonian System.		
Shale, black (Chattanooga)	50	465
Limestone, blue	55	520
Silurian System.		
Limestone, gray	81	601
Limestone ''sand,'' (oil)	9	610
Limestone, white	15	625
Limestone, blue	10	635
Limestone, white	15	650
Total depth		650

Log No. 1028

J. E. Hagan, No. 3, lessor. Location: 7½ miles east of Franklin, off Gold City Road. Authority: B. W. Lightburn.

Strata.

Mississippian System.	Thickness	Depth
Soil	35	35
Limestone, blue	15	50
Limestone, brown	12	62
Limestone ''sand,'' (oil)	18	80
Limestone, blue	105	185
Limestone, white, flint	80	265
Limestone, blue	10	275
Limestone, white	105	380
Shale, green	35	415
Devonian System.		
Shale, black (Chattanooga)	50	465
Limestone, blue	15	480
Limestone, white	10	490
Silurian System.		
Limestone, brown	10	590
Limestone, brown, flint	96	596
Limestone ''sand,'' (oil)	10	606
Total depth		606

Log No. 1029

J. E. Hagan, No. 4, lessor. Corinne Oil & Gas Co., Joplin, Mo., lessee. Location: 7½ miles east of Franklin, off Gold City Road. Authority: B. W. Lightburn, Field Manager.

Strata.

Mississippian System.	Thickness	Depth
Clay and boulders	7	7
Limestone, soft	43	50
Limestone, brown	25	75
Limestone, white	25	100
Limestone, blue	105	205
Limestone, white	140	345
Limestone, hard, yellow	5	350
Shale, green (New Providence)	60	410

Devonian System.	Thickness	Depth
Shale, black (Chattanooga)	50	460
Limestone, blue	65	525
Limestone, dark blue	65	590
Limestone ''sand,'' soft, (oil and sulphur water)	10	600
Total depth		600

NOTE— The Devonian-Silurian contact occurs about midway in the 65 feet of limestone above 525 feet. The oil is therefore Silurian.

Log No. 1030

Fowler Mitchell, No. 1, lessor. The Florida-Kentucky Oil Co., lessee. Location: 7½ miles north of Franklin, I. & N. Pike, 100 yards from the Warren-Simpson County line. Completed: Feb. 15, 1920. Authority: E. L. Reep.

Strata.

Mississippian System.	Thickness	Depth
Limestone	863	863

Devonian System.		
Shale, black (Chattanooga)	56	919
Limestone (cap rock), (oil)	12	931
Limestone ''sand,'' (pay)	9	940
Limestone ''sand,'' hard	4	944
Limestone ''sand,'' dark, (second pay)	9	953

Silurian System.		
Limestone ''sand,'' white	67	1,020
Total depth		1,020

Log No. 1031

Anderson, No. 1, lessor. Lick Creek Oil & Gas Co., lessee. Location: 3 miles northeast of Franklin. Production: Gas. Authority: Brady Perdue.

Strata.

Mississippian System.	Thickness	Depth
Limestone, white and gray	490	490
Limestone ''sand,'' brown and red	58	548

Devonian System.		
Shale, black (Chattanooga)	60	608
Limestone (cap rock)	4	612
Limestone	122	734
Total depth		734

Log No. 1032

Chas. Anglea, No. 1, lessor. Location: 1½ miles southeast of Franklin. Commenced: Aug. 19, 1919. Completed: Nov. 12, 1919. Production: Estimated production, 25,000 cu. ft. gas. Authority: Brady Perdue.

Strata.

Mississippian System.	Thickness	Depth
Soil	23	23
Limestone	382	405
Shale, green (New Providence)	40	445
Devonian System.		
Shale, black (Chattanooga)	47	492
Limestone (cap rock)	3	495
Limestone "sand," (gas)	7	502
Limestone "sand"	358	860
Total depth		860

NOTE—The contract between the Silurian and Devonian Systems occurs in the limestone 358 feet thick.

Log No. 1033

Ward Brown, No. 1, lessor. Location:——— Commenced: Nov. 10, 1919. Completed: Nov. 20, 1919. Sulphur water at 70 feet; show of oil and a little gas at 140 feet.

Strata.

Mississippian System.	Thickness	Depth
Limestone and shale	295	295
Devonian System.		
Shale, black (Chattanooga)	56	351
Limestone, very white	4	355
Limestone, very white dark brown	4	359
Limestone, dark brown	4	363
Limestone, lead color	16	379
Limestone, light brown	17	396
Silurian System.		
Limestone, light brown and lead color	4	400
Limestone, lead color	12	412
Limestone, light brown, fine, very hard	8	420
Limestone, gray	4	424
Limestone, gray	4	428
Total depth		428

Log No. 1034

Chas. Butt, No. 1, lessor.   Location: 4 miles southwest of Franklin.   Commenced: May 12, 1920.

Strata.

Mississippian System.	Thickness	Depth
Limestone and shale	438	438
Devonian System.		
Shale, black (Chattanooga)	58	496
Limestone	62	558
Limestone ''sand''	30	588
Limestone	6	594
Total depth		594

Water at 70 and 90 feet.

NOTE—The Devonian-Silurian contact occurs midway within the 62 feet of limestone above 558 feet in depth.

Log No. 1035

Chas. F. Butt, No. 3, lessor.   Location: 4 miles southwest of Franklin.   Commenced: July 24, 1920.   Completed: Aug. 18, 1920.   Authority: J. H. Buettner.

Strata.

Mississippian System.	Thickness	Depth
Limestone and shale	460	460
Devonian System.		
Shale, black (Chattanooga)	51	511
Limestone, salt and pepper	4	515
Limestone, gray	12	527
Limestone, sandy	24	551
Silurian System.		
Limestone, gray	8	559
Limestone ''sand,'' light gray	8	567
Limestone ''sand,'' light brown	8	575
Limestone, light brown, sandy	8	583
Limestone, blue	4	587
Limestone	17	604
Total depth		604

Water at 90 feet.

Log No. 1036

Dunn, No. 1, lessor. Location: 5 miles south of Franklin. Commenced: June 1, 1920. Completed: June 28, 1920. Authority: J. H. Buettner.

Strata.

Mississippian System.	Thickness	Depth
Limestone and shale	460	460

Devonian System.

Shale, black (Chattanooga)	54	514
Limestone	62	576
Limestone, 'sand''	33	609
Total depth		609

Water at 75 and 95 feet.

NOTE—The Devonian-Silurian contact occurs about midway within the 62 feet of limestone above 576 feet in depth.

Log No. 1037

O. Harris, No. 1, lessor. Location: 1½ miles southeast of Franklin. Commenced: July 8, 1919. Completed: Aug. 12, 1919. Authority: Brady Perdue.

Strata.

Mississippian System.	Thickness	Depth
Soil and gravel	31	31
Limestone	399	430
Shale, green (New Providence)	48	478

Devonian System.

Shale, brown (Chattanooga)	42	520
Limestone (cap rock)	10	530
Limestone ....:.....................	163	693
Total depth		693

The casing was pulled and the well abandoned.

Log No. 1038

Hughes, No. 1, lessor. Moore & Enders, lessees. Location: 4 miles southeast of Franklin, ¼ mile east of I. & N. Railroad. Commenced: July 23, 1920. Completed: Aug. 19, 1920. Shot: Aug. 23, 1920, 80 quarts. Authority: Walter Moore.

Strata.

Mississippian System.	Thickness	Depth
Limestone and shale	460	460
Devonian System.		
Shale, black (Chattanooga)	55	515
Limestone	5	520
Limestone ''sand''	20	540
Limestone, blue	50	590
Total depth		590

Log No. 1039

Tom Lewis, No. 1, lessor. Prestonsburg Oil & Gas Co., lessee. Location: 4½ miles northeast of Franklin, and 1 mile from Reeder pool. Production: Small sulphur gasser. Authority: Tom Lewis.

Strata.

Mississippian System.	Thickness	Depth
Soil	18	18
Limestone, (fresh water 125)	107	125
Limestone	375	500
Shale, green (New Providence)	40	540
Devonian System.		
Shale, black (Chattanooga)	50	590
Limestone	10 (plus)	600 (plus)
Total depth		600 (plus)

Log No. 1040

Meador, No. 1, lessor. Lick Creek Oil & Gas Co., lessee. Location: 7 miles east of Franklin. Production: Fine gas well. Authority: Brady Perdue.

Strata.

Mississippian System.	Thickness	Depth
Limestone, variable	225	225
Limestone ''sand,'' (gas)	4	229
Limestone ''sand,'' soft	6	235

Mississippian System.
	Thickness	Depth
Limestone, white	45	280
Limestone, pink	5	285
Limestone, white	45	330
Limestone, blue, and shells (New Providence)	72	402

Devonian System.
	Thickness	Depth
Shale, black, Chattanooga)	48	450
Fire clay	13	463
Limestone (cap rock)	2	465
Limestone, variable	220	685
Total depth		685

NOTE—The Devonian-Silurian contact is about 20 feet down in the last 220 feet of limestone. This well finished in the Silurian, or perhaps the top of the Ordovician.

Log No. 1041

Pearson, No. 1, lessor. Location: 6 miles northeast of Franklin, on Lick Creek. Authority: Brady Perdue.

Strata.

Mississippian System.
	Thickness	Depth
Limestone, (oil 100-108)	135	135
Limestone rock	163	298
Limestone "sand," (water)	7	305
Limestone "sand," dark, (water)	6	311
Limestone, blue and hard	79	390
Limestone, white	20	410
Limestone, blue and soft	5	415
Limestone, blue and hard	10	425
Limestone, brown and soft, (New Providence)	10	435
Limestone, white and green, (New Providence)	12	447

Devonian System.
	Thickness	Depth
Shale, black (Chattanooga), (some oil)	61	508
Limestone (cap rock) blue, (strong showing of oil)	6	514
Limestone "sand," (some showing of oil)	7	521
Limestone "sand," light, (oil)	2	523
Limestone, light gray	37	560
Limestone, gray	2	562
Limestone (cap rock), dark gray	12	574
Shale, blue, some gumbo	21	595
Limestone, dark blue	6	601

Devonian System.	Thickness	Depth
Limestone, light blue	4	605
Limestone, dark gray	10	615
Limestone, gray	12	627
Limestone, dark gray	6	633
Total depth		633

NOTE—The Devonian-Silurian contact is within the 37 feet of limestone above 560 feet in depth.

Log No. 1042

Pearson, No. 4, lessor. Location: 6 miles east of Franklin, on Lick Creek. Authority: Brady Perdue.

Strata.

Mississippian System.	Thickness	Depth
Soil	14	14
Limestone, variable	461	475
Shale, green	5	480
Devonian System.		
Shale, black (Chattanooga)	64	544
Limestone "sand," (oil strong showing)'	9	553
Limestone, variable	72	625
Limestone "sand," (oil, strong showing)....	10	635
Limestone	60	695
Total depth		695

NOTE—The Devonian-Silurian contact is within the 72 feet above 625 feet in depth.

Log No. 1043

Pearson, No. 6, lessor. Location: 8 miles of Franklin, on Lick Creek. Authority: Brady Perdue.

Strata.

Mississippian System.	Thickness	Depth
Soil	31	31
Limestone, blue	97	128
Limestone, blue	59	187
Limestone, white and fine	18	205
Limestone "sand," (oil, small showing)	5	210
Limestone, gray, blue and white	324	534

THE SEBREE SANDSTONE

This oil sand, productive in Union and Henderson counties, Ky., is here shown in the type locality between Sebree and the Green River, north of the Steamport Ferry Road. This is an ideal "sand" course, thick and mediumly cemented.

Devonian System.	Thickness	Depth
Shale, black (Chattanooga)	61	595
Limestone (cap rock)	5	600
Limestone "sand," (oil)	15	615
Limestone, white	60	675
Limestone "sand," (oil, good show)	15	690
Shale	7	697
Total depth		697

NOTE—The Devonian-Silurian contact is in the upper half of the 60 feet of limestone above 675 feet in depth.

Log No. 1044

O. M. Stringer, No. 2, lessor. Location: About 7 miles west of Franklin, on Sulphur Fork Creek. Commenced: Nov. 22, 1919. Completed: Dec. 22, 1919. Authority: Irvin J. Brown Oil Co.

Strata.

Mississippian System.	Thickness	Depth
Limestone and shale	320	320

Devonian System.		
Shale, black (Chattanooga)	56	376
Limestone, pepper and salt brown	8	384
Limestone, light gray	8	392
Limestone, lead color	8	400
Limestone, muddy brown	8	408
Shale, hard, lead colored	4	412
Limestone, light brown	16	428
Limestone, gray	6½	434½
Total depth		434½

Log No. 1045

O. M. Stringer, No. 6, lessor. Location: 8½ miles east of Franklin, on Middle Fork Creek. Commenced: April 6, 1920. Completed: April 18, 1920. Authority: Glen Neaville.

Strata.

Mississippian System.	Thickness	Depth
Soil	31	31
Limestone	280	311

Devonian System.

	Thickness	Depth
Shale, black (Chattanooga)	50	361
Limestone, pepper and salt	8	369
Limestone, muddy and gray	20	389
Limestone ''sand,'' muddy and brown	8	397

Silurian System.

Limestone ''sand,'' light brown	12	409
Limestone ''sand,'' dark brown, (oil show) ..	16	425
Limestone ''sand,'' lead color	4	429
Limestone, light gray and white	4	433
Limestone ''sand,'' brown sugar, (oil show)..	16	449
Limestone ''sand,'' brown and gray	8	457
Total depth		457

Log No. 1046

O. M. Stringer, No. 7, lessor. Location: 8½ miles east of Franklin, on Middle Fork Creek. Commenced: April 15, 1920. Completed: April 23, 1920. Authority: Glen Neaville.

Strata.

Mississippian System.

	Thickness	Depth
Limestone and shale	220	220

Devonian System.

Shale, black (Chattanooga)	55	275
Limestone, pepper and salt	12	287
Limestone, gray and brown	8	295
Shale, hard, lead color	4	299
Limestone, gray, brown and dark	4	303

Silurian System.

Limestone ''sand,'' brown, (oil show)	12	315
Limestone ''sand,'' brown and fine	4	319
Shale, hard, muddy, lead color	8	327
Shale, dark and clean	4	331
Limestone, gray and brown	4	335
Limestone, dingy brown	16	351
Limestone, muddy	4	355
Total depth		355

Water at 25 and 62 feet.

Log No. 1047

O. M. Stringer, No. 8, lessor.   Location: 8½ miles east of Franklin, on Middle Fork Creek.   Commenced: April 23, 1920.   Completed: May 20, 1920.   Authority: Glen Neaville.

Strata.

Mississippian System.	Thickness	Depth
Limestone and shale	230	230
Devonian System.		
Shale, black (Chattanooga)	47	287
Limestone, pepper and salt	4	291
Limestone, coarse and brown	8	299
Limestone, white and fine	12	311
Limestone, gray and brown	12	323
Silurian System.		
Shale, hard, muddy	8	331
Shale, muddy and brown	4	335
Sand, dark brown, (rainbow)	24	359
Shale, hard, light colored	4	363
Limestone, light brown, coarse	7	370
Limestone, dark	8	378
Shale, hard, dark	2	380
Total depth		380

Log No. 1048

O. M. Stringer, No. 9, lessor.   Location: 8½ miles east of Franklin, on Middle Fork Creek.   Commenced: April 30, 1920.   Completed: May 10, 1920.   Authority: Glen Neaville.

Strata.

Mississippian System.	Thickness	Depth
Limestone and shale	224	224
Devonian System.		
Shale, black (Chattanooga)	53	277
Limestone, pepper and salt	4	281
Limestone, light gray, blue, (gas)	4	285
Limestone, pepper and salt	8	293
Shale, hard and muddy	8	301
Limestone, light gray	4	305
Limestone, grayish brown	4	309

Silurian System.   Thickness Depth

    Limestone ''sand,'' brown and coarse   8   317
    Limestone ''sand,'' fine   4   321
    Shale, hard and muddy   12   333
    Limestone, fine and brown   16   349
    Limestone ''sand,'' (oil show)   15   364
        Total depth   364

Log No. 1049

    Stringer Bros., No. 4, lessors. Commenced: Dec. 29, 1919. Completed: Jan. 14, 1920. Authority: Irvin J. Brown Oil Co.

    Strata.

Mississippian System.   Thickness Depth

    Limestone and shale   340   340

Devonian System.

    Shale, black (Chattanooga)   55   395
    Limestone, pepper and salt   8   403
    Limestone, gray   4   407
    Limestone, muddy   8   415
    Limestone, medium, dark brown   8   423
    Limestone, gray   431

Silurian System.

    Limestone, whitish brown   435
    Limestone, (rainbow)   439
    Limestone, muddy gray   8   447
    Limestone, dark brown   4   451
    Limestone, light   4   455
    Limestone, little darker, (fair show of oil) ...   459
    Limestone, brown and gray   4   463
    Limestone, brown and gray   8   471
        Total depth   471

Log No. 1050

Stringer Bros., No. 5, lessors. Commenced: Dec. 30, 1919. Completed: Jan. 13, 1920. Authority: Irvin J. Brown Oil Co.

Strata.

Mississippian System.	Thickness	Depth
Limestone and shale	338	338
Devonian System.		
Shale, black (Chattanooga)	55	393
Limestone, pepper and salt	4	397
Limestone, black and gray	4	401
Limestone, muddy	8	409
Limestone, muddy, gray	4	413
Limestone, whitish	4	417
Limestone, light, (oil show)	8	425
Silurian System.		
Limestone ''sand,'' muddy, gray	4	429
Limestone ''sand,'' light brown gray	4	433
Limestone ''sand,'' (good oil show)	8	441
Limestone ''sand,'' darker	4	445
Total depth		445

Log No. 1051

Stringer Bros., No. 10, lessors. Irving J. Brown Oil Co., lessee. Location: 8½ miles east of Franklin, on Middle Fork Creek. Commenced: March 24, 1920. Completed: April 15, 1920. Authority: Glen Neaville.

Strata.

Mississippian System.	Thickness	Depth
Limestone and shale	345	345
Devonian System.		
Shale, black (Chattanooga)	54	399
Limestone, pepper and salt	8	407
Limestone, muddy gray	12	419
Limestone, muddy and brown	8	427
Shale, hard, lead color, dark	8	435
Silurian System.		
Limestone ''sand,'' brown	8	443
Shale, hard and muddy	4	447
Limestone ''sand,'' light brown, (rainbow) ..	24	471
Limestone ''sand,'' dark brown	10½	481½
Total depth		481½

Log No. 1052

Stringer Bros., No. 11, lessor. Location: 8½ miles east of Franklin, on Middle Fork Creek. Commenced: April 23, 1920. Completed: May 14, 1920. Authority: Glen Neaville.

Strata.

Mississippian System.	Thickness	Depth
Limestone and shale	360	360
Devonian System.		
Shale, black (Chattanooga)	54	414
Limestone, pepper and salt	8	422
Limestone, dark, muddy, gray	8	430
Limestone, muddy and brown	8	438
Shale, hard and muddy	8	446
Silurian System.		
Limestone "sand," light brown	6	452
Shale, hard and muddy	7	459
Limestone "sand," light brown (rainbow)	32	491
Shale	4½	495½
Total depth		495½

Log No. 1053

Chas. White, No. 1, lessor. Location: 5 miles east of Franklin, on Lick Creek. Authority: Moran Oil Refining Co.

Strata.

Mississippian System.	Thickness	Depth
Clay	3	3
Clay and limestone boulders	47	50
Limestone, gray	27	77
Shale, caving, (water)	3	80
Limestone and flint	42	122
Limestone, sandy	8	130
Limestone, gray, (sulphur water 165)	35	165
Limestone, crystallized	15	180
Limestone, dark and soft	55	235
Limestone, hard and gray	65	300
Limestone, white	20	320
Limestone, gray, and flint	35	355
Limestone, white, very hard	95	450
Limestone, green, and shale (New Providence)	17	467

Devonian System.	Thickness	Depth
Shale, black (Chattanooga)	59	526
Limestone, white	9	535
Limestone ''sand''	6	541
Limestone, blue	49	590
Limestone, gray	27½	617½
Limestone, white	12½	630
Limestone, white and blue	37	667
Limestone, white	10	677
Limestone ''sand,'' (little oil)	23	700
Limestone and sand, (salt water)	5	705
Total depth		705

NOTE—The Devonian-Silurian contact occurs within the upper half of the 49 feet of limestone above 590 feet in depth.

Log No. 1054

Pugh, No. 1, lessor. Location: 2 miles southeast of South Union. Drilled: June 21, 1921. Production: Orig. open flow 200 bbls. oil per day. Authority: C. A. Phelps.

Strata.

Mississippian and Devonian Systems.	Thickness	Depth
Limestone and shale	464	464
Limestone (cap rock)	8	472
Limestone ''sand,'' (gas)	8	480
Limestone, (oil show)	20	500
Limestone ''sand,'' (pay) (excellent)	2	522
Total depth		522

NOTE—No. 2 well same as No. 1, except larger.

TAYLOR COUNTY.

Production: Oil and Gas.   Producing Sands: Corniferous (Devonian); ''Second'' or ''Deep'', (Niagaran-Silurian).

Log No. 1055

J. R. Bailey, No. 1, lessor. Cash dollar, et al., lessees. Location: Just south of Sulphur Well P. O. Production: 2,470,000 cu. ft. gas. Casinghead el. above sea level, 790 feet.

Strata.

Mississippian System.	Thickness	Depth
Soil	3	3
Limestone, gray	140	143
Shale, blue	2	145
Limestone, white	2	147
Limestone, gray	84	231
Limestone, broken	60	291

Devonian System.   Thickness Depth

	Thickness	Depth
Shale, black (Chattanooga)	52	343
Limestone (cap rock)	3	346

Ordovician System.

Limestone, (gas)	30	376
Shale, blue and pink	55	431
Limestone, gray	2	433
Limestone, brown sand	4	437
Sand, shaly	5	442
Total depth		442

Log No. 1056

W. A. Russell, No. 1, lessor. Cashdollar, et. al., lessees. Location: ¾ mile southeast of Sulphur Well P. O. Production: 321,000 cu. ft. gas. Casing head el. above sea level, 690 feet.

Strata.

Mississippian System.

	Thickness	Depth
Limestone, (gas)	115	115
Limestone, (gas 160)	50	165
Shale, blue	5	170
Limestone	40	210
Shale, gray	10	220

Devonian System.

Shale, black (Chattanooga)	40	260
Limestone, soft	10	270

Ordovician System.

Limestone (pay) (gas)	30	300
Shale, blue and red	50	350
Sand, brown	44	394
Total depth		394

Log No. 1057

W. L. Hall, No. 4, lessor. Kenney Oil Co., lessee. Location: 1 mile northeast of Saloma P. O. Production: Dry. Casing head el. above sea level, 954 feet.

Strata.

Mississippian System.	Thickness	Depth
Soil	20	20
Limestone, black, and shale	50	70
Limestone, white, and flint	10	80
Limestone, brown, and flint	5	85
Limestone and flint	20	105
Limestone, gray	5	110
Limestone and shale	20	130
Sand, gray	5	135
Shale, black	15	150
Limestone, gray	5	155
Limestone, black, and shale	20	175
Limestone, gray and white	25	200
Limestone, gray	25	225
Limestone, brown, and shale, (gas)	40	265
Limestone, brown, and shale	100	365
Devonian System.		
Shale, black (Chattanooga)	10	375
Shale, black	35	410
Limestone, gray	8	418
Ordovician System.		
Limestone "sand," (neither oir or gas)	12	430
Limestone, brown and gray	7	437
Total depth		437

Log No. 1058

C. M. Hill, No. 3, lessor. Kenney Oil Co., lessee. Location: ¾ mile S. W. Saloma P. O. Production: Dry. Casing head el. above sea level, 884 feet.

Strata.

Mississippian System.	Thickness	Depth
Soil	20	20
Limestone, brown, and sand	180	200
Limestone, brown, and shale	45	245
Soapstone, white	25	270
Shale, green (New Providence)	80	350

Devonian System.	Thickness	Depth
Shale, black (Chattanooga)	52	402
Limestone, brown	5	407

Ordovician System.		
Limestone ''sand,'' white, (dry)	7	414
Limestone, brown, and flint	16	430
Total depth		430

Log No. 1059

Annie Campbell, No. 1, lessor. Kenney Oil Co., lessee. Location: 2 miles S. W. of Saloma P. O. Production: 250,000 cu. ft. gas. Casing head el. above sea level, 904 feet.

Strata.

Mississippian and Devonian Systems.	Thickness	Depth
Soil	7	7
Limestone, brown	11	18
Limestone and shale	52	70
Limestone, gray, (gas)	10	80
Limestone, white	5	85
Limestone, gray	55	140
Limestone, brown, and flint	107	247
Limestone, brown	285	532
Limestone ''sand,'' (oil) (good sand and gas)	28	560
Limestone and shale	25	585
Fire clay, red	35	620
Fire clay, red	65	685
Limestone, white	25	710
Limestone ''sand,'' (oil) (good and dry)	20	730
Limestone, sandy	15	745
Limestone, brown	30	775
Total depth		775

NOTE—The record of this well is not detailed enough to permit the showing of the Mississippian-Devonian contact. It is, however, close to 532 feet in depth. The Ordovician is close to this point, since the Devonian limestone is thin.

Log No. 1060

J. H. Hill, No. 1, lessor. Kenney Oil Co., lessee. Location: 1 mile S. W. of Saloma P. O. Production: 1,400,000 cu. ft. gas. Tests: 1/26 gallon gas to 1,000 feet. Casing head el. above sea level, 864 feet.

Strata.

Mississippian System.	Thickness	Depth
Soil	10	10
Limestone, white, (gas and water)	2	12
Limestone, brown, and shale	173	185
Limestone and shale, green	15	200
Limestone and shale, brown	100	300
Shale, green (New Providence)	30	330
Devonian System.		
Shale, black (Chattanooga)	70	400
Limestone, white	10	410
Ordovician system.		
Sand, white, (strong gas)	10	420
Limestone, brown	50	470
Shale, green	15	485
Limestone and shale	15	500
Limestone, white	20	520
Limestone "sand," (good and dry)	15	535
Limestone, white	5	540
Total depth		540

Log No. 1061

J, W. Wayne, No. 1, lessor. Cash dollar, et. al., lessees. Location: ¼ mile N. W. of Sulphur Well P. O. Production: 2,470,000 cu. ft. gas. Casing head el. above sea level, 790 feet.

Strata.

Mississippian System.	Thickness	Depth
Soil	5	5
Limestone, gray	15	20
Shale, blue	7	27
Limestone, gray	113	140
Flint rock	15	155
Limestone, gray	46	201
Limestone, broken	89	290
Shale, blue (New Providence)	7	297

Devonian System.

	Thickness	Depth
Shale, black (Chattanooga)	50	347
Limestone (cap rock)	5	352
Limestone ''sand,'' (pay) (gas)	17	369
Limestone, hard	1	370
Total depth		370

Log No. 1062

J. W. Cloyd, No. 1, lessor. Location: 2½ miles S. W. of Campbellsville. Commenced: Sept. 1, 1920. Completed: November, 1920. Drillers: Walter Hobson and Finn Litrell. Authority: F. L. Parrott, contractor.

Strata.

Mississippian System.

	Thickness	Depth
Clay, sandy	3	3
Limestone, hard, brown	100	103
Limestone, gray	25	128
Limestone, soft, brown	132	260
Shale, blue, and gumbo (New Providence) ...	30	290

Devonian System.

	Thickness	Depth
Shale, black (Chattanooga)	32	322
Shale, dark brown (Chattanooga)	11	333
Limestone (cap rock), (show of oil)	½	333½
Limestone, white	2½	336

Ordovician System.

	Thickness	Depth
Shale, blue, soft	27½	363½
Shale, brown, soft (pink)	3¾	367¼
Limestone ''sand,'' brown	4¼	371½
Limestone, brown	33½	405
Sand, pale yellow	10	415
Limestone, white	3	418
Limestone, broken	22	440
Limestone, brown	57	497
Total depth		497

Log No. 1063

W. B. Hill, No. 1, lessor. Location: 1 mile N. W. of Saloma. Commenced: May 16, 1921. Completed: June 18, 1921. Production: 500,000 cu. ft. gas. Authority: Green River Gas Co.

Strata.

Mississippian System.

	Thickness	Depth
Clay	6	6
Sand, yellow	24	30
Limestone, hard, gray	34	64
Limestone, brown, (water at 85)	101	165
Limestone, gray, flinty	40	205
Limestone, soft, brown, (gas 306)	115	320
Shale, green (New Providence)	60	380

Devonian System.

	Thickness	Depth
Shale, black (Chattanooga)	55	435
Limestone (cap rock) "sand," medium, (gas show)	20	455
Limestone "sand," light gray	10	465
Limestone, hard, gray, (no more gas)	5	470
Limestone, sandy with crystals	5	475
Limestone, gray, fine, fossils	10	485
Shale, hard, blue, muddy	11	496
Total depth		496

136 feet of 6¼" casing.

NOTE—The Devonian-Silurian contact is within the 20 feet above 455 feet in depth.

Log No. 1064

W. E. Stone, No. 1, lessor. Location: 1½ miles west of Campbellsville. Commenced: March 4, 1921. Production: 500,000 cu. ft. gas. Authority: Green River Gas Co.

Strata.

Mississippian System.

	Thickness	Depth
Clay, red	3	3
Limestone, gray	17	20
Cavity, mud and water	½	20½
Limestone, gray	15½	36
Cavity, water	1	37
Limestone, flinty	18	55

Mississippian System.	Thickness	Depth
Limestone, gray	12	67
Limestone, blue	13	80
Limestone, white, blue	35	115
Limestone, blue, very hard	130	245
Limestone, soft, dark, black	10	255
Shale, hard, blue	51	306

Devonian System.		
Shale, black (Chattanooga)	45	351
Limestone (cap rock), dark gray	49	400
Limestone ''sand,'' gray, fine, (gas)	2	402
Limestone ''sand,'' blue, gray, coarse, (gas)	8	410
Limestone ''sand,'' blue, gray, very coarse ..	5	415
Limestone ''sand,'' bluish gray, (no gas)	5	420
Limestone ''sand''	5	425
Shale, hard, blue	21	446
Total depth		446

NOTE—With the exception of a few feet (5-10) at the top of the 49 feet of limestone above 400, all of this strata is probably Ordovician. The Devonian-Ordovician contact is a few feet below the black (Chattanooga) shale.

Log No. 1065

T. E. Claycomb, No. 1, lessor. Location: 2 miles southeast of Saloma, 4 miles northwest of Campbellsville. Completed: June 25, 1920. Production: 962,000 cu. ft. gas. Contractor: William Claycomb.

Strata.

Mississippian System.	Thickness	Depth
Clay	17	17
Limestone, broken	173	190
Limestone, blue	105	295
Limestone and shale, blue (New Providence) (gas show)	33	328

Devonian System.		
Shale, black (Chattanooga)	52	380
Limestone (cap rock), dark gray, hard	5	385

Ordovician System.	Thickness	Depth
Limestone, hard, gray, sandy, (gas show 387)	10	395
Limestone, coarse pebbles, and sand, (large flow of gas)		403
Limestone ''sand,'' blue, soft, muddy, (show of salt water)	4½	407½
Total depth		407½

TODD COUNTY.

Production: **Oil and Gas.** Producing Sands: ''Shallow'' (Mississippian); Corniferous (Devonian); ''Deep'' (Silurian).

Log No. 1066

Tom Mimms, No. 1, lessor. Rogers & Wilson, lessees. Location: ¾ mile northwest of Guthrie. Authority: H. E. Wilson.

Strata.

Mississippian System.	Thickness	Depth
Limestone and shale	1,001	1,001
Devonian System.		
Shale, black (Chattanooga)	79	1,080
Sand, (pay)	20	1,100
Shale (red rock)	5	1,105
Total depth		1,105

Shot 240 quarts. No good.

Log No. 1067

T. C. Slack, No. 1, lessor. Rogers & Wilson, lessees. Location: About ¾ mile north of Guthrie. Authority: H. E. Wilson. Formation same as Mimms, No. 1, except no pay.

Log No. 1068

Bob Sydnor, No. 1, lessor. Location: 5 miles west of Guthrie. Authority: H. E. Wilson.

Strata.

Mississippian System.	Thickness	Depth
Limestone, gray	672	672
Limestone (flint), blue	350	1,022
Limestone ''sand,'' (small pay 1,162)	140	1,162
Total depth		1,162

Log No. 1069

"Bus" Terrell, No. 1, lessor. Elkton Oil Co., lessee. Location: About 150 feet from the north line and 150 feet from the east line of the Terrell Farm, 1 mile north of Elkton. Commenced: June 19, 1919. Completed: Feb. 1, 1920. Production: Volumes of salt water. Drilling contractors: Shaw Drilling Co., Inc., Oklahoma, Okla. Authority: Elkton Oil Co.

Strata.

Mississippian System.	Thickness	Depth
Clay, red, soft	13	13
Clay, yellow, hard	2	15
Limestone, gray, hard	6	21
Shale, gray, soft	2	23
Limestone, gray, hard	9	32
Shale, gray, soft, (water)	5	37
Limestone, blue, hard	9	46
Limestone, white, soft	21	67
Shale, gray, soft, (water)	1	68
Limestone, white, soft	56	124
Shale, blue, soft, (sulphur water)	4	128
Limestone, gray, hard	4	132
Limestone, white, soft	24	156
Shale, white, soft	2	158
Limestone, gray, hard	14	172
Limestone, white, soft	29	201
Limestone, gray, hard	4	205
Limestone, white, soft	24	229
Limestone, gray, hard	4	233
Limestone, gray, very hard	19	252
Limestone, gray, spar, soft, coarse	6	258
Limestone, gray, hard	4	262
Limestone, gray, very hard	2	264
Limestone, brown, decomposed, coarse	6	270
Limestone, brown, hard, fine	20	290
Limestone, gray, hard	5	295
Limestone and shale, soft, (gas show)	15	310
Limestone, gray, hard	5	315
Limestone, gray (crystalline oolitic)	5	320
Limestone, gray, hard	22	342
Limestone, gray, light, hard	12	354
Limestone, brown and gray, soft	40	394
Limestone, brown, decomposed	13	407
Limestone, gray, hard	55	462
Limestone, brown, decomposed, (sulphur water)	43	505
Limestone, white, soft	20	525

Mississippian System.   Thickness Depth

    Limestone, black 65   590
    Limestone, gray, and red rock 6   596
    Shale and lime shell, (casing 6-5/8) 8   604
    Limestone, black 66   670
    Shale (break) 2   672
    Sand, brown 8   680
    Sand, white 12   692
    Shale (break) 2   694
    Limestone, black 8   702
    Limestone, gray 13   715
    Pebble sand, brown 6   721
    Shale and lime shell 9   730
    Limestone, white 10   740
    Shale and lime shell 12   752
    Limestone, black 18   770
    Limestone, white 7   777
    Limestone, white 73   850
    Limestone, blue 351   1,201
    Shale and lime shell 44   1,245

Devonian System.

    Shale, brown, (Chattanooga) 50   1,295
    Lime shell, black (Chattanooga) 5   1,300
    Shale, black (Chattanooga) 50   1,350
    Limestone "sand" 20   1,370
    Limestone "sand," brown, (oil show) 8   1,378
    Limestone "sand,". white 2   1,380
    Limestone "sand," brown, (oil show) 10   1,390

Silurian System.

    Limestone "sand," white 13   1,403
    Limestone "sand," brown 5   1,408
    Limestone "sand," white 37   1,445
    Limestone "sand," white, soft 85   1,530

Ordovician System.

    Limestone "sand," brown, (water) 20   1,550
        Total depth 1,550

Casing record: 140 ft. 10 in. 8¼" casing; 604 ft. 6-5/8" casing.

NOTE—The "sands" referred to from 1,350 to the bottom of the well are not true silicious sands, but are either soft granular limestones, or sandy limestones.

UNION COUNTY.

Log No. 1069-A.

George Proctor, No. 1, lessor. Mt. Carmel Syndicate, Mt. ·Carmel, Ill., lessee. Completed: February, 1922. Production: 20 bbls. oil approximately. Authority: Ivyton Oil & ·Gas Co., Louisville, Ky.

Strata.

Pennsylvanian System.	Thickness	Depth
Clay	7	7
Clay, blue	20	27
Shale, ,blue	30	57
Shale, sandy	60	117
Limestone rock	14	131
Shale (fire clay)	2	133
Coal	1	134
Shale, blue	2	136
Limestone rock	3	139
Shale	5	144
Shale, dark	4	148
Coal	5	153
Shale (fire clay)	2	155
Shale, sandy	5	160
Shale, soft	15	175
Shale, gray,	10	185
Shale, dark (Conemaugh and Allegheny Series)	30	215
Shale, blue	15	230
Shale, gray	20	250
Shale, black	5	255
Coal	4	259
Shale (fire ·clay)	5	264
Limestone, blue	2	266
Shale, sandy	10	276
Shale, gray	10	286
Shale, dark	30	316
Shale, gray	40	356
Shale, dark	35	391
Shale, sandy	20	411
Shale, dark	30	441
Shale, black	4	445
Shale (fire clay)	3	448
Shale, sandy	10	458
Shale, dark	20	478
Coal	1	479
Shale (fire clay)	2	481
Shale, dark	10	491

A WESTERN KENTUCKY OIL "SAND" OF PROMISE

The Sebree Sandstone, basal formation, in the Allegheny series, is thick and coarse grained. It has recently become recognized as an oil producer in Union County and undoubtedly has an important future. Outcrop type locality

Pennsylvanian System.	Thickness	Depth
Limestone, blue	2	493
Shale, dark	40	533
Shale, white	10	543
Shale, dark	46	589
Sand,	48	637
Sand (oil, 20 bbls), } Sebree Sandstone	9	646
Total depth		646

NOTE—This record stops at the base of the Alleghany Series.

CHAPTER X.

WARREN COUNTY.

Production: Oil and Gas. Producing Sands: "Shallow," "Beaver," and "Amber Oil Sand" (Mississippian); Corniferous (Devonian); "Deep" (Niagaran age) (Silurian).

Log No. 1070

Graham, No. 1, lessor. Location: 3 miles northeast of Bowling Green, Richardsville, Pike. Completed: January, 1920. Authority: E. W. Cooper, contractor.

Strata.

	Thickness	Depth
Mississippian System.		
Soil, gravel and boulders	30	30
Limestone	652	682
Devonian System.		
Shale, black (Chattanooga)	85	767
Limestone (cap rock)	16	783
Limestone, white, oil odor	12	795
Limestone, brown, light show	24	819
Silurian System.		
Limestone, gray	37	856
Limestone, soft, fair show of oil	20	876
Limestone, light gray	26	902
Ordovician System.		
Limestone, streaks of oil sands	173	1,075
Limestone (cap rock)	11	1,086
Limestone, light brown, strong oil odor	35	1,121
Limestone, dark	9	1,130
Total depth		1,130

Fresh water from 40 to 60 feet.
Sulphur water from 210 to 225 feet.

Log No. 1071

W. B. Anderson, No. 1, lessor. Completed: November 5, 1919.
Strata.

	Thickness	Depth
Mississippian System.		
Soil	65	65
Limestone, gray	60	125
Limestone, blue (sulphur water)	15	140
Limestone, blue	5	145
Limestone, blue	5	150
Limestone, bluer (showing of oil)	15	165

Mississippian System.	Thickness	Depth
Limestone (lot of gas, 215-20)	55	220
Limestone, dark	40	260
Limestone, dark	20	280
Limestone, lighter	30	310
Limestone, white	30	340
Limestone, white	40	380
Limestone, clear white	30	410
Devonian System.		
Shale, black	65	475
Limestone (cap rock), white, sandy, (showing of oil at 483)	8	483
Oil ''sand,'' brown	32	515
Limestone, gray	18	533
Total depth		533

Drilled by J. S. Garretson & Son, drilling contractors, Bowling Green, Ky. Commenced spudding on September 29, 1919. Amount of casing used, 654 feet, 8¼ and 180 feet 6¼.

Log No. 1072

Chandler, No. 1, lessor. Location: Moulder Pool. Authority: W. N. Thayer.

Strata.

Mississippian System.	Thickness	Depth
Limestone	295	295
Devonian System.		
Shale, black	52	347
Limestone, ''sand,'' (dry)	18	365
Limestone	37	402
Limestone, ''sand,'' (oil show)	13	415
Total depth		415

Log No. 1073

Chandler, No. 2, lessor. Location: Moulder Pool.

Strata.

Mississippian System.	Thickness	Depth
Limestone	291	291
Devonian System.		
Shale, black	52	343
Limestone, ''sand,'' (dry)	18	361
Limestone	27	388
Total depth		388

Log No. 1074

W. A. Hewitt, No. 1, lessor. Location: Martin Precinct. Completed: July 3, 1920. Authority: The New Domain Oil & Gas Company.

Strata.

Mississippian System.	Thickness	Depth
Clay, red	22	22
Limestone, gray	123	145
Limestone, blue	295	440
Devonian System.		
Shale, brown	50	490
Limestone, cap rock	5	495
Limestone, white	10	505
Total depth		505

Log No. 1075

W. A. Hewitt, No. 4, lessor. Completed: April 8, 1920. Production: Estimated at 4 barrels. Authority: New Domain Oil & Gas Company.

Strata.

Mississippian System.	Thickness	Depth
Clay, red	25	25
Limestone, gray, hard	417	442
Devonian System.		
Shale, black	53	495
Limestone, dark	10	505
Limestone, gray	6	511
Total depth		511

Log No. 1076

W. A. Hewitt, No. 6, lessor. Completed: May 18, 1920. Authority: New Domain Oil & Gas Company.

Strata.

Mississippian System.	Thickness	Depth
Clay, red	17	17
Limestone, gray, light	137	154
Limestone, gray, dark	272	426
Devonian System.		
Shale, black	52	478
Limestone (cap rock), black	9	487
Limestone "sand"	6	493
Total depth		493

Log No. 1077

W. A. Hewitt, No. 7, lessor. Completed: June 10, 1920. Authority: New Domain Oil & Gas Company.

Strata.

Mississippian System.	Thickness	Depth
Clay, red	24	24
Limestone, gray	129	153
Limestone, white	288	441

Devonian System.		
Shale, black (Chattanooga)	53	494
Limestone (cap rock), black	6	500
Limestone, ''sand,'' gray (Corniferous)	9	509
Total depth		509

Log No. 1078

J. C. Cole, No. 1, lessor. Completed: September 29, 1919. Authority: The Swiss Oil Corporation.

Strata.

Mississippian System.	Thickness	Depth
Clay and gravel (water)	32	32
Limestone, gray, dark (water at 60)	33	65
Limestone, white	65	130
Limestone, blue, and flint	28	158

Devonian System.		
Shale, black	45	203
Shale, black, and limestone	6	209
Limestone, brown	13	222
Limestone, oil ''sand,'' rainbow	3	225
Limestone, gray	5	230
Total depth		230

Log No. 1079

J. C. Cole, No. 2, lessor. Commenced: September 30, 1919. Completed: October 8, 1919. Production: Dry. Authority: The Swiss Oil Corporation.

Strata.

Mississippian System.	Thickness	Depth
Clay and gravel	33	33
Limestone, gray, dark	32	65
Limestone, white	60	125
Limestone, blue	30	155
Limestone and shale, blue	10	165
Devonian System.		
Shale, black (Chattanooga)	47	212
Limestone, black	5	217
Limestone (cap rock)	5	222
Limestone, oil ''sand,'' rainbow and stain	5	227
Limestone, blue (salt water at 268)	41	268
Limestone, salty	3	271
Total depth		271

Log No. 1080

Brunson, No. 1, lessor. Authority: The Swiss Oil Corporation.

Strata.

Mississippian System.	Thickness	Depth
Clay, soft	24	24
Limestone	36	60
Mud cave (fresh water at 75)	15	75
Limestone, black, hard (sulphur water at 110)	45	120
Limestone, white, medium	40	160
Limestone and shell, dark, soft (show of oil at 215)	100	260
Limestone, sandy, light	40	300
Limestone, sandy, white	75	375
Shale, green, soft	40	415
Devonian System.		
Shale, black, soft (Chattanooga)	45	460
Limestone (cap rock)	8	468
Limestone, ''sand,'' (first)	7	475
Limestone, break	25	500
Limestone, ''sand,'' (second)	8	508
Shale (break)	17	525
Limestone, ''sand,'' (third)	6	531
Total depth		531

Log No. 1081

Brunson, No. 5, lessor. Commenced: May 31, 1920. Completed: June 15, 1920. Authority: The Swiss Oil Corporation.

Strata.

Mississippian System.	Thickness	Depth
Limestone, gray, hard	30 ·	30
Mud cave, soft (fresh water at 45)	15	45
Limestone, black, hard (sulphur water at 80)	35	80
Limestone, white, medium	62	142
Limestone, gritty, white, hard, shells	108	250
Limestone, gritty, white, hard	50	300
Limestone, white, medium	55	355
Shale, green, medium	55	410

Devonian System.		
Shale, black, soft (Chattanooga)	50	460
Limestone (cap rock), brown, hard (gas at 469)	9	469
Limestone, "sand," white, soft (oil at 470)	8	477
Shale, break, brown, hard	1	478
Total depth		478

Log No. 1082

Brunson, No. 7, lessor. Commenced: June 18, 1920. Completed: July 9, 1920. Authority: The Swiss Oil Corporation.

Strata.

Mississippian System.	Thickness	Depth
Clay and gravel, soft	18	18
Limestone, gray, hard (fresh water at 80, sulphur water at 110)	92	110
Limestone, white, hard	50	160
Limestone and shells, dark, medium	140	300
Limestone, white, gritty	50	350
Limestone, white, hard	50	400
Shale, green, gritty	50	450

Devonian System.		
Shale, black (Chattanooga)	50	500
Limestone (cap rock), brown, hard	6	506
Limestone, "sand," white, soft (first)	9	515
Shale, break, brown, hard	1	516
Total depth		516

Log No. 1083

Goodnight, No. 3, lessor. Authority: The Swiss Oil Corporation.

Strata.

Mississippian System.	Thickness	Depth
Clay and gravel	50	50
Limestone, brown	90	140
Limestone, brown	15	155
Limestone, dark	33	188
Limestone, gray	75	263
Limestone, yellow and brown	30	293
Limestone and shale, green	50	343
Devonian System.		
Shale, black (Chattanooga)	50	393
Limestone, brown and black	11	404
Limestone (cap rock), white	3	407
Limestone, oil and water	8	415
Limestone, soft, dark	33	448
Limestone, ''sand,'' brown (oil showing)	15	463
Silurian System.		
Limestone and shale, hard, dark	75	538
Limestone, ''sand,'' (oil odor)	15	553
Limestone, blue	3½	556½
Total depth		556½

Log No. 1084

J. E. Moulder, No. 9, lessor. Commenced: July 24, 1919. Completed: Aug. 23, 1919. Contractor: J. D. Turner, Bowling Green. Authority: The Swiss Oil Corporation.

Strata.

Mississippian System.	Thickness	Depth
Soil and gravel	7	7
Limestone boulders, hard	18	25
Limestone, soft	15	40
Limestone, gray, and flint, white, hard	10	50
Limestone, brown, soft	20	70
Limestone and flint, white, hard	20	90
Flint, blue, white	10	100
Limestone, white, soft	30	130
Flint, white, hard	20	150
Limestone and flint, white	20	170
Flint, white, shelly	30	200
Shale, green, soft (New Providence)	35	235

Devonian System.

	Thickness	Depth
Shale, black, soft (Chattanooga)	40	275
Shale, brown	10	285
Limestone, hard	8	293
Limestone, oil ''sand,'' gray, soft	3	296
Total depth		296

Log No. 1085

J. E. Moulder, No. 10, lessor. Commenced: August 9, 1919. Completed: September 3, 1919. Contractor: L. D. Turner. Production: Dry. Authority: The Swiss Oil Corporation.

Strata.

Mississippian System.

	Thickness	Depth
Clay, red	22	22
Limestone, hard, gray	158	180
Limestone and flint, gray, white, hard	105	285
Limestone, gray, hard	70	355
Shale, green, hard, (New Providence)	10	365

Devonian System.

	Thickness	Depth
Shale, black, hard (Chattanooga)	45	410
Limestone and shale, gray, hard	11	421
Limestone (cap rock), mixed, hard	6	427
Limestone and flint, black, hard	6	433
Limestone, oil ''sand''	6	439
Limestone, salt water, hard	8	447
Limestone, gray, soft	4	451
Total depth		451

Log No. 1086

J. E. Moulder, No. 11, lessor. Authority: The Swiss Oil Corporation.

Strata.

Mississippian System.

	Thickness	Depth
Clay, red, soft	23	23
Limestone, blue, medium (water at 123)	137	160
Sand, fine, white, hard	10	170
Shale, hard, white	86	256
Sand, coarse, gray, soft	12	268
Limestone, white, hard	17	285

Mississippian System.   Thickness Depth

	Thickness	Depth
Sand, brown, soft (gas at 286)	15	300
Flint, blue, hard	5	305
Shale, green, soft	41	346

Devonian System.

Shale, black, soft	54	400
Limestone, brown, soft	5	405
Shale, black, hard	3	408
Limestone (cap rock), brown, gas	6	414
Limestone, "sand," light, soft	4	418
Total depth		418

Log No. 1087

J. E. Moulder, No. 12, lessor.   Authority: The Swiss Oil Corporation.

Strata.

Mississippian System.   Thickness Depth

	Thickness	Depth
Unrecorded	79	79
Limestone, black, hard	5	84
Limestone, gray, hard	28	112
Limestone and flint, white and hard	8	120
Limestone, white, hard	30	150
Limestone, gray, soft	12	162
Limestone, gray, soft	18	180
Limestone spur, gray, soft	5	185
Shale, green, soft (New Providence)	29	214
Limestone and black shale, hard	6	220

Devonian System.

Shale, black, hard (Chattanooga)	47	267
Shale, brown, soft	6	273
Limestone (cap rock), soft	$2\frac{2}{3}$	$275\frac{2}{3}$
Limestone "sand"	$2\frac{5}{6}$	$278\frac{1}{2}$
Total depth		$278\frac{1}{2}$

Log No. 1088

J. E. Moulder, No. 13, lessor.   Completed: November 7, 1919.
Authority: The Swiss Oil Corporation.

Strata.

Mississippian System.	Thickness	Depth
Gravel and clay, yellow, soft	2	2
Limestone, hard, gray	10½	12½
Clay, boulders and gravel, yellow, soft	10	22½
Limestone, dark gray, medium	40½	63
Limestone and shale, blue, medium	97	160
Limestone, white, hard	85	245
Shale, green, hard, flinty (New Providence) ..	25	270

Devonian System.		
Shale, black, soft (Chattanooga)	58	328
Limestone (cap rock), gray, hard	5	333
Limestone, oil "sand," hard	9	342
Total depth		342

Log No. 1089

J. E. Moulder, No. 14, lessor.   Authority: The Swiss Oil Corpora_
tion.

Strata.

Mississippian System.	Thickness	Depth
Soil, yellow, soft	25	25
Limestone, gray, hard	6	31
Mud, blue, soft	5	36
Limestone, hard, gray	72	108
Sand, light, hard	5	113
Limestone, hard, gray	27	140
Limestone, white, hard	75	215
Limestone, green, soft	25	240
Limestone, yellow, hard, (New Providence) ..	19	259
Shale, green, soft, (New Providence)	13	272

Devonian System.		
Shale, black, soft (Chattanooga)	55	327
Limestone (cap rock), gray, hard	12	339
Limestone, "sand," brown, soft	11	350
Limestone, "sand," light, hard	2	352
Total depth		352

Log No. 1090

J. E. Moulder, No. 15, lessor. Commenced: September 12, 1919. Completed: September 23, 1919. Production: Dry. Authority: The Swiss Oil Corporation.

Strata.

Mississippian System.	Thickness	Depth
Soil, soft	3	3
Clay, yellow, soft	13	16
Limestone, blue, hard	32	48
Mud, blue, soft (fresh water)	4	52
Limestone, blue, hard	26	78
Water sand, gray, soft	6	84
Shale, blue, soft	5	89
Limestone, blue, hard	21	110
Limestone, yellow, soft (sulphur water)	8	118
Limestone, blue, hard	20	138
Limestone, white, hard	52	190
Limestone, gray, soft	7	197
Limestone, blue, hard	33	230
Shale, gray, soft	47	277
Devonian System.		
Shale, brown, soft	58	335
Limestone (cap rock), gray, hard	4	339
Limestone, "sand," gray, hard	5	344
Limestone, white, hard (salt water)	1	345
Total depth		345

Log No. 1091

J. E. Moulder, No. 16, lessor. Production: Dry. Authority: The Swiss Oil Corporation.

Strata.

Mississippian System.	Thickness	Depth
Clay and gravel, soft	18	18
Limestone, hard, (fresh water, top; sulphur water, bottom	117	135
Limestone, white	55	190
Limestone, gray	30	220
Limestone, blue	40	260
Limestone, green, hard (little gas)	20	280
Devonian System.		
Shale, black (Chattanooga)	58	338
Limestone (cap rock)	3	341
Limestone "sand," (small show of oil)	10	351
Limestone, light (water)	5	356
Total depth		356

Log No. 1092

J. E. Moulder, No. 17, lessor.   Completed: October 29, 1919. Production: Dry.   Authority: The Swiss Oil Corporation.

Strata.

Mississippian System.	Thickness	Depth
Clay boulders, red	40	40
Limestone, blue	110	150
Limestone, white	110	260
Shale, green	5	265
Devonian System.		
Shale, black (Chattanooga)	50	315
Shale, brown	13	328
Limestone (cap rock)	5	333
Sand	20	353
Silurian System.		
Shale, blue	21	374
Limestone, blue	9	383
Total depth		383

Log No. 1093

J. E. Moulder, No. 18, lessor.   Authority: The Swiss Oil Corporation.

Strata.

Mississippian System.	Thickness	Depth
Soil	8	8
Limestone boulders, gray, soft	3	11
Limestone, white, extra hard	9	20
Limestone, black, soft	4	24
Limestone and mud, yellow, soft (water at 40)	21	45
Limestone, black, hard	40	85
Limestone and flint, white, hard	5	90
Limestone, gray, soft	30	120
Limestone, white, hard	30	150
Limestone, gray, bard	25	175
Spar, light, and shale, gray, soft	40	215
Devonian System.		
Shale, black, hard	47	262
Shale, brown, hard	12	274
Limestone (cap rock), soft	3	277
Limestone, gray, soft	2	279
Limestone, black	1	280
Total depth		280

Log No. 1094

J. E. Moulder, No. 19, lessor. Production: Dry. Authority: The Swill Oil Corporation.

Strata.

Mississippian System.	Thickness	Depth
Surface, yellow, soft	30	30
Limestone, gray, hard	4	34
Mud, blue, soft	1	35
Limestone, gray, hard	70	105
Sand, light, hard	7	112
Limestone, gray, hard	32	144
Limestone, white, hard	80	224
Limestone, green, soft	25	249
Limestone, yellow, hard	20	269
Shale, green, soft	16	285
Devonian System.		
Shale, black, soft (Chattanooga)	54	339
Shale, brown, soft	4	343
Limestone (cap rock), gray, hard	8	351
Limestone, brown, hard	12	363
Limestone, dark, hard	7	370
Silurian System.		
Limestone, blue, soft	23	393
Total depth		393

Log No. 1095

Joe Shipley, No. 1, lessor. Commenced: June 30, 1919. Completed: August 26, 1919. Authority: The Swiss Oil Corporation.

Strata.

Mississippian System.	Thickness	Depth
Clay	32	32
Limestone, yellow	6	38
Limestone, white (mud seam 2 feet)	45	83
Limestone, gray (water)	12	95
Limestone, brown (sulphur water at 162), (black sulphur water at 181)	144	239
Limestone, brown, sandy, very hard	19	258
Shale, brown, hard	9	267
Limestone, brown	43	310
Soapstone	6	316
Limestone, blue	17	333
Limestone, blue, shelly (gas at 365)	43	376
Limestone, white, sandy, (show of oil at 384)	14	390
Limestone, blue, shelly (gas at 408 and 550)	248	638

	Thickness	Depth
Devonian System.		
Shale, black (Chattanooga)	66½	704½
Limestone (cap rock)	4½	709
Limestone, white	11	720
Limestone and sand, gray, very hard	22	742
Silurian System.		
Limestone, ''sand,'' (show of oil)	3	745
Limestone, blue (show of oil)	4	749
Limestone, gray, and sand, gritty	22	771
Limestone, soft	4	775
Limestone, blue, rotten (show of oil at 780) ..	31	806
Limestone, gray, and sand	14	820
Limestone, blue	19	839
Total depth		839

Log No. 1096

Joe Shipley, No. 2, lessor. Commenced: September 1, 1919. Completed: October 28, 1919. Authority: The Swiss Oil Corporation. Strata.

	Thickness	Depth
Mississippian System.		
Clay, yellow	10	10
Clay and gravel, yellow	30	40
Limestone, white (water at 95)	60	100
Limestone, gray	15	115
Limestone, brown (white sulphur water at 135 black sulphur water at 175)	60	175
Limestone, brown, sandy, very hard	25	200
Limestone, brown (black sulphur water at 250)	50	250
Limestone, blue, flinty	50	300
Limestone (show of oil)	6	306
Limestone, white	32	338
Limestone, broken	22	360
Limestone, gray	90	450
Limestone, brown	202	652
Devonian System.		
Shale, brown (Chattanooga)	76	728
Limestone (cap rock)	6	734
Limestone, white :	19	753
Limestone, brown, sandy	24	777
Silurian System.		
Limestone, blue (show of oil 776-780)	4	781
Limestone, gray, gritty	22	803
Limestone, rotten	22	825
Limestone, brown	6	831
Shale, soft and slick ....:............	4	835
Limestone, gray	35	870
Limestone, blue	13	883
Total depth		883

Jeffersonville limestone

Contact

Louisville limestone

DEVONIAN AND SILURIAN LIMESTONE

The Jeffersonville is of Devonian age, and the Louisville Limestone is of Silurian age. The reason the driller frequently cannot note the change is at once apparent. Photo in eastern Louisville Quarry, by Charles Butts.

Log No. 1097

Bryan, No. 1, lessor. Tampa-Kentucky Oil Co., lessee. Location: On the Simpson-Warren County line. Completed: July 10, 1920. Authority: Mr. Reep.

Strata.

Mississippian System.	Thickness	Depth
Lime, variable	848	848
Devonian System.		
Shale, black (Chattanooga)	61	909
Limestone (cap rock), oil	5½	914½
Limestone, "sand"	25	939½
Limestone, harder	6½	946
Total depth		946

Log No. 1098

Bryan, No. 2, lessor. Tampa-Kentucky Oil Co., lessee. Location: On the Simpson-Warren County line. Drilled: In 1920. Authority: Mr. Reep.

Strata.

Mississippian System.	Thickness	Depth
Limestone, variable	842½	842½
Devonian System.		
Shale, black (Chattanooga)	61½	904
Limestone (cap rock), oil	3	907
Limestone, "sand" (first pay)	8	915
Limestone, brown	4	919
Limestone, white, hard (second pay)	24	943
Total depth		943

Log No. 1099

Widow of George Nye, No. 1, lessor. Shrout and Wright, lessee. Completed: October, 1919. Authority: The Big Dipper Oil Company.

Strata.

Mississippian System.	Thickness	Depth
Soil, dark, soft	4	4
Limestone, light and dark, hard (sulphur)	150	154
Limestone (gas)	126	280
Limestone, light and dark, hard	62	342
Limestone, brown (sand), (good show of oil)	17	359
Limestone, dark, light, hard	315	674

Devonian System.   Thickness Depth

	Thickness	Depth
Shale, dark, hard	62	736
Limestone (cap rock), light, dark, hard	15	751
Limestone, white, hard	8	759
Limestone, brown, sandy, hard	15	774

Silurian System.

	Thickness	Depth
Limestone, gray, hard	30	804
Flint, brown, hard	5	809
Limestone, brown, sandy, hard	8	817
Limestone, brown, dark, hard	19	836
Limestone, brown, hard	48	884
Limestone, white, soft	20	904
Limestone and salvage, blue, soft	101	1,005
Total depth		1,005

Log No. 1100

S. Purdue, No. 1, lessor. Completed: In 1920. Authority: The Big Dipper Oil Company.

Strata.

Mississippian System.   Thickness Depth

	Thickness	Depth
Limestone	904	904

Devonian System.

	Thickness	Depth
Shale, black (Chattanooga)	76	980
Limestone	35	1,015

Silurian System.

	Thickness	Depth
Limestone, "sand," first	11	1,026
Limestone, red rock	9	1,035
Limestone, "sand," second	11	1,046
Total depth		1,046

Log No. 1101

William Stone, No. 1, lessor.   Completed: In 1920.   Authority: The Big Dipper Oil Company.

Strata.

Mississippian System.	Thickness	Depth
Limestone	1,005	1,005

Devonian System.		
Shale, black (Chattanooga)	78	1,083
Limestone	102	1,185
Oil sand, first	15	1,200

Ordovician System.		
Limestone	15	1,215
Limestone, "sand"	8	1,223
Limestone and red rock	9	1,232
Limestone, "sand"	13	1,245
Limestone	5	1,250
Total depth		1,250

NOTE—The base of the Devonian and the top of the Silurian is within the 102 feet of limestone above 1,185 feet.

Log No. 1102

J. T. Hunter, No. 1, lessor.   Completed: September 11, 1919. Show of oil: At 424 feet.   Oil and gas from 941 to 947 feet.   Authority: The Big Dipper Oil Company.

Strata.

Mississippian System.	Thickness	Depth
Soil	2	2
Limestone, gray	43	45
Limestone, white	35	80
Limestone, blue	16	96
Limestone, gray	12	108
Limestone, brown	38	146
Limestone, gray	22	168
Limestone, shelly	20	188
Limestone, white	20	208
Limestone, brown	122	330
Limestone, gray	70	400
Limestone, brown	15	415

Mississippian System.	Thickness	Depth
Limestone, gray	38	453
Limestone, brown	70	523
Limestone, gray	77	600
Limestone, black	195	795

Devonian System.		
Shale, black (Chattanooga)	75	870
Limestone	94	964
Limestone (red rock)	2	966
Total depth		966

NOTE—The base of the Devonian and the top of the Silurian is found in the 94 feet of limestone above 964 feet.

Log No. 1103

John Thomas, No. 1, lessor. Commenced: January 12, 1920. Authority: The Big Dipper Oil Company.

Strata.

Mississippian System.	Thickness	Depth
Limestone	410	410

Devonian System.		
Shale, black (Chattanooga)	63	473
Limestone	7	480
Limestone, "sand"	15	495
Limestone	12	507
Total depth		507

Log No. 1104

Robert Lawrence, No. 1, lessor. Completed in January, 1920. Authority: The Big Dipper Oil Company.

Strata.

Mississippian System.	Thickness	Depth
Limestone	339	339

Devonian System.		
Shale, black (Chattanooga)	60	399
Limestone	11	410
Limestone, "sand"	7	417
Limestone	3	420
Total depth		420

Log No. 1105

Tom Lawrence, No. 2, lessor. Commenced: November 20, 1919. Completed: December 31, 1919. Production: Dry. Authority: The Big Dipper Oil Company.
Strata.

	Thickness	Depth
Mississippian System.		
Limestone	380	380
Devonian System.		
Shale, black (Chattanooga)	58	438
Limestone	6	444
Limestone, first "sand"	26	470
Silurian System.		
Limestone	32	502
Limestone, second "sand"	16	518
Limestone	12	530
Total depth		530

Log No. 1106

Henry Lawrence, No. 2, lessor. Commenced: October 16, 1919. Completed: November 15, 1919. Production: Pumping 25 barrels daily, flush. Authority: The Big Dipper Oil Company.
Strata.

	Thickness	Depth
Mississippian System.		
Limestone	345	345
Devonian System.		
Shale, black (Chattanooga)	35	380
Limestone	5	385
Limestone, first "sand"	10	395
Limestone	45	440
Total depth		440

NOTE—The Devonian-Silurian contact is within the last 45 feet.

Log No. 1107

Henry Lawrence, No. 3, lessor. Commenced: November 18, 1919. Completed: December 5, 1919. Production: Started pumping 40 barrels daily. Authority: The Big Dipper Oil Company.
Strata.

	Thickness	Depth
Mississippian System.		
Limestone	339	339
Devonian System.		
Shale, black (Chattanooga)	58	397
Limestone, first "sand"	10	407
Limestone	13	420
Total depth		420

Log No. 1108

Henry Lawrence, No. 4, lessor. Commenced: December 18, 1919. Completed: January 15, 1920. Production: Pumped 200 barrels daily for 6 days, then pumped 100 barrels daily for 5 days. Authority: The Big Dipper Oil Company.

Strata.

Mississippian System.	Thickness	Depth
Limestone	350	350
Devonian System.		
Shale, black (Chattanooga)	61	411
Limestone	6	417
Limestone, ''sand''	12	429
Limestone	16	445
Total depth		445

Log No. 1109

Lydia Miller, No. 1, lessor. Authority: The Bertram Developing Company.

Strata.

Mississippian System.	Thickness	Depth
Limestone	573	573
Devonian System.		
Shale, black (Chattanooga)	57	630
Limestone, white	23	653
Limestone, soft	12	665
Limestone	22	687
Silurian System.		
Limestone, second ''sand''	8	695
Limestone	39	734
Total depth		734

Log No. 1110

Kister, No. 1, lessor. Completed: February 17, 1920. Authority: The Bertram Developing Company.

Strata.

Mississippian System.	Thickness	Depth
Soil, red	10	10
Limestone, hard (little water)	66	76
Limestone, white	94	170

Mississippian System.	Thickness	Depth
Limestone, brown (water at 260)	130	300
Limestone, blue	25	325
Limestone, dark	105	430
Limestone, gray (water at 540)	125	555
Limestone, blue	18	573
Limestone, brown	39	612
Limestone, blue	62	674
Limestone, dark	76	750
Limestone, blue	50	800
Limestone, white and black	70	870
Shale, brown	5	875
Limestone	90	965
Shale, variable in color	139	1,104
Total depth		1,104

NOTE—This is a poorly kept record. The base of the Mississippian System and the top of the Devonian (Chattanooga Shale) is evidently within the 70 feet above 870. The change was not noted by the driller. The base of the Devonian and top of the Silurian is within the last 139 feet of the well.

Log No. 1111

J. P. Lowe, No. 1, lessor.   Authority: The Bertram Developing Company.

Strata.

Mississippian System.	Thickness	Depth
Soil	70	70
Limestone	320	390
Shale, green (New Providence)	30	420
Devonian System.		
Shale, brown (Chattanooga)	60	480
Limestone (cap rock)	7	487
Limestone, "sand"	18	505
Shale, hard	55	560
Limestone, blue	20	580
Limestone, "sand," white	25	605
Limestone, blue	45	650
Ordovician System.		
Limestone	21	671
Limestone, salt water	16	687
Total depth........................		687

NOTE—The base of the Devonian and the top of the Silurian is within the 55 feet above 560 feet.

Log No. 1112

Tarrants, No. 3, lessor. Commenced: June 10, 1920. Completed: July 13, 1920. Authority: Stein, Johnson and Kersetter.

Strata.

Mississippian System.	Thickness	Depth
Soil	8	8
Boulders and clay	22	30
Limestone, black and white	419	449
Limestone, gas "sand"	4	453
Limestone oil "sand"	14	467
Limestone, black	6	473
Total depth		473

Log No. 1113

Ben F. Hewitt, No. 1, lessor. Commenced: August 9, 1919. Completed: August 27, 1919. Production: 48 hours after shot, well pumped 12 bbls. oil. Authority: The Swiss Oil Corporation.

Strata.

Mississippian System.	Thickness	Depth
Clay, red, soft	20	20
Limestone and caves, hard	60	80
Limestone, gray, hard	100	180
Limestone, light gray, sandy, soft (gas)	5	185
Limestone, white, medium hard	15	200
Limestone, sandy, soft (gas)	5	205
Limestone, white, hard	140	345
Limestone, white, sandy	55	400
Limestone, green, soft (New Providence)	40	440
Devonian System.		
Shale, black, soft (Chattanooga)	52	492
Limestone (cap rock), black, hard	8	500
Limestone, "sand," gray, hard	3	503
Limestone, "sand," white, hard, (puff of gas)	7	510
Limestone, "sand," brown, medium hard	12	522
Limestone, gray, hard	3	525
Silurian System.		
Limestone, gray, coarse, soft	7	532
Limestone, "sand," brown, soft (second)	10	542
Limestone, gray, coarse, soft	12	554
Limestone, "sand," brown, soft (third)	9	563
Limestone, "sand"	5	568
Total depth		568

Log No. 1114

B. F. Hewitt, No. 2, lessor.   Commenced: September 8, 1919.
Authority: The Swiss Oil Corporation.

Strata.

Mississippian System.	Thickness	Depth
Clay, red, soft	10	10
Limestone, dark	90	100
Limestone, white, sandy, hard	20	120
Limestone, brown, hard (sulphur water at 145)	40	160
Limestone, white, medium	11	171
Limestone, white, soft (gas and oil at 200) ..	104	275
Limestone, gray, hard (gas at 300)	25	300
Limestone, white, sandy, hard	50	350
Limestone, white, hard	50	400
Shale, green, soft (New Providence)	45	445
Devonian System.		
Shale, black, soft (Chattanooga)	50	495
Limestone (cap rock), gray, hard	11	506
Limestone, ''sand,'' white, medium (first) ...	8	514
Limestone, ''sand,'' brown, medium (oil) ...	13	527
Silurian System.		
Limestone (break), gray, medium		535
Limestone, ''sand,'' brown, soft (second) (oil)	10	545
Limestone (break), gray, soft	12	557
Limestone, ''sand,'' brown, medium (third oil)	8	565
Total depth		565

Log No. 1115

B. F. Hewitt, No. 3, lessor.   Commenced: September 29, 1919.
Completed: October 23, 1919.   Authority: The Swiss Oil Corporation.

Strata.

Mississippian System.	Thickness	Depth
Clay, red, soft	30	30
Limestone, black, hard	50	80
Mud cave, soft	10	90
Limestone, black, hard (fresh water at 120) ..	30	120
Limestone, white, medium	20	140
Limestone, black, hard (sulphur water at 145)	10	150
Limestone, white, medium	15	165

Mississippian System.   Thickness Depth

    Limestone, black, hard (sulphur water at 170)   10   175
    Limestone, white, medium   25   200
    Limestone, dark, medium   100   300
    Limestone (shells), dark, hard   50   350
    Limestone, white, medium   50   400
    Limestone, green, soft (New Providence)   60   460

Devonian System.

    Shale, black, soft (Chattanooga)   51   511
    Limestone, black, hard   6   517
    Shale, black, hard   3   520
    Limestone, white, hard   8   528
    Limestone, ''sand,'' brown, medium (first oil)   20   548

Silurian System.

    Limestone (break), gray, soft   8   556
    Limestone, ''sand,'' brown, medium (second)
        (oil)   10   566
    Limestone (break), gray, soft   12   578
    Limestone, ''sand,'' brown, medium (third)
        (oil)   586

        Total depth   586

Log No. 1116

    B. F. Hewitt, No. 4, lessor.   Commenced: September 28, 1919.
Completed: October 30, 1919.   Authority: The Swiss Oil Company.

    Strata.

Mississippian System.   Thickness Depth

    Clay, red   35   35
    Limestone, hard   70   105
    Mnd cave   15   120
    Limestone, black   50   170
    Limestone   40   210
    Shale   110   320
    Limestone, sandy   70   390
    Limestone, white   15   405
    Limestone, sandy   25   430
    Shale, green (New Providence)   40   470

Devonian System.	Thickness	Depth
Shale, black (Chattanooga)	56	526
Shale, brown	8	534
Limestone, ''sand''	12	546
Limestone (cap rock)	12	558

Silurian System.		
Limestone	11	569
Limestone, ''sand''	6	575
Limestone (break)	10	585
Limestone, ''sand''	7	592
Total depth		592

Log No. 1117

B. F. Hewitt, No. 5, lessor.   Commenced: December 25, 1919.
Completed: January 17, 1920.   Authority: The Swiss Oil Corporation.

Strata.

Mississippian System.	Thickness	Depth
Clay, red, soft	24	24
Limestone, and crevices, hard	46	70
Mud cave, soft (fresh water at 85)	15	85
Limestone, black, hard (sulphur water at 140)	55	140
Limestone, white	35	175
Limestone and shells, black	125	300
Limestone, white, sandy	95	395
Shale, green (New Providence)	40	435

Devonian System.		
Shale, black (Chattanooga)	50	485
Limestone (cap rock), brown	11	496
Limestone, white	9	505
Limestone ''sand'' (first)	15	520

Silurian System.		
Limestone ''sand'' (break)	10	530
Limestone ''sand'' (second), (oil)	8	538
Limestone ''sand'' (break)	14	552
Limestone ''sand'' (third) (oil)	12	564
Total depth		564

Log No. 1118

Hatcher, No. 1, lessor. Location: 1 mile northeast of Bowling Green. Commenced: November 25, 1919. Completed: January 30, 1920. Authority: The Bertram Developing Company.

Strata.

Mississippian System.	Thickness	Depth
Boulders, flint, yellow clay	26	26
Limestone	44	70
Boulders, limestone and clay	10	80
Limestone	25	105
Cavern	10	115
Limestone, gray (oil)	119	234
Limestone, dark (gas at 286)	40	276
Limestone, gray	54	330
Limestone, black	25	355
Limestone, white	65	420
Limestone, blue and green (New Providence)	40	460

Devonian System.		
Shale, brown (Chattanooga)	53	513
Limestone (cap rock)	8	521
Limestone "sand," white (little oil)	4	525
Limestone, brown	28	553

Silurian System.		
Limestone, gray	12	565
Total depth		565

Water at 140 feet.

Log No. 1119

Hobdy, No. 1, lessor. Authority: The Bertram Developing Company.

Strata.

Mississippian System.	Thickness	Depth
Soil and gravel	7	7
Cave	114	121
Cased	189	310
Limestone (light show of oil and gas)	20	330
Limestone	140	470
Limestone "sand" (oil)	10	480
Limestone	334	814

Devonian System.

	Thickness	Depth
Shale, black (Chattanooga)	77	891
Limestone "sand," brown	10	901
Limestone "sand," white	25.	926
Limestone, brown gray, (pay)	15	941

Silurian System.

Limestone, gray	30	971
Limestone "sand," brown, (third pay)	12	983
Limestone (red rock)	4	987
Limestone, gray black	235	1,222
Total depth		1,222

Log No. 1120

Slate well, No. 1. Authority: The Bertram Developing Company.

Strata.

Mississippian System.

	Thickness	Depth
Soil and gravel	45	45
Limestone, gray, white	479	524
Limestone, black	10	534

Devonian System.

Shale, black (Chattanooga)	68	602
Limestone (cap rock)	20	622
Limestone, oil "sand," first	20	642
Limestone, broken	40	682
Limestone, oil "sand," second	28	710
Limestone, shelly	49	759
Total depth		759

Fresh water at from 40 to 60 feet.

Sulphur water at 190 to 265 feet.

Log No. 1121

William Neale, No. 1. Location: 1 mile north of Woodburn. Drilled in May, 1920. Authority: Moran Oil & Refining Company.

Strata.

Mississippian System.

	Thickness	Depth
Clay, red	45	45
Limestone, red (water)	10	55
Limestone, gray	120	175
Limestone, brown (fresh water)	20	195
Limestone, gray	25	220

Mississippian System.	Thickness	Depth
Limestone, brown (sulphur water)	20	240
Limestone, light brown .. ,................	100	340
Limestone, gray	110	450
Sand, brown (salt water)	10	460
Limestone, gray	65	525
Limestone, blue, sharp	135	660
Devonian System.		
Shale, brown (Chattanooga, ...	67	727
Limestone (cap rock)	3	730
Limestone, white	45	775
Silurian System.		
Limestone, blue, sandy (gas)	6	781
Limestone, gray, fine	6	787
Limestone, blue	28	815
Limestone, soft, shaly	400	1,215
Shale, hard	25	1,240
Shale, light brown	35	1,275
Shale, black	5	1,280
Limestone, blue	20	1,300
Limestone, brown, fine	15	1,315
Limestone, rotten	30	1,345
Total depth		1,345

NOTE—The Devonian-Silurian contact is within the upper 50 feet of the 400 feet above 1,215 feet in depth.

Log No. 1122

Noah Manley, No. 1, lessor. Location: Oakland, R. F. D. No. 1. Commenced: March 24, 1920. Completed: April 9, 1920. Authority: The Kenco Oil Company.

Strata.

Mississippian System.	Thickness	Depth
Clay and gravel	18	18
Limestone (water at 50 and 165)	292	310
Shale, green	42	352
Devonian System.		
Shale, black (Chattanooga)	62	414
Limestone (cap rock)	4	418
Limestone "sand," brown, hard, (gas and oil)	14	432
Limestone (salt water)	5	437
Limestone and shale	48	485
Limestone "sand" (showing of oil)	9	494
Limestone, blue, hard	34	528
Total depth		528

NOTE—The Devonian-Silurian contact is midway in the 48 feet above 485 feet in depth.

Log No. 1123

Noah Manley, No. 3, lessor. Drilled in 1920. Authority: The Kenco Oil Company.

Strata.

Mississippian System.	Thickness	Depth
Limestone	314	314
Shale, green (New Providence)	44	358
Devonian System.		
Shale, black (Chattanooga)	52	410
Limestone (cap rock)	7	417
Limestone "sand" (gas)	3	420
Limestone "sand" (oil)	3½	423½
Limestone, gray	76½	500
Total depth		500

NOTE—This well finished in the Silurian.

Log No. 1124

Turner Farm, No. 3, lessor. Location: 3 miles from Bowling Green, Nashville Pike. Authority: A. B. Hughes and Son, drillers.

Strata.

Mississippian System.	Thickness	Depth
Limestone, hard (casing)	165	165
Limestone, variable	235	400
Limestone and flint, hard	200	600
Limestone	122	722
Devonian System.		
Shale, black (Chattanooga)	76	798
Limestone (cap rock)	16	814
Total depth		814

NOTE—Black sulphur water at 158 feet.

Log No. 1125

Perkins Lease, No. 2, lessor. Location: Davenport Oil Pool. Drilled in 1920.

Strata.

Mississippian System.	Thickness	Depth
Limestone	820	820
Devonian System.		
Shale, black (Chattanooga)	80	900
Limestone	105	1,005
Total depth		1,005

Oil at 940.

Black sulphur water at 185.

Cased off at 223.

NOTE—This well finished in the Silurian.

Log No. 1126

Fleenor Farm, No. 1, lessor.   Location: 3 miles south of Bowling Green.   Commenced: July 26, 1920.   Authority: Giles Overton, driller.

Strata.

Mississippian System.	Thickness	Depth
Clay	4	4
Limestone, gray	126	130
Limestone, brown	15	145
Limestone, gray, and flint, brown	75	220
Limestone, brown (good show of oil at 300)	80	300
Limestone, dark, flint, gray	100	400
Limestone, dark gray	47	447
Limestone, black (gas)	3	450
Limestone, oil "sand" (fair showing of oil)	14	464

NOTE—This well is entirely in the Mississippian.

Log No. 1127

Well in the Davenport Pool.   Authority: The Leon Oil Producers Company.

Strata.

Mississippian System.	Thickness	Depth
Limestone and cherty limestone	710	710
Devonian System.		
Shale, black	80	790
Limestone	20	810
Shale and "sand"	15	825
Limestone "sand"	5	830
Limestone, white	20	850
Silurian System.		
Limestone, on top of sand	10	860
Sand	10	870
Limestone, brown	20	890
Sand, blue	15	905
Limestone, white	5	910
Limestone (pay sand), (gas and oil)	10	920
Total depth		920

Log No. 1129

Henry S. Chapman, lessor. (Deep Test.) Location: On the Davenport Farm. Authority: M. L. Chenoweth.

Strata.

Mississippian System.	Thickness	Depth
Limestone, white	80	80
Cavern, mud	8	88
Limestone, white	932	1,020

Devonian System.

Shale, brown (Chattanooga)	90	1,110
Limestone, dark gray	90	1,200
Limestone, oil "sand," dry	10	1,210

Ordovician System.

Limestone, blue	90	1,300
Limestone, gray and white	100	1,400
Limestone, brownish gray, very hard	100	1,500
Limestone, gray	90	1,590
Limestone, red	12	1,602
Limestone, gray, soft	58	1,660
Limestone, white and gray	65	1,725
Limestone, blue gray, medium soft	375	2,100

Mississippian System.

Limestone, pale brown and white, hard	100	2,200
Limestone, brown, hard, flinty, with particles of black limestone mixed	50	2,250
Limestone, dark brown, very hard	150	2,400
Limestone, dark brown	100	2,500
Limestone, light and brown chertz, mixed	50	2,550
Limestone, gray and black, soft	50	2,600
Limestone, dark gray and brown, mixed, hard	100	2,700
Limestone, dark and light, hard	50	2,750
Limestone, dark gray, hard	50	2,800
Limestone, light gray, with black particles showing	50	2,850
Limestone, medium gray, very hard	75	2,925
Total depth		2,925

NOTE—Top of Trenton probably at 1,800 to 1,900. Trenton 700 to 900 feet. The Devonian-Silurian contact is within the 90 feet of Limestone above 1,200 feet in depth.

Log No. 1130

Edwin Willoughby, No. 1, lessor. Location: Near Sledge Pool and Bays Fork. Elevation: About 610 A. T.

Strata.

	Thickness	Depth
Mississippian System.		
Limestone	360	360
Devonian System.		
Shale, black (Chattanooga)	53	413 .
Limestone, blue	8	421
Limestone, brown	14	435
Silurian System.		
Limestone sand	17	452
Limestone	132	584
Limestone sand	29	613
Limestone	728	1,341
Limestone (Cap Rock), dark gray	5	1,352
Limestone (Trenton), (1st oil show)	65	1,417
Limestone, crystalized, hard	21	1,438
Limestone, (2nd show)	2	1,440

1st shot at 475 feet, 60 quarts.
2nd shot at 435 to 452 feet, 80 quarts.
Salt at 584.

NOTE—The Silurian-Devonian contact is within the 132 feet above 584 feet in depth.

Log No. 1131

J. W. McGuire, No. 1, lessor. Hoge Oil & Gas Co., lessee. Commenced: Aug. 14, 1919. Completed: Aug. 28, 1919. Contractor: Lloyd Roetramel. Rig: Steam star. Production: Oil. Plugged up well to 480 feet, and shot second sand with 20 quarts nitro glycerin.

Strata.

	Thickness	Depth
Mississippian System.		
Conductor	47	47
Limestone, (cased 175)	128	175
Limestone	142	317
Devonian System.		
Shale, black (Chattanooga)	68	385
Limestone (cap)	16	401
Limestone, 1st "sand"	12	413

Silurian System. Thickness Depth
    Limestone, 2nd "sand" 12 425
    Limestone, blue 50 475
    Limestone, 3rd "sand" 5 480
    Limestone, blue 18 498
    Limestone, green 22 520
        Total depth 520

	Thickness	Depth

Log No. 1132

J. W. McGuire, No. 2, lessor. Hoge Oil & Gas Co., lessee. Commenced: June 4, 1920. Completed: July 12, 1920. Contractor: Regal & Madison. Rig: New star No. 38.
    Strata.

Mississippian System.	Thickness	Depth
Conductor	14	14
Limestone, (cased 220)	206	220
Limestone	152	372
Devonian System.		
Shale, black (Chattanooga)	68	440
Limestone (cap and 1st "sand")	19	459
Limestone	6	465
Silurian System.		
Limestone, 2d "sand"	30	495
Limestone	39	534
Limestone, 3d "sand"	5	539
Limestone	5	544
Total depth		544

    Shot 2nd sand 467 to 482½ with 60 quarts nitro glycerin.
      6½ feet anchor on bridge.
    Cleaned out well to 515 feet.
    80 feet of fluid in hole on Aug. 6, 1920.

Log No. 1133

J. W. McGuire, No. 3, lessor. Hoge Oil & Gas Co., lessee. Commenced: July 15, 1920. Completed: July 28, 1920. Contractor: Regal & Madison. Rig: New gasoline star No. 38. Production: Bailed ½ bbl. oil 30 minutes after shot. Aug. 6, one week after shot, fluid stood 158'4" in hole.
    Strata.

Mississippian System.	Thickness	Depth
Conductor,	38	38
Limestone, (cased 226)	188	226
Limestone	160	386

Devonian System.

	Thickness	Depth
Shale, black (Chattanooga)	47	433
Limestone (cap)	13	446
Limestone, 1st "sand,"	2	448
Limestone	10	458
Limestone, 2nd "sand"	21	479

Silurian System.

Limestone	46	525
Limestone, 3d "sand"	15	540
Limestone	70	610

Ordovician System

Limestone, white, 4th "sand"	51	661
Limestone	8	669
Limestone, 5th "sand"	5	674
Limestone	30	704
Total depth		704

NOTE—Shot third sand with 60 qts. nitro glycerin, 525 to 540. Bridge stood at 579 after shot and cleaned out.

Log No. 1134

J. W. McGuire, No. 4, lessor. Hoge Oil & Gas Co., lessee. Completed: Aug. 10, 1920. Contractor: E. P. Meredith. Rig: Steam star. Production: Dry; oil shows only.

Strata.

Mississippian System.

	Thickness	Depth
Conductor	41	41
Limestone, (cased 217)	176	217
Limestone	185	402

Devonian System.

Shale, black (Chattanooga)	57	459
Limestone (cap)	9	468
Limestone, 1st "sand"	24	492

Silurian System.

Limestone, (includes 2nd "sand")	54	546
Limestone, 3d "sand"	9	555
Limestone	77	632

Ordovician System.

Limestone, 4th "sand"	26	658
Limestone	22	680
Limestone, 5th "sand"	15	695
Limestone, gray	42	737
Total depth		737

Log No. 1135

F. P. Tabor, No. 1, lessor. Hoge Oil & Gas Co., lessee. Completed: Nov. 12, 1919. Contractor: Russell & Gardner. Rig: Cyclone class D.

Strata.

Mississippian System.	Thickness	Depth
Conductor, (8¼ case.)	40	40
Limestone, (cased 235)	195	235
Limestone	138	373
Devonian System.		
Shale, black (Chattanooga)	55	428
Limestone, (cap)	8	436
Limestone, 1st "sand," (oil show)	5	441
Limestone	17	458
Silurian System.		
Limestone, 2nd "sand"	15	473
Limestone, gray	12	485
Total depth		485

NOTE—Shot second sand with 20 qts. nitro glycerin and made good shot.

Log No. 1136

F. P. Tabor, No. 2, lessor. Hoge Oil & Gas Co., lessee. Commenced: Aug. 19, 1920. Completed: Oct. 12, 1920. Contractor: E. P. Meredith. Rig: Steam star. Production: Oil, well shot.

Strata.

Mississippian System.	Thickness	Depth
Conductor	52	52
Limestone, (cased 252)	200	252
Limestone	65	317
Devonian System.		
Shale, black (Chattanooga)	55	372
Limestone (cap)	9	381
Limestone, 1st "sand"	4	385
Limestone	13	398
Silurian System.		
Limestone, 2nd "sand"	17	415
Limestone	69	484
Total depth		484

Log No. 1137

E. E. Buchanon, No. 1, lessor. Hoge Oil & Gas Co., lessee. Commenced: November 20, 1919. Completed: Dec. 28, 1919. Contractor: Reagle & Madison. Rig: Gasoline star.

Strata.

Mississippian System.	Thickness	Depth
Conductor, (case. 8¼)	27	27
Limestone	184	211
Limestone	164	375
Devonian System.		
Shale, black (Chattanooga,)	53	428
Limestone (cap)	5	433
Limestone, 1st ''sand''	13	446
Limestone	4½	450½
Silurian System.		
Limestone, 2nd ''sand''	22½	473
Limestone, gray	4½	477½
Total depth		477½

NOTE—Shot 16 feet second sand with 40 qts. nitro glyc. Dec. 19, 1919.

Log No. 1138

E. E. Buchanon, No. 2, lessor. Hoge Oil & Gas Co., lessee. Completed: Aug. 20, 1920. Contractor: Reagle & Madison. Rig: New gasoline star No. 38.

Strata.

Mississippian System.	Thickness	Depth
Conductor, (case. 8¼)	30	30
Limestone, (cased 210)	180	210
Limestone	171	.381
Devonian System.		
Shale, black (Chattanooga)	59	440
Limestone (cap)	5	445
Limestone, 1st ''sand''	10	.455
Limestone	10	465
Silurian System.		
Limestone, 2nd ''sand''	20	485
Limestone	40	525
Limestone, 3d ''sand''	15	540
Total depth		540

Driller's Note: Shot third sand with 60 qts. nitro glycerin. Good showing in second sand shot later 470 to 480.

Log No. 1139

E. E. Buchanon, No. 3, lessor. Hoge Oil & Gas Co. lessee, Completed: Aug. 24, 1920. Contractor: Regal & Madison. Rig: New gasoline star. Production: Good oil showing in all three sands.
Strata.

Mississippian System.	Thickness	Depth
Conductor, (case. 8¼)	27	27
Limestone	153	180
Limestone	162½	342½
Devonian System.		
Shale, black (Chattanooga)	56	398½
Limestone (cap)	10½	409
Limestone, 1st "sand"	7	416
Limestone	7	423
Siilurian System.		
Limestone, 2nd "sand"	19	442
Limestone, 3d "sand"	49	491
Limestone	8	499
Total depth		499

Shot first sand 407 to 414 with 20 qts. Shot second sand 427 to 442 with 30 qts. Shot third sand 491 to 499 with 20 qts.

WAYNE COUNTY.

Production: **Oil and Gas.** Producing Sands: **"Beaver"** (Mississippian); Sunnybrook and Trenton (Ordovician).

Log No. 1140

J. H. Duncan, No. 1, lessor. Location: Monticello Precinct. Completed: Nov. 23, 1903. Production: Dry. Authority: New Domain Oil & Gas Co.
Strata.

Mississippian System.	Thickness	Depth
Soil,	44	44
Limestone, hard, (gas 303)	319	363
Shale, hard, white, soft	60	423
Devonian System.		
Shale, black, soft (Chattanooga)	34	457
Ordovician System.		
Limestone, hard	493	950
Limestone "sand" (Sunnybrook), hard	120	1,070
Limestone, hard	88	1,158
Total depth		1,158

Log No. 1141

J. H. Duncan, No. 3, lessor.   Location: Monticello District.   Completed: July 30, 1904.   Production: Dry.   Authority: New Domain Oil & Gas Co.

Strata.

Mississippian System.	Thickness	Depth
Clay, red, soft	25	25
Shells, limy, soft	15	40
Limestone, white, hard, (water 65)	25	65
Limestone, gray, hard, (water & gas 400)	335	400
Limestone "sand," (Beaver) New Providence	10	410
Limestone, blue, (New Providence)	90	500
Total depth		500

Log No. 1142

J. A. Brown, No. 2, lessor.   Completed: Oct. 29, 1904.   Production: commenced producing 20 bbls.   Authority: New Domain Oil & Gas Co.

Strata.

Mississippian System.	Thickness	Depth
Limestone, white, hard	90	90
Limestone "sand," black, hard	50	140
Limestone, white, hard	20	160
Limestone, dark, soft	40	200
Limestone, dark, white, hard	225	425
Limestone "sand," (Stray) white, hard	13	438
Shale, hard, dark, soft	10	448
Limestone, blue, hard	32	480
Shale, hard, dark, soft	15	495
Limestone "sand," white, soft	52	547
Total depth		547

Log No. 1143

J. A. Brown No. 3, lessor.   Completed: Mar. 16, 1905.   Production: Dry; casing pulled, well plugged and abandoned.   Authority: New Domain Oil & Gas Co.

Strata.

Mississippian System.	Thickness	Depth
Soil, red, soft	43	43
Limestone, gray, hard	67	110
Cave and gravel, soft	63	173

Mississippian System.

	Thickness	Depth
Limestone, white, hard	197	370
Limestone, blue soft	125	495
Limestone, blue, hard	75	570
Shale, hard, blue, soft	92	662
Limestone "sand" (Beaver), gray, hard New Providence	10	672
Shale, hard, blue, soft, New Providence	7	679

Devonian System.

	Thickness	Depth
Shale, black, soft (Chattanooga)	40	719

Ordovician System.

	Thickness	Depth
Limestone, black, soft	479	1,198
Limestone "sand" (Sunnybrook), brown, hard	220	1,418
Limestone, blue, hard	82	1,500
Total depth		1,500

NOTE—A Silurian component is regarded as forming the upper portion of the 479 feet of limestone above 1,198 feet in depth.

Log No. 1144

J. A. Brown, No. 10, lessor. Completed: Feb. 5, 1910. Location: Slick Ford Precinct. Production: Dry. Authority: New Domain Oil & Gas Co.

Strata.

Mississippian System.

	Thickness	Depth
Clay, blue, soft	44	44
Limestone, white, medium	200	244
Limestone, gray, medium	215	459
Grit, gray, hard	200	659
Shale, hard, blue, soft	68	727
Limestone "sand" (Beaver), gray, medium, New Providence	12	739
Shale, hard, blue, soft, New Providence	13	752
Total depth		752

Log No. 1145

W. M. Hill, No. 1, lessor. Location: Little South Fork. Completed: Dec. 13, 1911. Production: Dry; oil at 70; salt water at 75 and 390; sulphur water at 300; gas at 545 and 550; well was plugged and abandoned. Authority: New Domain Oil & Gas Co.

Strata.

Mississippian System.

	Thickness	Depth
Sand, brown, soft	10	10
Clay, blue, soft	12	22

Mississippian System.	Thickness	Depth
Limestone, gray, hard	40	62
Shale, blue, soft	10	72
Limestone, hard, variable	470	542
Limestone ''sand,'' black, hard	20	562
Limestone, blue, hard	60	622
Limestone, ''grit,'' gray, hard	60	682
Shale, hard, blue, New Providence	18	700
Shale, hard, blue, soft, New Providence	18	718
Total depth		718

Log No. 1146

G. W. Roberts No. 1, lessor. Completed:; June 30, 1913. Production: Dry; show of gas at 215, 310 and 370 feet. Authority: New Domain Oil & Gas Co.
Strata.

Mississippian System.	Thickness	Depth
Clay	11	11
Limestone, hard, veriable	387	398
Shale, hard, mixed, soft	97	495
Limestone ''sand'' (Beaver), gray, hard New Providence	13	508
Shale, hard, blue, soft, New Providence	21	529
Total depth		529

Log No. 1147

J. L. Dobbs, No. 1, lessor. Completed: Oct. 1, 1914. Production: Dry; casing pulled and well abandoned. Authority: New Domain Oil & Gas Co.
Strata.

Pennsylvanian System.	Thickness	Depth
Clay	9	9
Sandstone	91	100
Mississippian System.		
Clay shale, blue	175	275
Shale	180	455
Limestone, gray	40	495
Limestone, white	300	795
Limestone, gray	50	845
Limestone, black	175	1,020
Shale, hard, mixed	110	1,130
Limestone ''sand'' (Beaver), New Providence	15	1,145
Shale, hard, blue, New Providence	7	1,152
Total depth		1,152

Log No. 1148

J. L. Dobbs, No. 3, lessor. Completed: Dec. 12, 1914. Production: showing for 15 bbls. Authority: New Domain Oil & Gas Co.

Strata.

Pennsylvanian System.	Thickness	Depth
Sandstone	125	125

Mississippian System.		
Clay and shale, blue and red	355	480·
Limestone, gray, white	390	870
Limestone, black	175	1,045
Shale, hard, mixed	.125	1,170
Limestone ''sand'' (Beaver), New Providence	12	1,182
Shale, hard, blue, New Providence	10	1,192
Total depth		1,192

Log No. 1149

J. L. Dobbs, No. 5, lessor. Completed: April 3, 1915. Production: Dry; casing pulled and well abandoned. Authority: New Domain Oil & Gas Co.

Strata.

Pennsylvanian System.	Thickness	Depth
Clay	10	10
Sandstone	90	100
Clay, shale, blue, and red	330	430

Mississippian System.		
Limestone, gray	50	480
Limestone, white	300	780
Limestone, gray	50	830
Limestone, black	175	1,005
Shale, hard, mixed	135	1,140
Limestone ''sand'' (Beaver), New Providence	10	1,150
Shale, hard, blue, New Providence	6	1,156
Total depth		1,156

Log No. 1150

J. L. Dobbs, No. 6. lessor. Completed: April 26, 1915. Production: production first day was 5 bbls. Authority: New Domain Oil & Gas Co.

Strata.

Pennsylvanian System.	Thickness	Depth
Clay, yellow	9	9
Sandstone, yellow	25	34
Shale, blue	150	184
Shale, red	50	234
Mississippian System.		
Shale, blue	50	284
Shale, red	20	304
Shale, blue	50	354
Limestone gray	50	404
Limestone, white	300	704
Limestone, gray	50	754
Limestone, black	175	929
Shale, hard, mixed	127	1,056
Sand (Beaver), brown, New Providence	15	1,071
Shale, hard, blue, New Providence	5	1,076
Total depth		1,076

Log No. 1151

Riley Correll, No. 1, lessor. Completed: Feb. 21, 1905. Production: Dry. Gas at 276 feet. Authority: New Domain Oil & Gas Co.

Strata.

Mississippian System.	Thickness	Depth
Shale, soft, red, loose	17	17
Limestone, white, hard	21	38
Gravel and cave, soft	10	48
Limestone, white, hard	50	98
Limestone, gray, hard	12	110
Gravel and cave, red, soft	30	140
Limestone, white, hard	210	350
Shale, hard, blue, soft New Providence	63½	413½
Limestone, white, hard, soft, New Providence	25	438½
Shale, hard, blue, soft, New Providence	20	458½
Devonian System.		
Shale, hard, black, soft (Chattanooga)	41½	500
Ordovician System.		
Limestone, hard, soft	1,001½	1,501½
Total depth		1,501½

NOTE—A Silurian component is regarded as present forming the upper portion of the 1001½ feet above 1,501½ feet in depth.

Log No. 1152

Jordan McGowan, No. 1, lessor. Completed: Jan. 24, 1905. Production: Dry. Authority: New Domain Oil & Gas Co.

Strata.

Mississippian System.	Thickness	Depth
Limestone, hard	320	320
Shale, hard, gray, soft, New Providence	32	352
Limestone "sand" (Beaver), hard, New Providence	3	355
Shale, hard, gray, soft, New Providence	2	357
Devonian System.		
Shale, black, soft (Chattanooga)	40	397
Ordovician System.		
Limestone, brown, gray, soft, hard	603	1,000
Shale (pencil cave), soft	3	1,003
Limestone, brown, gray, hard, soft	849	1,852
Limestone "sand," gray	35	1,887
Limestone "sand,"	24	1,911
Shale, limy	10	1,921
Total depth		1,921

NOTE—A Silurian component is regarded as present forming the upper portion of the 603 feet above 1000 in depth.

Log No. 1153

W. F. Dick, No. 1, lessor. Completed: Dec. 23, 1904. Production: Dry. Authority: New Domain Oil & Gas Co.

Strata.

Mississippian System.	Thickness	Depth
Clay and gravel, soft ....:................	11½	11½
Limestone	507½	519
Shale, hard, blue, soft, New Providence	8	527
Limestone, white (Beaver), sandy, New Providence	13	540
Shale, hard, blue, soft, New Providence	14	554
Devonian System.		
Shale, black, soft, Chattanooga	40	594
Shale, blue, soft, Chattanooga	10	604

Ordovician System.

	Thickness	Depth
Limestone, Sandy, white, hard	11	615
Limestone, blue, loose	200	815
Limestone, white, black, hard, shelly	150	965
Limestone, blue, open	135	1,100
Limestone, black, white, shelly	100	1,200
Limestone, white, very sandy	12	1,212
Limestone, blue, soft	20	1,232
Limestone, gray, very hard	100	1,332
Limestone, gray, very shelly	150	1,482
Limestone, gray, very hard	25	1,507
Total depth		1,507

NOTE—The upper portion of the 200 feet above 815 is regarded as Silurian.

Log No. 1154

H. C. Dobbs, No. 1, lessor. Completed: Mar. 16, 1916. Production: Dry; casing pulled and well abandoned. Authority: New Domain Oil & Gas Co.

Strata.

Mississippian System.

	Thickness	Depth
Limestone, gray	75	75
Limestone, white	90	165
Limestone, gray	55	220
Shale (red rock)	40	260
Limestone, black	65	325
Limestone, blue	170	495
Shale, hard, blue New Providence	10	505
Limestone "sand" (Beaver), white, New Providence	5	510
Shale, hard, blue, New Providence	9	519
Total depth		519

Water at 40 feet; sulphur water at 160 feet; gas at 330 and 425 feet.

Log No. 1155

E. R. Walker, No. 1. lessor. Completed: Oct. 7, 1904. Production: Dry; fresh water at 65 feet; oil show at 188 feet; small gas show at 360 feet. Authority: New Domain Oil & Gas Co.

Strata.

Mississippian System.

	Thickness	Depth
Limestone, white, blue, sandy	620	620
Devonian System.		
Shale, black, soft (Chattanooga)	60	680

Ordovician System.   Thickness Depth
    Limestone, hard, soft, white, blue   822   1,502
    Total depth   1,502
    NOTE—The upper portion of the 822 feet above 1502 feet in depth is regarded as Silurian.

Log No. 1156

Cyrus Brown, No. 7, lessor. Completed: June 14, 1915. Production: Dry; casing pulled and well abandoned. Authority: New Domain Oil & Gas Co.
    Strata.

Mississippian System.	Thickness	Depth
Clay	15	15
Limestone, gray	185	200
Limestone, white	200	400
Limestone, gray	50	450
Limestone, yellow	25	475
Limestone, black	101	576
Limestone, white	90	666
Limestone, blue	10	676
Limestone, white	94	770
Shale, blue, New Providence	8	778
Limestone, white, New Providence	8	786
Shale, blue, New Providence	8	794
Devonian System.		
Shale, black (Chattanooga)	2	796
Total depth		796

Log No. 1157

Frank Hurt, No. 1, lessor. Completed: Sept. 25, 1907. Production: showed for 10 bbls. before shot; dry after shot. Authority: New Domain Oil & Gas Co.
    Strata.

Mississippian System.	Thickness	Depth
Soil	20	20
Limestone, white, hard	140	160
Shale, blue, soft	10	170
Limestone, white, hard	150	320
Limestone, black, soft	20	340

Mississippian System.	Thickness	Depth
Limestone, white	60	400
Limestone, gritty, brown, hard	170	570
Limestone, black, soft	140	710
Shale, hard, blue, New Providence	10	720
Limestone "sand," (oil) New Providence ...	14	734
Shale, hard, blue, soft, New Providence	26	760
Devonian System.		
Shale, black, soft (Chattanooga)		761
Total depth		761

Log No. 1158

Frank Hurt, No. 2, lessor. Completed: Dec. 11, 1907. Production: Dry. Authority: New Domain Oil & Gas Co.

Strata.		
Mississippian System.	Thickness	Depth
Limestone, dark, hard	126	126
Limestone, white, hard	160	286
Limestone, dark, medium	110	396
Limestone, black, medium	200	596
Limestone, white, medium	54	650
Limestone, black, medium	75	725
Shale, hard, medium, New Providence	2	727
Sand (Beaver), New Providence	10	737
Shale, hard, blue, New Providence	9	746
Total depth		746

Log No. 1159

William Foster, No. 1, lessor. Completed: June 26, 1907. Production: commenced producing 5 bbls. Authority: New Domain Oil & Gas Co.

Strata.		
Pennsylvanian System.	Thickness	Depth
Limestone "sand"	200	200
Mississippian System.		
Limestone, light	650	850
Limestone "sand" Beaver Creek, light, New Providence	15	865
Shale, hard, blue, New Providence	2	867
Total depth		867

Log No. 1160

William Foster, No. 2, lessor. Completed: Aug. 17, 1907. Production: commenced producing 25 bbls. Authority: New Domain Oil & Gas Co.

Strata.

	Thickness	Depth
Pennsylvanian System.		
Sand, light, hard	10	10
Mississippian System.		
Limestone, light, hard	860	870
Limestone "sand" (Beaver), dark, medium ..	13	883
Total depth		883

Log No. 1161

B. Foster, No. 1, lessor. Completed: Aug. 19, 1913. Production: Dry. Authority: New Domain Oil & Gas Co.

Strata.

	Thickness	Depth
Mississippian System.		
Shale, blue, soft	275	275
Limestone, hard, variable	565	840
Shale, hard, black	83	923
Limestone "sand," white, hard, New Providence	28	951
Shale, hard, blue, soft, New Providence	6	957
Total depth		957

Log No. 1162

B. Foster, No. 2, lessor. Location: Bell Hill Precinct. Production: commenced producing 10 bbls., Dec. 16, 1913. Authority: New Domain Oil & Gas Co.

Strata.

	Thickness	Depth
Mississippian System.		
Shale, soft	136	136
Limestone, gray, hard	30	166
Limestone, white, hard	320	486
Limestone, gray, hard	40	526
Limestone, black	200	726
Limestone and shale, hard	73	799
Limestone "sand" (Beaver), New Providence	21	820
Shale, hard, blue	5	825
Total depth		825

Log No. 1163

B. Foster, No. 3, lessor. Completed: Jan. 26, 1914. Production: Dry. Authority: New Domain Oil & Gas Co.
Strata.

Mississippian System.	Thickness	Depth
Shale, soft	130	130
Limestone, gray	30	160
Limestone, white	310	470
Limestone, black	200	670
Limestone "sand," white	122	792
Shale, hard, blue, New Providence	21	813
Shale, hard, blue, New Providence	44	857
Total depth		857

NOTE—The Berea sand (limestone) occurring within the New Providence formation was not recognized by the driller of this well.

Log No. 1164

T. T. Davis, No. 10, lessor. Location: Turkey Rock Pool, near Slickford. Commenced: Oct. 11, 1919. Completed: Nov. 8. 1919. Drilled by the Vulcan Oil Co. Authority: The Vulcan Oil Co.
Strata.

Pennsylvanian System.	Thickness	Depth
Shale, soft	19	19
Sandstone,	11	30
Shale, soft	10	40
Sandstone	160	200
Shale, hard	180	380
Mississippian System.		
Limestone, gray	70	450
Shale, hard	20	470
Limestone	30	500
Shale, hard	5	505
Limestone, (gas)	385	890
Limestone, black	195	1,085
Shale, hard	42	1,127
Limestone "sand" (Beaver)	20½	1.147½
Shale, (oil)	28	1,175½
Total depth		1,175½

Log No. 1165

T. T. Davis, No. 11, lessor. Location: Turkey Rock Pool, near Slickford. Commenced: Nov. 27, 1919. Completed: Jan. 15, 1920. Authority: The Vulcan Oil Co., drillers.

Strata.

Pennsylvanian System.	Thickness	Depth
Shale, soft	15	15
Sandstone	10	25
Shale, soft	15	40
Sandstone	180	220
Shale soft, and shale, hard	180	400

Mississippian System.		
Limestone, gray	75	475
Shale, hard	20	495
Limestone	380	875
Limestone, black	175	1,050
Limestone, blue	50	1,100
Shale, hard, New Providence	48	1,148
Limestone ''sand'' (Beaver), New Providence	12	1,160
Shale, hard, New Providence	3	1,163

Devonian System.		
Shale, black (Chattanooga)	29	1,192
Total depth	.	1,192

Log No. 1166

T. T. Davis, No. 12, lessor. Location: Turkey Rock Pool, near Slickford. Commenced: Feb. 21, 1920. Completed: April 16, 1920. Authority: The Vulcan Oil Co., drillers.

Strata.

Pennsylvanian System.	Thickness	Depth
Shale, soft	15	15
Shale, hard	15	30
Shale, soft	20	50
Sandstone	150	200
Shale, hard and soft	200	400

Mississippian System.		
Limestone	85	485
Shale, hard	12	497
Limestone	403	900
Limestone, black	250	1,150
Shale, hard, New Providence	37	1,187
Limestone ''sand'' (Beaver), New Providence	14½	1,201½
Shale, hard, New Providence	½	1,202

Devonian System.		
Shale, black (Chattanooga), (oil)	27	1,229
Total depth		1,229

WEBSTER COUNTY.

Production: Small oil and gas.   Producing Sands of commercial importance not recognized to date.

Log No. 1167

Jim Trice, No. 1, lessor.   Noon Oil & Gas Co., lessee.   Location: 1½ miles northeast of Dixon, Ky.   Spudded Dec. 13, 1918.   Production: Dry.   Driller: Morarity.

Strata.

Pennsylvanian System.	Thickness	Depth
Soil, (conductor 16 in. case.)	20	20
Shale	80	100
Shale	100	200
Limestone	50	250
Sand, (water 270)	30	280
Shale	70	350
Limestone	50	400
Sand	50	450
Shale	150	600
Sand	50	650
Shale	100	750
Limestone	50	800
Shale, blue	100	900
Sand	25	925
Shale, white	75	1,000
Limestone, very hard	50	1,050
Shale	50	1,100
Limestone	25	1,125
Shale	35	1,160
Limestone, very hard	10	1,170
Sand	70	1,240
Shale	60	1,300
Sand	60	1,360
Shale	40	1,400
Limestone	50	1,450
Shale	150	1,600
Limestone	10	1,610
Sand	40	1,650
Shale	80	1,730
Sand, (water 1760)	45	1,775
Shale	25	1,800
Limestone	150	1,950
Shale	50	2,000
Sand	200	2,200

Mississippian System.	Thickness	Depth
Shale	50	2,250
Sand	50	2,300
Shale	100	2,400
Sand	100	2,500
Shale	25	2,525
Limestone	25	2,550
Unrecordel sedments	190	2,740
Total depth		2,740

NOTE—A very poor record indeed. The full Pennsylvanian section, Conemaugh, Carbondale (Alleghany) and Pottsville are here represented.

WHITLEY COUNTY.

Production: Oil and gas.   Producing Sands.   Pottsville (Pennsylvanian);   Maxton and Big Lime (Mississippian).

Log No. 1168

S. M. Brown, No. 1, lessor.   Completed: Jan. 16, 1905.   Production: Dry; casing pulled, well abandoned.   Authority: New Domain Oil & Gas Co.

Strata.

Pennsylvanian System.	Thickness	Depth
Soil, yellow, soft	15	15
Shale, hard, black	85	100
Sandstone, gray, hard	10	110
Shale, hard, black, soft	25	135
Sandstone, white, hard	4	140
Sand, white, medium	12	152
Shale, brown, soft	33	185
Shale, black, soft	10	195
Coal, black, soft	2	197
Shale, brown, hard, limy	5	202
Shale, brown, soft	108	310
Sand, gray, soft	90	400
Shale, white, soft, limy, (gas 415)	15	415
Sand, white, soft	85	500
Shale, hard, black, soft	20	520
Sandstone, white, hard, shaly	15	535
Sand, gray, hard	165	700
Shale, hard, brown	20	720
Sand, yellow, hard	50	770
Shale, hard, yellow	5	775
Sand, white, hard, (oil show 800)	25	800

Pennsylvanian System.

	Thickness	Depth
Sand, gray, hard	100	900
Shale, hard, black	54	954
Sandstone, gray hard	29	983
Coal, black, soft	2	985
Shale, hard, white, soft	10	995

Mississippian System.

Shale (red rock), hard	18	1,013
Limestone and shells,. red, very hard	37	1,050
Shale (red rock), very hard	50	1,100
Limestone (Big Lime), white, very hard	140	1,240
Limestone, blue, hard	5	1,245
Limestone, blue, hard	45	1,290
Shale, black, soft	6	1,296
Limestone, white, hard, (gas 1470)	234	1,530
Sand, white, hard	60	1,590
Shale, blue, soft	230	1,820
Total depth		1,820

Log No. 1169

John Foley, No. 1, lessor. Iroquois Oil Co., Knoxville, Tenn., lessee. Location: ¾ mile west of Williamsburg, ½ mile above mouth on Briar Creek. Commenced: September, 1920. Incomplete record secured July 1, 1921. Driller and authority: Tom Langton. Casinghead elevation: 1036 A. T. 2 feet above Lily coal. Production: Dry. Structural position: South flank of Williamsburg Anticline ¾ mile from crest.

Strata.

Pennsylvanian System.

	Thickness	Depth
Clay and soil	6	6
Sandstone	104	110
Shale, hard	30	140
Sand, hard	290	430
Shale, hard	10	440
Sand, hard	200	640
Shale, hard	20	660
Sand, hard, (oil show 728)	103	763
Shale, hard	7	770
Sand hard	42	812
Shale, hard	10	822
Sand, hard, (Williamsburg oil sand 865)	43	865
Shale, hard	25	890
Sand, hard, (oil, gas water 1122)	255	1,145

Mississippian System.	Thickness	Depth
Lime shell and shale, hard, (Little Lime)	20	1,165
Limestone, hard (Big Lime)	80	1,245
Shale	10	1,255
Limestone (Big Lime)	245	1,500
Shale, hard	10	1,510
Flint rock	40	1,550
Shale, hard	10	1,560
Lime shell	20	1,580
Shale, hard	20	1,600
Lime shell	10	1,610
Red rock	15	1,625
Shale, hard	75	1,700
Devonian System.		
Shale, brown (Chattanooga)	90	1,790
Lime shell	20	1,810
Silurian System.		
Shale, hard	40	1,850
Limestone, brown	850	2,700
Limestone, black	15	2,715
Limestone, brown	635	3,350
Total depth, July 1, 1921		3,350

NOTE—The Silurian-Ordovician contact occurs within the upper quarter of the 850 feet of brown limestone above 2,700 feet.

Log No. 1170

Rose, No. 1, lessor. Iroquois Oil Co., lessee. Location: 1½ miles west of Williamsburg. Completed in the spring of 1920. Production: about 4,000,000 cu. ft. gas. Authority: E. C. Dicel.

Strata.

Pennsylvanian System.	Thickness	Depth
Soil	2	2
Sand	148	150
Shale, hard	120	270
Sand	320	590
Shale, hard	20	610
Sand	95	705
Shale, hard	25	730
Sand	85	815
Shale, hard	83	898
Sand, (some oil)	35	933
Shale, hard	30	963

Mississippian System.	Thickness	Depth
Shale (red rock), sandy	12	975
Shale, hard	15	990
Limestone (Little Lime)	20	1,010
Shale, hard	90	1,100
Limestone (Big Lime), (gas 1,265)	205	1,305
Total depth		1,305

Log No. 1171

Baptist Educational Society, No. 1, lessor. Empire Oil & Gas Co., lessee. Location: First left hand branch of Dog Slaughter Creek, 1 mile north of Dog Slaughter Creek. Completed: 1918. Contractors: J. H. Wilt Drilling Co. Authority: E. C. Dicel.

Strata.

Pennsylvanian System.	Thickness	Depth
Drift, yellow, soft	6	6
Rock, very firm, yellow, hard	180	186
Shale, hard, black and shells	35	221
Shale, gritty, white, firm, and sand	85	306
Shale, black, soft, and coal	15	321
Sand, gray, hard	100	421
Shale, hard, black and soft	10	431
Shale, black, hard	7	438
Shale, hard, blue, sticky	30	468
Shale, hard, blue, and shells	70	538
Shale shell, gray, very hard, limy	20	558
Shale, sticky, red, soft	15	573
Sand, red, firm	10	583
Sand, pink, hard	5	588
Coal, black, soft	7	595
Sand, white, hard	15	610

Mississippian System.		
Shale, hard, red, soft	5	615
Shell, hard, dark	10	625
Shale, hard, white, soft	15	640
Sand, white, hard	20	660
Shale, hard, white, eaving	10	670
Shale, hard, dark, limy	10	680
Shell, dark, hard .....:..............	7	687
Limestone, hard, dark	12	699
Shale, hard, white, soft	5	704
Sand, white, hard, very close (Maxon)	40	744
Limestone, white, soft (Little Lime)	4	748

Pennsylvanian System.	Thickness	Depth
Sand, white, hard	7	755
Limestone, white, soft, Big Lime (184)	34	789
Limestone, white, hard, Big Lime (184)	45	834
Limestone, brown, soft, Big Lime (184)	10	844
Limestone, brown, firm, Big Lime (184)	20	864
Limestone, brown, hard, Big Lime (184)	25	889
Limestone, brown, soft, Big Lime (184)	14	903
Flint, hard, dark, Big Lime (184)	15	918
Limestone, brown, soft, Big Lime (184)	6	924
Limestone flint, hard, dark, Big Lime (184) ..	15	939
Sandstone, gray, firm, limy, (odor of gas)	35	974
Shell, hard, dark	10	884
Limestone, white, soft	30	1,014
Sand, pink, soft, shaly	5	1,019
Sand, gray, hard	161	1,180
Shale, hard, white, soft	30	1,210
Shell, hard, black	10	1,220
Shale, hard, green, firm (New Providence)	80	1,300
Devonian System.		
Shale, black, soft (Chattanooga)	90	1,390
Shale, hard, white, soft	60	1,450
Limestone, gray, hard	20	1,470
Silurian System.		
Shale, hard, white, soft	10	1,480
Limestone, gray, hard	12	1,492
Shale, hard, white, soft	10	1,502
Limestone, gray, hard	10	1,512
Shale, hard, blue, firm	8	1,520
Limestone, black, hard	60	1,580
Total depth		1,580

Log No. 1172

H. M. Young, No. 1, lessor. Empire Oil & Gas Co., lessee. Location: about 13 miles from Williamsburg, and about 6 miles from Cumberland Falls, on the road from Williamsburg to Cumberland Falls. Head of Dog Slaughter Creek. Completed: Feb. 5, 1919. Authority: E. C. Dicel.

Strata.

Pennsylvanian System.	Thickness	Depth
Drift, yellow, soft, (little water)	15	15
Sand, rock, yellow, hard	45	60
Sand, gritty, blue, firm, (hole full of water) ..	90	150
Shale, hard, black, soft	35	185
Sandstone, hard, white	10	195
Sand, gray, soft	55	250

Pennsylvanian System.   Thickness Depth

	Thickness	Depth
Shale, hard, black, soft	10	260
Shale, white, hard, sandy	65	325
Shale, hard, brown, soft	40	365
Sandstone, gray, hard	10	375
Sand, white, medium, (settled quickly)	80	455
Shale, hard, black, soft	3	458
Shale, gray, hard, limy	14	472
Shale, hard, brown, soft	4	476
Sand, white, hard, medium	52	528
Coal, black, soft	1	529
Shale, hard, brown, soft	22	551
Sandstone, hard, gray, (little gas 555)	10	561
Shale, hard, blue, soft, and shells	74	635
Sandstone, hard, gray	10	645
Shale, hard, brown, soft	15	660
Sandstone, dark, very hard	10	670
Shale, hard, brown, soft	15	685
Sandstone, gray, hard	10	695
Shale, hard, brown	15	710
Limestone, black, very hard	5	715
Shale, brown soft	10	725
Sandstone, gray, hard	20	745
Shale and shells, hard, white, soft	45	790
Sandstone, gray, hard	10	800
Shale, hard, red, soft	40	840
Shale, hard, gray, soft	15	855
Sand, gray, medium	22	877
Shale, hard, brown, soft	11	888
Shale, hard, gray, soft	14	902
Sand, gray, hard	10	912

Mississippian System.

Limestone, brown, hard	28	940
Shale, hard, gray, soft	5	945
Limestone, brown, hard	5	950
Shale, hard, gray, soft	10	960
Limestone, white, hard	25	985
Shale, hard, white, soft	5	990
Limestone, hard, white, Big Lime (180)	10	1,000
Limestone, hard, brown, Big Lime (180)	70	1,070
Limestone, hard, gray, Big Lime (180)	5	1,075
Limestone, hard, brown, Big Lime (180)	18	1,093
Sand, soft, gray, (light oil show), Big Lime (180)	5	1,098
Limestone, brown, hard, Big Lime (180)	7	1,105

Mississippian System.	Thickness	Depth
Limestone, hard, gray, Big Lime (180)	35	1,140.
Limestone, gray, hard, Big Lime (180)	30	1,170
Sand, brown, limy, (50,000 cu. ft. gas at 1185)	31	1,201
Total depth		.1,201

Log No. 1173

Nelson, No. 1, lessor. The Cumberland Bend Oil Co., lessee. Location: 1½ miles southeast of Williamsburg. Completed: in 1907. Shot Nov. 15, 1907, 90 qts. Production: about 1½ bbls. Authority: E. C. Dicel, Williamsburg.
Strata.

Pensylvanian System.	Thickness	Depth
Sandstone, shale and coal	455	455
Sand, white	10	465
Sand, white	30	495
Limestone, white	10	505
Sand, white	70	575
Shale, hard	7	582
Shale, hard	3	585
Sand, white, (gas 595; oil 605)	55	640
Sand, gray	20	660
Shale, hard, shelly	55	715
Sand, gray	20	735
Shale, hard	31	766
Sand, (gas)	1	767
Sand, white, (oil 767 to 780)	13	780
Sand, white	25	805
Sand, white, (oil and gas 809)	6	811
Shale, hard	4	815
Shale, hard	10	825
Flint rock	1½	826½
Total depth		826½

Log No. 1174

G. W. Rains, No. 3, lessor. Location: mouth of Clear Fork Creek, 1½ miles southeast of Williamsburg. Cased Feb. 14, 1919. Shot with 10 qts. glycerine by the Ky. Glycerine Co. Recommenced drilling at 939 feet. Authority: E. C. Dicel.
Strata.

Pennsylvanian System.	Thickness	Depth
Sandstone, shale and coal	939	939
. Shale, hard	3	942

Mississippian System.	Thickness	Depth
Shale and shells, hard, pink, (Mauch Chunk) ..	68	1,010
Sand, gray	15	1,025
Shale, hard, red	2	1,027
Limestone, light	51	1,078
Shale, hard, pink, light (Mauch Chunk)	48	1,126
Limestone, blue	5	1,131
Shale, hard, dark	46	1,177
Limestone, black, (show of oil)	14	1,191
Limestone, black	24	1,215
Shale, hard, light	6	1,221
Limestone, light, Big Lime (253)	63	1,284
Sand, light, Big Lime (253)	35	1,319
Limestone, light softer, Big Lime (253)	33	1,352
Limestone, light, dark, Big Lime (253)	31	1,383
Limestone, light, Big Lime (253)	20	1,403
Limestone, dark, Big Lime (253)	4	1,407
Limestone, light, Big Lime (253)	7	1,414
Limestone, light, Big Lime (253)	10	1,424
Limestone, gray, Big Lime (253)	11	1,435
Limestone, dark, Big Lime (253)	10	1,445
Limestone, gray, (oil show 1460-1463) Big Lime (253)	18	1,463
Limestone, light, Big Lime (253)	4	1,467
Limestone, gray, dark, Big Lime (253)	7	1,474
Sand, dark, gray	8	1,482
Limestone, dark, gray, & pebbles (gas 1515)	35	1,517
Limestone, dark gray, pebbles & crystal rock	5	1,522
Sand, pebble and crystal rock	4	1,526
Sand, gray, and limestone	6½	1,532½
Total depth		1,532½

Log No. 1174-A.

Well at Saxton, ½ mile S. E. Saxton, between L. & N. and Southern R. R., 20 feet or more above railroad.
Strata.

Pennsylvanian System.	Thickness	Depth
Shale	130	130
Sandstone	30	160
Sandstone	10	170
Sandstone	10	180
Sandstone	30	210
Sandstone	35	245
Sandstone	30	275
Shale	20	295

Pennsylvanian System.	Thickness	Depth
Shale	5	300
Shale	30	330
Shale	50	380
Sandstone	45	425
Sandstone	35	460
Sandstone	35	495
Sandstone	30	525
Sandstone	30	555
Sandstone	30	585
Sandstone	25	610
Sandstone	25	635
Sandstone	20	655
Sandstone	20	675
Sandstone	15	690
Sandstone	15	705
Sandstone	15	720
Sandstone	15	735
Sandstone	10	745
Shale, (cased at 747, 6¼" casing)	10	755
Shale	55	810
Sandstone	30	840
Shale	55	895
Shale and sandstone (salt water 935)	40	935
Sandstone	10	945
Sandstone (more water, rerimmed 8¾")	15	960
Sandstone	10	970
Sandstone	15	985
Sandstone	10	995
Sandstone (8" hole to 1000, cased)	10	1,005
Sandstone	15	1,020
Sandstone	20	1,040
Sandstone	15	1,055
Sandstone	10	1,065
Sandstone	15	1,080
Sandstone (oil show)	12	1,092
Sandstone and shale	33	1,125
Sandstone	30	1,155
Mississippian System.		
Limestone	20	1,175
Pink rock	35	1,210
Shale, red	50	1,260
Sandstone	40	1,300
Limestone and shale	35	1,335
Limestone and shale	35	1,370
Shale and shell	30	1,400

THE BASAL POTTSVILLE OIL SAND

Cliffs or "rock houses" are common in the heavy bedded Pottsville conglomerate of Whitley and McCreary counties. This strata outcropping on Eagle Creek, below Cumberland Falls, goes under cover with the normal southeast dip

Mississippian System.	Thickness	Depth
Shale and shell	25	1,425
Limestone	25	1,450
Limestone	25	1,475
Limestone	30	1,505
Limestone	25	1,530
Shale and limestone	30	1,560
Limestone	25	1,585
Limestone	25	1,610
Limestone	6	1,616
Total depth		1,616

Log No. 1174-B.

R. N. Adkins, No. 1, lessor. Location: 1½ miles S. W. Williamsburg, on 1st right-hand branch of Briar Creek, Whitley County, Ky. Production: Gas. Completed: 1920. Casing head elevation: 1042 A. T. Structural position: Nose of anticline, south flank near crest. Authority: C. E. Dicel, Williamsburg, Ky.

Strata.

Pennsylvanian System.	Thickness	Depth
Soil	5	5
Sandstone	85	90
Shale	175	265
Sandstone	162	427
Shale and coal	3	430
Sandstone	163	593
Shale (cased 600 6¼" casing)	32	625
Sandstone	115	740
Shale	10	750
Sandstone	55	805
Shale (coal due at 809)	60	865
Sandstone	35	900
Shale	23	923
Sandstone (oil)	20	943
Sandstone, broken	22	965
Shale	13	978
Mississippian System.		
Shale, pink (Mauch Chunk)	72	1,050
Shale	50	1,100
Limestone	10	1,110
Shale	45	1,155
Limestone (Big Lime)	25	1,180
Shale	5	1,185
Limestone (Big Lime)	40	1,225

Mississippian System.	Thickness	Depth
Shale	5	1,230
Limestone (gas 1365)	135	1,365
Limestone (more gas at 1370)	19	1,384
Total depth		1,384

WOLFE COUNTY.

Production: Oil and gas. Producing Sands: Corniferous (Devonian), Niagaran (Silurian).

Log No. 1175

J. T. Day, No. 1, lessor. High Gravity Oil Co., lessee. Location: on Red River, 1 mile north of intersection of Wolfe, Breathitt and Magoffin County lines. Commenced: June 10, 1920. Completed: Aug. 27 1920. Drillers: A. A. Wolfe, J. C. Gibson.

Strata.

Pennsylvanian System.	Thickness	Depth
Conductor	20	20
Shale, (cased 8¼, 215 ft.)	460	480
Limestone	10	490
Sand	50	540
Shale	60	600
S. Sand	75	675
Shale	90	765
Limestone	10	775
Sand	100	875
Shale	5	880
Sand	60	940
Shale	5	945
Limestone, sandy	10	955
Shale	5	960
Mississippian System.		
Limestone (Little Lime)	5	965
Shale (pencil cave)	5	970
Limestone (Big Lime), (cased 6-5/8, 995) ..	105	1,075
Shale, white	415	1,490
Sandstone (Wier)	35	1,525
Shale, brown (Sunbury)	10	1,535
Sandstone (Berea)	30	1,565
Shale, gray	35	1,600
Devonian System.		
Shale, brown (Chattanooga)	155	1,755
Shale (boulder), very hard	5	1,760
Shale, brown	115	1,875

Devonian System.	Thickness	Depth
Shale, white	35	1,910
Limestone (Corniferous in part)	190	2,100
Limestone ''sand,'' sharp, (Big 6)	55	2,155
Shale, black	5	2,160
Sand, limy	15	2,175
Shale, black	13	2,188
Shale (red rock)	2	2,190
Total depth		2,190

Water at 40 feet.
Gas to light, 250 feet.
Gas at 500 feet.
Water at 625 feet.
Finished in Red Rock, 2 feet.

NOTE—The upper part only of the 190 feet of limestone above 2,100 feet in depth is Corniferous. The Devonian-Silurian contact comes at the base of the Corniferous here.

Log No. 1176

• J. D. Spencer, No. 1, lessor. Commenced: Sept. 3, 1918. Completed: Sept. 21, 1918. Gas from 1,227 to 1,231 feet; water from 1,231 to 1237 feet. Authority: The Ohio Oil Co.
Strata.

Pennsylvanian System.	Thickness	Depth
Soil, yellow, soft	14	14
Shale, hard, black, soft	16	30
Sand, brown, soft	170	200
Shale, hard, black, soft	30	230
Sand, hard, white	15	245
Shale, light, soft	170	415
Mississippian System.		
Limestone (Little Lime) white, soft, broken ..	20	435
Limestone (Big Lime), hard, white	80	515
Shale, hard, and shells, blue, broken	20	535
Limestone, hard, white	35	570
Shale and limestone, hard, blue, soft	40	610
Shale, hard, blue, soft	335	945
Rock, pink, soft	10	955
Shale, hard, Limestone, broken, soft	25	980
Devonian System.		
Shale, brown, soft (Chattanooga)	190	1,170
Fire clay, white, soft	10	1,180
Shale, black, soft	5	1,185
Limestone (cap rock), hard, blue	4	1,189

Devonian System.	Thickness	Depth
Limestone ''sand,'' hard, brown	16	1,205
Limestone ''sand,'' hard, brown, (little gas)..	4	1,209
Limestone ''sand,'' hard, gray, (water)	8	1,217
Total depth		1,217

Log No. 1177

Dr. A. Congleton, No. 1, lessor. Commenced: July 6, 1918. Completed: Aug. 9, 1918. Production: 1,000 cu. ft. gas was gotten from this well. Casing pulled and well plugged. Authority: The Ohio Oil Co.

Strata.

Pennsylvanian System.	Thickness	Depth
Soil, brown, soft	7	7
Sandstone, gray, soft	·143	150
Shale and shells, brown and soft	185	335
Sand, hard, white, watery	20	355
Mississippian System.		
Limestone (Big Lime), hard, white	100	455
Shale, hard, and shells, soft	145	600
Shale, hard, blue, soft	375	975
Shale, hard, and shells	20	995
Devonian System.		
Shale, brown, soft (Chattanooga)	165	1,160
Shale, hard, blue, soft	26	1,186
Limestone (cap rock), hard, gray	10	1,196
Limestone ''sand,'' hard, brown	13	1,209
Limestone ''sand,'' gray, hard, (pay)	12	1,221
Limestone, hard, black	15	1,236
Silurian System.		
Limestone ''sand,'' hard, brown	6	1,242
Limestone ''sand,'' white, hard, (water)	7	1,249
Limestone, hard black	17	1,266
Limestone ''sand,'' hard, white, (water)	9	1,275
Total depth		1,275

Log No. 1178

A. Rose, No. 1, lessor. Location: Lee City. Commenced: Nov. 15, 1918. Completed: Dec. 26, 1918. Well was dismantled on Jan. 4, 1919. 6¼ inch casing used at 835 feet. Authority: L. Beckner.

Strata.

Pennsylvanian System.	Thickness	Depth
Soil and shale	105	105
Shale, hard	96	205
Coal	4	109

Pennsylvanian System.

	Thickness	Depth
Coal	5	210
Shale, hard	130	340
Sand	210	550
Shale, hard	45	595
Sand and shell	10	605
Shale, hard	15	620
Sand and shell	15	635
Sand	80	715
Shale, hard	25	740

Mississippian System.

Limestone (Little Lime)	10	750
Shale, hard	5	755
Sand	10	765
Limestone (Big Lime)	85	850
Sand, broken	540	1,390
Shale, black (Sunbury)	10	1,400
Shale, hard, blue (New Providence)	25	1,425

Devonian System.

Shale, brown (Chattanooga)	245	1,670
Shale	25	1,695
Limestone (cap rock)	4	1,699
Limestone ''sand''	6	1,705
Limestone, brown	121	1,826
Total depth		1,826

NOTE—The Devonian-Silurian contact occurs within the upper quarter of the 121 feet of limestone above 1,826 feet in depth.

Log No. 1179

W. L. Hobbs, No. 7, lessor. Commenced: Oct. 9, 1919. Completed: Nov. 25, 1919. Production: commenced producing Nov. 29, 1919. Authority: The Superior Oil Corporation.

Strata.

Pennsylvanian System.

	Thickness	Depth
Clay, yellow, soft,	20	20
Shale, dark, soft	85	105
Sand (mountain), yellow, soft	200	305
Shale, hard, white, medium	105	410

Mississippian System.

Limestone, white, hard, (Big Lime)	90	500
Shale, hard, gray, medium	500	1,000

Devonian System.	Thickness	Depth
Shale, brown, hard (Chattanooga)	160	1,160
Fire clay, white, soft	43	1,203
Limestone (cap rock), gray, hard	6	1,209
Limestone "sand," gray, soft, (oil)	11	1,220
Limestone "sand," gray, hard	4	1,224
Total depth		1,224

Log No. 1180

W. L. Hobbs, No. 8, lessor. Location: Township in the fourth precinct. Commenced: Dec. 22, 1919. Completed: Jan. 14, 1920. Production: commenced producing Jan. 15, 1920. Authority: The Superior Oil Corporation.

Strata.

Pennsylvanian System.	Thickness	Depth
Soil and shale, hard	100	100
Sand (mountain)	110	210
Shale, hard, and soft	150	360
Mississippian System.		
Limestone (Big Lime)	140	500
Shale, hard, and lime shells	530	1,030
Devonian System.		
Shale, black (Chattanooga)	160	1,190
Fire clay	16	1,206
Limestone (cap rock)	6	1,212
Limestone "sand," (oil pay 1222)	19	1,231
Total depth		1,231

Log No. 1181

W. L. Hobbs, No. 9, lessor. Commenced: Mar. 17, 1920. Completed: Mar. 31, 1920. Production: commenced producing April 2, 1920. Authority: The Superior Oil Corporation.

Strata.

Pennsylvanian System.	Thickness	Depth
Soil and clay	10	10
Sand (mountain)	170	180
Shale, hard and soft	140	320

Mississippian System.	Thickness	Depth
Limestone (Big Lime)	105	425
Shale, hard, and shells	505	930 .

Devonian System.		
Shale, black (Chattanooga)	160	1,090
Fire clay	15	1,105
Limestone (cap rock), (Corniferous)	12	1,117
Limestone "sand," (Corniferous), (oil pay 1120-1130)	20	1,137
Total depth		1,137

Log No. 1182

W. L. Hobbs, No. 10, lessor. Commenced: April 14, 1920. Completed: April 30, 1920. Production: Dry. Authority: The Superior Oil Corporation.

Strata.

Pennsylvanian System.	Thickness	Depth
Soil and clay	20	20
Shale, hard	60	80
Sand (Mountain)	180	260
Shale, hard, and soft	100	360

Mississippian System.		
Limestone (Big Lime)	120	480
Shale, hard, and lime shells	450	930

Devonian System.		
Shale, brown (Chattanooga)	190	1,120
Fire clay	20	1,140
Limestone (cap rock)	5	1,145
Limestone "sand," (small show of oil)	12	1,157
Limestone, gray	33	1,190
Limestone "sand," (water)	24	1,214
Total depth		1,214

NOTE—The Devonian-Corniferous contact occurs toward the base of the 33 feet of limestone above 1190 feet. The lower part of the last 24 feet of the record is probably Silurian.

Log No. 1183

W. L. Hobbs, No. 11, lessor. Commenced: May 11, 1920. Completed: May 26, 1920. Production: Dry. Authority: The Superior Oil Corporation.

Strata.

Pennsylvanian System.	Thickness	Depth
Soil, soft	10	10
Shale, hard	15	25
Sand (Mountain), soft	165	190
Shale, hard	115	305
Mississippian System.		
Limestone (Big Lime), hard	125	430
Shale, hard, green, soft	30	460
Shale, hard	455	915
Devonian System.		
Shale, brown, soft (Chattanooga)	180	1,095
Fire clay	13	1,108
Limestone (cap rock), Corniferous	9	1,117
Sand, (dry) Corniferous	40	1,157
Total depth		1,157

Log No. 1184

A. C. Creech, No. 5, lessor. Location: 1 mile from Torrent. Completed: Feb. 2, 1920. Authority: The Sable Oil & Gas Co.

Strata.

Pennsylvanian & Mississippian Systems.	Thickness	Depth
Sandstone, shale, and limestone	1,068	1,068
Limestone (cap rock)	15	1,083
Limestone "sand," (Corniferous-Devonian)	10	1,093
Limestone	14	1,107
Total depth		1,107

Log No. 1185

A. C. Creech, No. 5, lessor. Location: 1 mile from Torrent. Completed and shot Feb. 27, 1920. First pay was from 1106 to 1121 feet. Second pay was from 1106 to 1132 feet. Authority: The Sable Oil & Gas Co.

Strata.

Pennsylvanian & Mississippian Systems.	Thickness	Depth
Sandstone, shale, and limestone	1,092	1,092
Limestone "sand," (Corniferous-Devonian)	40	1,132
Total depth		1,132

Log No. 1186

A. C. Creech, No. 6, lessor. Location: 1 mile from Torrent. Completed: June 1, 1920. Authority: The Sable Oil & Gas Co.

Strata.

Pennsylvanian & Mississippian Systems.	Thickness	Depth
Conductor	10	10
Sandstone, shale and limestone	1,030	1,040
Limestone (cap rock), Corniferous	15	1,055
Limestone, (first pay) Corniferous	15	1,070
Limestone	10	1,080
Silurian System.		
Limestone, (second pay)	25	1,105
Total depth		1,105

Log No. 1187

A. F. Johnson, No. 1, lessor. Completed: July 12, 1910. Production: Dry; water at 100 and 220 feet; showing of oil and gas at 160 feet. Authority: New Domain Oil & Gas Co.

Strata.

Pennsylvanian System.	Thickness	Depth
Soil and sand, white, soft	7	7
Sand and shale, hard, light, soft	43	50
Shale, hard, blue, soft	30	80
Sand, white, loose	8	88
Shale, hard, blue, soft	5	93
Sand, white, hard	127	220
Sand and shale hard, light, blue, loose	50	270
Shale, hard, light, blue	55	325
Shale, blue, soft	50	375
Mississippian System.		
Sand and shale, white, hard	25	400
Limestone (Big Lime), white, hard	80	480
Shale, variable in color and hardness	770	1,250
Limestone ''sand,'' light, extra hard, (oil)	80	1,330
Limestone ''sand,'' light, hard	10	1,340
Limestone ''sand,'' blue, hard	30	1,370
Limestone ''sand,'' light, hard	40	1,410
Limestone, light, hard	5	1,415
Limestone and shale, hard, light, soft	4	1,419
Shale (red rock), pink, soft, limy	5	1,424
Total depth		1,424

NOTE—The Mississippian-Devonian contact occurs in the lower part of the 770 feet of colored shale above 1,250 feet in depth.

Log No. 1188

George Spencer, No. 3, lessor. Commenced: Jan. 28, 1920. Completed: Mar. 4, 1920. Authority: The Superior Oil Corporation.

Strata.

Pennsylvanian System.	Thickness	Depth
Soil and clay	16	16
Shale, hard	30	46
Sand (mountain)	155	201
Shale	25	226
Sand (water)	30	256
Shale, hard	90	346
Mississippian System.		
Limestone	104	450
Shale, hard and soft	470	920
Shale, hard, red	10	930
Shale, hard	15	945
Devonian System.		
Shale, brown, and fire clay	200	1,145
Limestone (cap rock)	29	1,174
Limestone ''sand,'' (water 1197)	23	1,197
Total depth		1,197

Log No. 1189

Spencer Heirs, No. 12, lessors. Location: The fourth precinct. Commenced: Nov. 7, 1919. Completed: Jan. 17, 1920. Authority: The Superior Oil Corporation.

Strata.

Pennsylvanian System.	Thickness	Depth
Soil	6	6
Sandstone	9	15
Shale, hard, blue	85	100
Mississippian System.		
Limestone (Big Lime)	95	195
Shale, hard, green	205	400
Shale, hard, blue	320	720
Devonian System.		
Shale, brown (Chattanooga)	173	893
Fire clay	30	923
Limestone (cap rock)	5	928
Limestone ''sand,'' top	2½	930½
Limestone ''sand,'' salt	20	950½
Limestone, black	18½	969
Total depth		969

Log No. 1190

Spencer Heirs, No. **13**, lessors. Completed: Jan. **14, 1920.** Authority: The Superior Oil Corporation.

Strata.

Pennsylvanian System.	Thickness	Depth
Sandstone, shale and limestone	410	410
Mississippian System.		
Limestone (Big Lime)	135	545
Shale, hard, and lime shells	495	1,040
Devonian System.		
Shale (Chattanooga)	175	1,215
Fire clay	15	1,230
Limestone (cap rock)	6	1,236
Limestone ''sand,'' (Corniferous)	9	1,245
Total depth		1,245

There was some salt water under pay.

Log No. 1191

Spencer Heirs, No. **14**, lessors. Commenced: Feb. **2, 1920.** Completed: Mar. **2, 1920.** Authority: The Superior Oil Corporation. Production: Dry; well plugged and casing pulled.

Strata.

Pennsylvanian System.	Thickness	Depth
Soil and clay	15	15
Shale, hard	25	40
Sand (mountain)	160	200
Shale	30	230
Sand, white, (a little water)	30	260
Shale, hard	92	352
Shale, black	3	355
Limestone, (Big Lime)	85	440
Break, (Big Lime)	10	450
Limestone, (Big Lime)	10	460
Shale, hard and soft	465	925
Shale, hard, red	10	935
Limestone shells	2	937
Shale, hard	13	950
Devonian System.		
Shale, brown (Chattanooga)	165	1,115
Fire clay	24	1,139
Limestone (cap rock)	11	1,150
Limestone ''sand,'' (oil show 1,155, salt water 1,160 & 1,200)	59	1,209
Total depth		1,209

NOTE—The Devonian-Silurian contact occurs within the lower half of the last **59** feet of limestone.

Log No. 1192

Spencer Heirs, No. 15, lessors. Commenced: Mar. 15, 1920. Completed: Apr. 1, 1920. Production: commenced producing Apr. 3, 1920. Authority: The Superior Oil Corporation.

Strata.

Pennsylvanian System.	Thickness	Depth
Soil and clay	20	20
Shale, hard and soft	110	130
Sand (mountain)	140	270
Shale, hard, and soft	130	400
Mississippian System.		
Limestone (Big Lime) ....:	115	515
Shale, hard, and lime shells	500	1,015
Shale	165	1,180
Fire clay	20	1,200
Limestone (cap rock)	14	1,214
Limestone "sand," (oil pay 1,220-1,230) ..	16	1,230
Total depth		1,230

Log No. 1193

Spencer Heirs, No. 16, lessors. Commenced: April 19, 1920. Completed: April 24, 1920. Authority: The Superior Oil Corporation.

Strata.

Pennsylvanian Systetm.	Thickness	Depth
Soil and clay	20	20
Shale, hard, and soft	110	130
Sand (Mountain)	160	290
Shale, hard and soft	80	370
Mississippian System.		
Limestone (Big Lime)	105	475
Shale, hard and soft	470	945
Devonian System.		
Shale (Chattanooga)	180	1,125
Fire clay	30	1,155
Limestone (cap rock)	14	1,169
Limestone "sand," (oil pay 1171-1180)	18	1,187
Total depth		1,187

Log No. 1194

Spencer Heirs, No. 17, lessors. Commenced: May 5, 1920. Completed: May 20, 1920. Production: Dry; casing pulled and well abandoned. Authority: The Superior Oil Corporation.

Strata.

Pennsylvanian System.	Thickness	Depth
Soil, soft	10	10
Shale, hard, white, soft	130	140
Sand (Mountain), yellow, soft	150	290
Shale, hard	125	415
Mississippian System.		
Limestone (Big Lime), white, hard	105	520
Shale, hard, green, soft	30	550
Shale, hard	450	1,000
Devonian System.		
Shale, brown, soft (Chattanooga)	170	1,170
Fire clay, soft	18	1,188
Limestone (cap rock), hard	12	1,200
Limestone "sand," hard	49	1,249
Total depth		1,249

NOTE—The Devonian-Silurian contact occurs within the lower half of the last 49 feet of limestone.

Log No. 1195

Hall and Burke, No. 29, lessors. Commenced: Dec. 30, 1919. Completed: Jan. 17, 1920. Authority: The Superior Oil Corporation.

Strata.

Pennsylvanian System.	Thickness	Depth
Soil and sand	18	18
Sand, gray, hard	218	236
Shale, dark, soft	60	296
Sand, white, hard	80	376
Mississippian System.		
Limestone (Big Lime), light, hard	120	496
Shale, light, medium	440	936

Devonian System. Thickness Depth

	Thickness	Depth
Shale, black, medium (Chattanooga)	140	1,076
Fire clay, light, medium	20	1,096
Limestone "sand," salt, gray, hard, (salt water)	29	1,125
Limestone "sand," gray, hard, and medium, (oil) (pay)	40	1,165
Limestone, rotten, gray, soft	5	1,170
Limestone "sand," gray, medium, (oil (pay)	15	1,185
Shale, hard, light, medium	4	1,189
Total depth		1,189

NOTE—The Devonian-Silurian contact occurs within the 40 feet of limestone above 1,165 feet in depth.

Log No. 1196

William Adams, No. 1, lessor. Location: Torrent District. Completed: May 16, 1917. Initial production: 65 bbls. oil. Authority: The Superior Oil Corporation.

Strata.

Pennsylvanian System. Thickness Depth

	Thickness	Depth
Sandstone, shale and coal	335	335

Mississippian System.

	Thickness	Depth
Limestone (Big Lime)	100	435
Shale, gray	30	465
Shale, white	460	925

Devonian System.

	Thickness	Depth
Shale, black (Chattanooga)	155	1,080
Fire clay	15	1,095
Limestone "sand," (oil show 1,114, water 1,124)	70	1,165
Total depth		1,165

NOTE—The base of the Devonian occurs at the top of the last one-third of the last 70 feet of limestone.

Log No. 1197

William Adams, No. 3, lessor. Location: Torrent District. Completed: Jan. 12, 1918. Initial production: 40 bbls. oil. Authority: The Superior Oil Corporation.

Strata.

Pennsylvanian System.	Thickness	Depth
Sandstone, shale and coal	340	340
Mississippian & Devonian Systems.		
Limestone (Big Lime)	100	440
Rock, green	30	470
Sandstone and shale and fire clay	624	1,094
Limestone (cap rock)	3	1,097
Limestone "sand," (salt water 1,127)	65	1,162
Total depth		1,162

The well showed lots of gas.

NOTE—The base of the Devonian occurs at about the top of the last one-third of the last 65 feet of limestone.

Log No. 1198

William Adams, No. 4, lessor. Location: Torrent District. Commenced: Nov. 25, 1917. Completed: Dec. 22, 1917. Initial production: about 90 bbls. oil. Commenced producing: Dec. 23, 1917. Authority: The Superior Oil Corporation.

Strata.

Pennsylvanian System.	Thickness	Depth
Sandstone, shale and coal	344	344
Mississippian System.		
Limestone (Big Lime)	146	490
Shale	470	960
Devonian System.		
Shale, brown, and fire clay (Chattanooga)	171	1,131
Limestone "sand," (oil)	59	1,190
Total depth		1,190

NOTE—The base of the Devonian occurs toward the base of the last 59 feet of recorded limestone.

Log No. 1199

R. H. Taulbee, No. 1, lessor. Federal Oil Corp., lessee. Location: 3½ mi. south of Campton on Upper Devil's Creek. Commenced: Aug. 22, 1921. Completed: Sept. 19, 1921. Production: Gas, 50,-000 cu. ft. est. & ½ bbl. natural. Rig: 28 star. Driller: Glenn Mc-Coun, Campton.

Strata.

Pennsylvanian System.	Thickness	Depth
Soil	5	5
Sandstone & Shale (Mountain Sand)	285	290
Mississippian System.		
Limestone, (Little Lime)	35	325
Shale, blue, soft	30	355
Limestone (Big Lime), (cased 445)	90	445
Sandstone (shaly), gray-green	502	947
Devonian System.		
Shale, black (Chattanooga)	200	1,147
Shale, white (fire clay)	20	1,167
Shale, brown	10	1,177
Limestone (cap)	1	1,178
Limestone (gas "sand")	10	1,188
Limestone	5½	1,193½
Limestone (oil "sand")	11	1,204½
Limestone	27½	1,232
Limestone, (oil show)	5½	1,237½
Limestone	16½	1,254
Total depth		1,254

WOODFORD COUNTY.

Production: **Neither oil or gas. Producing Sand: None recognized.**

Log No. 1200

United Phosphate & Chemical Co., No. 4, owners and operators. Location: at Wallace Station. Completed: Dec. 1, 1920. Authority: W. R. Golson, mgr.

Strata.

Ordovician System.	Thickness	Depth
Soil, (10 in.)	16	16
Limestone, (10 in.)	780	796
Limestone, (8 in.)	12	808
Limestone, (6 in.)	389	1,197
Total depth		1,197

NOTES—Small water flow 1½ gal. per min. at 90 ft. Small dry cavity at 796 ft. Small wet cavity at 1,143 feet, which pumped 30 gal. fresh water per minute. No recognizable oil or gas show.

(THE END)

APPENDIX

List of Commercially Important Oil and Gas Pools in Kentucky.

(Corresponds to numbering of map on page 20).

No. 1, Meade County (old) Gas Field; No. 2, Cloverport (old) Gas Field; No. 3, Hartford Oil Pool; No. 4, Caneyville Oil Pool; No. 5, Leitchfield Oil and Gas Field; No. 6, Bear Creek Gas Field; No. 7, Diamond Springs Gas Field; No. 8, Warren County Oil and Gas Fields; No. 9, Allen County Oil and Gas Fields; No. 10, Barren County Oil and Gas Fields; No. 11, Green River Gas Field; No. 12, Lincoln County Oil Pools; No. 13, Wayne County Oil Pools; No. 14, Knox County Oil and Gas Field; No. 15, Clay County Gas Field; No. 16, Island Creek Gas Field; No. 17, Station Camp Oil Pool; No. 18, Irvine Oil Pool; No. 19, Big Sinking Oil Pool; No. 20, Ross Creek Oil Pool; No. 21, Menifee County Gas Field; No. 22, Menifee County Oil Pool; No. 23, Ragland Oil Pool; No. 24, Campton Oil Pool; No. 25, Stillwater Oil Pool; No. 26, Breathitt County Gas Field; No. 27, Cannel City Oil and Gas Pool; No. 28, Knott County Oil Pool; No. 29, Beaver Creek Oil and Gas Fields; No. 30, Prestonsburg Oil and Gas Fields; No. 31, Burning Fork Gas Field; No. 32, Paint Creek Oil and Gas Field; No. 33, Laurel Creek Oil and Gas Fields; No. 34, Martin County Gas Field; No. 35, Bussevville Oil Pool; No. 36, Fallsburg Oil Pool.

INDEX

D

E

F

G

INDEX—Continued.

H

I

J

K

L

Lightning Source UK Ltd.
Milton Keynes UK
UKHW020124220119
335965UK00008B/349/P